Programming Language Theory and Formal Methods

Programming Language Theory and Formal Methods

Edited by:

Zoran Gacovski

ARCLER
PRESS

www.arclerpress.com

Programming Language Theory and Formal Methods

Zoran Gacovski

Arcler Press
224 Shoreacres Road
Burlington, ON L7L 2H2
Canada
www.arclerpress.com
Email: orders@arclereducation.com

ISBN: 978-1-77469-447-3 (Hardcover)

DECLARATION

Some content or chapters in this book are open access copyright free published research work, which is published under Creative Commons License and are indicated with the citation. We are thankful to the publishers and authors of the content and chapters as without them this book wouldn't have been possible.

ABOUT THE EDITOR

Dr. Zoran Gacovski's current position is a full professor at the Faculty of Technical Sciences, "Mother Tereza" University, Skopje, Macedonia. His teaching subjects include Software engineering and Intelligent systems, and his areas of research are: information systems, intelligent control, machine learning, graphical models (Petri, Neural and Bayesian networks), and human-computer interaction. Prof. Gacovski has earned his PhD degree at Faculty of Electrical engineering, UKIM, Skopje. In his career he was awarded by Fulbright postdoctoral fellowship (2002) for research stay at Rutgers University, USA. He has also earned best-paper award at the Baltic Olympiad for Automation control (2002), US NSF grant for conducting a specific research in the field of human-computer interaction at Rutgers University, USA (2003), and DAAD grant for research stay at University of Bremen, Germany (2008 and 2012). The projects he took an active participation in, are: "A multimodal human-computer interaction and modelling of the user behaviour" (for Rutgers University, 2002-2003) - sponsored by US Army and Ford; "Development and implementation of algorithms for guidance, navigation and control of mobile objects" (for Military Academy – Skopje, 1999-2002); "Analytical and non-analytical intelligent systems for deciding and control of uncertain complex processes" (for Macedonian Ministry of Science, 1995-1998). He is the author of 3 books (including international edition "Mobile Robots"), 20 journal papers, over 40 Conference papers, and he is also a reviewer/ editor for IEEE journals and Conferences.

TABLE OF CONTENTS

LIST OF CONTRIBUTORS

Shagufta Shafiq
UIIT-PMAS Arid Agriculture University, Rawalpindi, Pakistan

Nasir Mehmood Minhas
UIIT-PMAS Arid Agriculture University, Rawalpindi, Pakistan

Saiqa Bibi
UIIT-PMAS Arid Agriculture University, Rawalpindi, Pakistan

Saira Mazhar
UIIT-PMAS Arid Agriculture University, Rawalpindi, Pakistan

Nasir Mehmood Minhas
UIIT-PMAS Arid Agriculture University, Rawalpindi, Pakistan

Irfan Ahmed
UIIT-PMAS Arid Agriculture University, Rawalpindi, Pakistan

Monika Singh
Faculty of Engineering & Technology (FET), Mody University of Science & Technology, Sikar, India

Ashok Kumar Sharma
Faculty of Engineering & Technology (FET), Mody University of Science & Technology, Sikar, India

Ruhi Saxena
Computer Science & Engineering, Thapar University, Patiala, India

Nasir Mehmood Minhas
University Institute of Information Technology, PMAS-University Institute of Information Technology, Rawalpindi, Pakistan

Asad Masood Qazi
University Institute of Information Technology, PMAS-University Institute of Information Technology, Rawalpindi, Pakistan

Sidra Shahzadi
University Institute of Information Technology, PMAS-University Institute of Information Technology, Rawalpindi, Pakistan

Shumaila Ghafoor
University Institute of Information Technology, PMAS-University Institute of Information Technology, Rawalpindi, Pakistan

Lorina Negreanu
POLITEHNICA University of Bucharest, Splaiul Independentei 303, 060042 Bucharest, Romania

Zheng Yang
School of Information and Software Engineering, University of Electronic Science and Technology of China, No.4 Section 2 North Jianshe Road, Chengdu 610054, China

Hang Lei
School of Information and Software Engineering, University of Electronic Science and Technology of China, No.4 Section 2 North Jianshe Road, Chengdu 610054, China

Ekaterina M. Lavrischeva
Moscow Physics-Technical Institute, Dolgoprudnuy, Russia

Muthu Ramachandran
Faculty of Arts, Environment and Technology, School of Computing and Creative Technologies, Leeds Metropolitan University, Leeds, UK.

Emdad Khan
College of Computer and Information Sciences, Al-Imam Muhammad Ibn Saud Islamic University, Riyadh, KSA.

Mohammed Alawairdhi
College of Computer and Information Sciences, Al-Imam Muhammad Ibn Saud Islamic University, Riyadh, KSA.

Jielan Zhang
Department of Information Technology, Yingtan Vocational and Technical College, Yingtan, China

Zhongsheng Qian
School of Information Technology, Jiangxi University of Finance and Economics, Nanchang, China.

Yalu Li
School of Mathematics and Statistics, Shandong Normal University, Jinan 250014, China

Wenhui Dou
School of Mathematics and Statistics, Shandong Normal University, Jinan 250014, China

Haitao Li
School of Mathematics and Statistics, Shandong Normal University, Jinan 250014, China
Institute of Data Science and Technology, Shandong Normal University, Jinan 250014, China

Xin Liu
School of Mathematics and Statistics, Shandong Normal University, Jinan 250014, China

Rui Wang
College of Computer Science, National University of Defense Technology, Changsha, Hunan 410073, China

Wanwei Liu
College of Computer Science, National University of Defense Technology, Changsha, Hunan 410073, China

Tun Li
College of Computer Science, National University of Defense Technology, Changsha, Hunan 410073, China

Xiaoguang Mao
College of Computer Science, National University of Defense Technology, Changsha, Hunan 410073, China

Ji Wang
College of Computer Science, National University of Defense Technology, Changsha, Hunan 410073, China

Ali Sever
Pfeiffer University, Misenheimer, USA

Bin Wen
State Key Lab of Software Engineering, Wuhan University, Wuhan, China.

Keqing He
State Key Lab of Software Engineering, Wuhan University, Wuhan, China.

Jian Wang
State Key Lab of Software Engineering, Wuhan University, Wuhan, China.

Ning Huang
Beihang University, Beijing, China

Xiao Juan Wang
Beihang University, Beijing, China

Camilo Rocha
University of Illinois at Champaign Urbana, USA

Elodie Marie Gontier
Professor of French and History, Paris, France

Islam Elgedawy
Computer Engineering Department, Middle East Technical University, Northern Cyprus Campus, Guzelyurt, Mersin 10, Turkey

LIST OF ABBREVIATIONS

ABox	Assertion Box
ADLs	Architectural Description Languages
AI	Artificial Intelligence
BMC	Bounded Model Checking
BOP	Base of the Pyramid People
CBSE	Components Based Software Engineering
CM	Conceptual Model
CO	Connecting Ontologies
CSEG	Concepts Substitutability Enhanced Graph
CSG	Concept Substitutability Graph
DAO	Decentralized Autonomous Organization
DFA	Deterministic Finite Automata
DFD	Data Flow Diagram
DLs	Description Logics
DPO	Domain Problem Ontology
DSL	Domain Specific Language
DSSA	Domain-Specific Software Architectures
EVM	Ethereum Virtual Machine
FODA	Feature-Oriented Domain Analysis
FOL	First Order Logic
GADTs	Generalized Algebraic Datatypes
GAP	Goal Achievement Pattern
GDP	Gross Domestic Products
GSE	Guidelines Based Software Engineering
IA	Intelligent Agent
ICT	Information and Communication Technologies
IFDS	Integrated Formal Development Support
IoS	Internet of Services

ITP	Inductive Theorem Prover
KB	Knowledgebase
MBPN	Modeling Biasness Process Notation
MDD	Model Driven Development
NFA	Non-deterministic Finite Automata
NLP	Natural Language Processing
NLU	Natural Language Understanding
ODM	Organizational Domain Modeling
ODSD	Ontology-Driven Software Development
OWL	Web Ontology Language
OWL-S	Web Ontology Language for Services
PIM	Platform Independent Model
PP	Program Products
PS	Program Systems
PSM	Platform Specific Models
QoE	Quality of Experience
RAA	Requirements Acquiring & Analysis
RE	Requirements Engineering
RML	Requirement Modeling Language
RoI	Return on Investment
SA	Structured Analysis
SAAS	Software as a Service
SAWSDL	Semantic Annotations for Web Services Description Language
SD	Structured Design
SEBLA	Semantic Engine using Brain-Like Approach
SMP	Sequence Mediation Procedure
SOA	Service Oriented Architecture
SQL	Structured Query Language
STP	Semitensor Product
TBox	Terminology Box
URI	Uniform Resource Identifier
W3C	World Wide Web Consortium
WS	Web Service

WS-BPEL	Web Services Business Process Execution Language
WSCI	Web Services Choreography Interface
WSMO	Web Services Modelling Ontology
XFM	Extreme Formal Modeling
XML	Extensible Markup Language

PREFACE

In informatics, particularly in software and hardware engineering, formal methods are a special type of mathematically-defined techniques that perform specification, development and verification of software and hardware systems. The use of formal methods for software and hardware design is motivated by the expectation that, as in other engineering disciplines, performing appropriate mathematical analysis can contribute to the reliability and robustness of the design.

Formal methods can be described as an application of a fairly wide range of theoretical informatics fundamentals, especially: logical methods, formal languages, automata theory, dynamic system of discrete events and program semantics, but also type systems and algebraic data types on software and hardware specification and verification problems. Formal methods provide the basic methods of symbolic logic in the application of software development, both classical and modern.

They define the elements of syntax and semantics of classical court calculus, and methods of automatic deduction, based on the rule of resolution for court calculus and its modifications (semantic resolution, linear resolution, hyper-resolution), or the Davis-Putnam method. The adopted formal language for reasoning with its subsystems (Horn logic) and supersystems (quantified court accounts), as well as automatic deduction methods developed for them - can serve as a means of modeling and solving a range of problems: artificial intelligence planning, strategic modeling problems (chess), combinatorial (e.g. "four in a row" games), and propositional information and expert systems.

Formal specification and verification methods are widely used during software systems development. Theoretical background for these methods include: process algebras, Petri nets and temporal logic, and finite discrete automata. Formal models of communication between processes are used during the model verification, testing and verification of reactive competing systems. Practical application of formal methods is for language for specification, testing and verification. On the market - there are many model verification tools and software testing tools.

This edition covers different topics from: formal grammars in programming, programming languages semantics, finite automata, and formal methods and semantics in distributed software.

Section 1 focuses on formal methods in programming, describing integrating formal methods in XP (extreme programming) - a conceptual solution, formal methods for commercial, applications issues vs. solutions, why formal methods are considered for safety critical systems, and integration of UML sequence diagram with formal specification methods-a formal solution based on Z.

Section 2 focuses on programming languages semantics, describing declarative programming with temporal constraints, in the language CG, Lolisa: formal syntax and semantics for a subset of the solidity programming language in mathematical tool coq, ontology of domains. ontological description software engineering domain - the standard life cycle, guidelines based software engineering for developing software components, intelligent agent based mapping of software requirement specification to design model.

Section 3 focuses on finite automata, describing the equivalent conversion between regular grammar and finite automata, controllability, reachability, and stabilizability of finite automata: a controllability matrix method, bounded model checking of ETL cooperating with finite and looping automata connectives, an automata-based approach to pattern matching, tree automata for extracting consensus from partial replicas of a structured document.

Section 4 focuses on formal methods and semantics in distributed software, describing building requirements semantics for networked software interoperability, formal semantics of OWL-s with rewrite logic, web semantic and ontology, web services conversation adaptation using conditional substitution semantics of application domain concepts.

SECTION 1: FORMAL METHODS IN PROGRAMMING

Chapter

INTEGRATING FORMAL METHODS IN XP—A CONCEPTUAL SOLUTION

1

Shagufta Shafiq and Nasir Mehmood Minhas

UIIT-PMAS Arid Agriculture University, Rawalpindi, Pakistan

ABSTRACT

Formal methods can be used at any stage of product development process to improve the software quality and efficiency using mathematical models for analysis and verification. From last decade, researchers and practitioners are trying to establish successful transfer of practices of formal methods into industrial process development. In the last couple of years, numerous analysis approaches and formal methods have been applied in different settings to improve software quality. In today's highly competitive software development industry, companies are striving to deliver fast with low cost and improve quality solutions and agile methodologies have proved their

Citation: Shafiq, S. and Minhas, N. (2014), "Integrating Formal Methods in XP (Extreme Programming) - A Conceptual Solution". Journal of Software Engineering and Applications, 7, 299-310. doi: 10.4236/jsea.2014.74029.

efficiency in acquiring these. Here, we will present an integration of formal methods, specifications and verification practices in the most renowned process development methodology of agile i.e. extreme programming with a conceptual solution. That leads towards the development of a complete formalized XP process in future. This will help the practitioners to understand the effectiveness of formal methods using in agile methods that can be helpful in utilizing the benefits of formal methods in industry.

Keywords: Formal Methods, Specification, Verification, Agile, Extreme Programming

INTRODUCTION

Formal methods have proved as a powerful technique to ensure the correctness of software. The growth in their use has been slow but steady and FMs are typically applied in safety critical systems. Use of formal methods requires both expertise and efforts, but this is rewarded if they are applied wisely. It must be seen as good prudently news that Microsoft products increasingly use formal methods in key parts of their software development, particularly checking the interoperability of third-party software with Windows. We believe formal methods are here to stay and will gain further traction in the future [1] .

Formal methods are used for developing software/hardware systems by employing mathematical analysis and verification techniques and often supported by the tools [2] . Mathematical model's steadiness enables developers to analyse and verify these models in any phase of the development process i.e., requirements engineering, specification, design and architecture, implementation, testing and maintenance [2] .

Since their inception and use in the domain of real-time, critical systems, now these methods are finding their way to other widens area of industrial applications especially developing high quality software products [2] .

Traditional software development process can be categorised into three phases: 1) requirement gathering. Sometime, specifications are also incorporating in requirement to get more precise and accurate requirements; 2) phase of design, modelling and implementation; 3) the late phase involves verification and validation process activities.

It can be suggest that formal methods can be effectively used in traditional software development process to get accurate system specifications using the formal specification methods like ASM, B, Z and VDM even these can be effectively used for representation and management of complex system specifications. Formal methods can also be used in software system design by defining formal models to refine the data, abstract function to represent system functionality [2] and to implement. Formal methods can be used for automated code generation and verification from formal models [2] .

Formal methods are always perceived as highly mathematical based processes and can only be used by mathematicians and specialist software experts. This inclination leads towards the limited usage in industry-based software development processes. To change this misconception, a much wider industrial research has to be performed to get the true benefits of formal methods in industry [3] .

In today's fast growing software industries, software industries make every effort to produce fast delivery, with better quality and low cost software solutions [2] . With Lightweight iterative approach with the focus on communication between client and developing team, family of agile methods has turned out as solution to achieve all these goals. Agile methods have a wide range of approaches from development process methods like extreme programming to complete project management process like scrum [4] . These methods have been effectively used in the software industry to develop systems on time and within budget with improved software quality and customer satisfaction [3].

A main reason of not using agile approaches for the development of safety critical systems is the lack of more formal evaluation techniques in agile methods where as safety critical systems require more rigorous development and evaluation techniques to ensure quality products [3] .

As agile approaches less focus on documentation over processes with informal techniques which are often insufficient in determining the quality of safety critical systems [3] , agile methods are still not effectively used to create systems which require more formal development and testing techniques for development [3] .

It has been observed in literature that combination of agile and formal methods can bring best features of both the worlds [5] which can lead towards a better software development solution. In [6] , authors present an evaluation of agile manifesto and agile development principles to show that how formal and agile approaches can be integrated and identify the challenges and issues in doing so. In [3] , authors suggest that agile software development can used light weight formal analysis techniques effectively to bring potential difference in creating system, with formally verified techniques, on time and within budget.

Motivation

It has been observed through literature that application of formal techniques in early phases of software development improves the quality of software artefacts and as a result ensure precise and error free requirement details to the later phases of the development process. As a result the overall cost of a software project is significantly lower because of the minimized error rate. After that formal specifications transformed into concrete models to verify its consistency with the specification that lead towards the implementation.

Till date formal methods couldn't be effectively used in industry based product engineering but it has potential of widespread effectiveness for application development in different domains, whereas agile approaches lack precise techniques for planning and evaluation. A combination of formal methods and agile development processes can significantly encourage the use of formal techniques in industry base software development solutions [3] .

Here in this article, in Section 2 we first describe the use of formal specification and verification techniques with frequently used formal specification languages. We then present an overview of extreme programming in Section 3. Section 4 contains the related research work which shows the integration of formal methods with traditional software development and agile methodologies to support our main concept and the reason of choosing agile process method for our proposed approach. Section V describes our proposed approach.

FORMAL METHODS IN PRACTICE

Formal methods can be generally categorized into two basic techniques and practices i.e. formal specifications and verification [7] .

Formal Specifications can be described as the technique that uses a set of notations derived from formal logic to explicitly specify the requirements that the system is to achieve with the design to accomplish those requirements and also the context of the stated requirements with assumptions and constraints to specify system functions and desired behaviour explicitly [7] .

In design specifications, a set of hierarchical specifications with a high-level abstract representation of the system to detailed implementation specifications are designed, Figure 1 shows that hierarchy of specification levels [7] .

Formal Verification is the use of verification methods from formal logic to examine the specifications for required consistency and completeness to ensure that the design will satisfy the requirements, assumptions and constraints that system required [7] .

There are several techniques available for formal specifications with automated tool support. These automated tools can perform rigorous verification that can be a tedious step in formal methods [7] .

There are many different types of formal methods techniques used in different settings; following are the most commonly used examples of formal specifications, i.e. VDM, B and Z [3] .

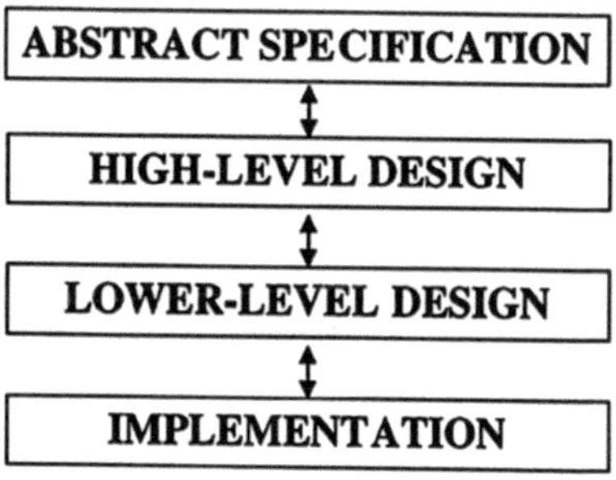

Figure 1. Hierarchy of formal specifications [7] .

VDM

VDM stands for "The Vienna Development Method" and consider as one of the oldest formal methods. VDM is a collection of practices for the formal specification and computational development [8] . It consists of a specification language called VDM-SL. Specifications in VDM-SL based on the mathematical models develop through simple data types like sets, lists and mappings, and the operations, causes the state change in the model [8] .

B-Methods

B-method is another formal specification method consists of abstract notations and uses set theory for system modelling, and mathematical proof for consistency verification between the different refinement phases [7] .

Z

Another most commonly used formal specification language for critical system development, using mathematical notations and schemas to provide exact descriptions of a system. System is described in a number of small Z modules called schemas which can cross refer each other as well as per system required. Each module is expected to have some descriptive informal language text to help users to understand it.

The selection of formal specification language made on the basis of developer's past experience with the selected method or the suitability of any model with respect to the system under develop and its application domain [3] .

EXTREME PROGRAMMING AN AGILE APPROACH

An agile development methodology extreme programming can be define as light weight iterative approach for small and medium size development teams having incomplete or continuously changing requirements. XP works in small iterations with simple practices which focus on close collaboration, simple design to produce high quality products with continuous testing.

Extreme programming created by K. Beck in 1990's, is a set of twelve key practices [9] applied with four core values including communication, simplicity, courage and feedback.

Extreme programming [9] provides a complete solution for product development process and widely accepted for development of industry based and as well as object oriented software systems [10] . With the principles of agile methodology XP proves as novel approach in the family of agile methods that significantly increase productivity that produce high quality error free code [10] -[12] . Figure 2 shows the complete XP development process in traditional settings.

Extreme programming is a Test driven approach. In TDD each user story converted to a test and at the time of release code developed during iteration will be verified though these pre develop tests. This regression testing technique provides a better test coverage at every phase of software development which involves writing of unit tests for each individual task in the system domain [11] .

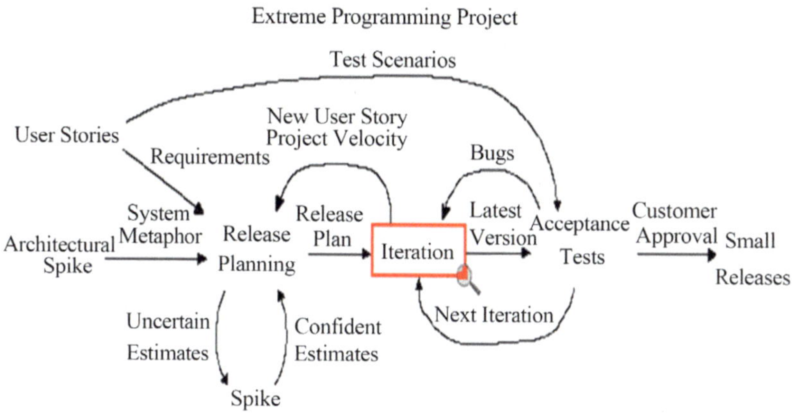

Figure 2. Extreme programming development process. [11]

AGILE APPROACHES TOWARDS FORMAL METHODS

Many efforts have been made to integrate the rapidly growing agile practices, having wide industrial acceptance for producing quality products, with the formal methods having limited industrial acceptance but with strong background, distinctive features and benefits. Table 1 shows the categorization of formal and agile methods.

Richard Kemmerer has made first attempt to investigate the integration approach for conventional development process using formal methods [15] . Kemmerer's work was related to the addition of formal techniques into the different stages of conventional waterfall model, our work is a step towards the integration of formal specification and verification within the agile software development process i.e. extreme programming.

Another study [16] proposed an integration approach for agile formal development approach with the name of XFun, proposed integration of formal notation using X-machine within the unified process.

In [17] author suggests an agile approach, combines the agile practice of tests driven development with formal approach design by contract within XP [17] .

There is another study [18] that proposes the integration of formal method techniques into the traditional Vmodel for refinement of critical aspects of system [18] . The V-model representing the structure of activities providing guideline for software engineers to follow during the process development, while our study focuses on suggesting a complete solution as software development methodology which can be used by the system developers effectively.

In another study [19] authors have made an effort to develop a light weight approach using formal methods with the industry development standards, SOFL [19] . They have used a more graphical notation instead of pure mathematical syntax to define the high level architecture of the system. Later on author refine his proposed approach by developing agile based SOFL method [20] .

Table 1. Agile and formal methods.

Characterizations of Formal and Agile Methods	
Agile Methods	**Formal Methods**
Validation	Verification
Pleasantness	Correctness
Refactoring	Refinement
Concrete	Abstract
Particular	General
Tests	Roofs
Design evolve with code	upfront design
Cowboy coding	Analysis paralysis
Team	Programmer
Beck [9] [10]	Dijkstra [13] [14]

In [21] , authors proposed an extreme Formal Modeling (XFM) (agile formal methodology) to design the specifications from an informal description into a more formal language uses extreme programming approach.

Recently [3] presented an integration approach of formal specification and agile. Suggest a theoretical agile approach using scrum methodology and integrating formal specifications for safety-critical systems. In the proposed method formal specifications are applied within iteration phase and having a developing team that consists of both conventional as well as formal modelling engineers [3] .

Most industrially accepted agile methods i.e. extreme programming [9] and scrum [22] have been used as emergent trends in dealing with core challenges and issues in software development process [23] such as: increase time, with low quality and increase cost at delivery time. [24] . although, it has been observed that agile software development practices are also effectively applied in different development settings for safety critical systems as well [25] [26] . In [26] author argued that Plan driven approaches are better suited for these types of systems. Whereas further studies suggested that the integration of agile approaches with the company's existing plan driven software development activities can be more effective and beneficial for producing safety critical systems [26] -[28] . Another study [29] suggests that integration of agile and CMMI can produce significant difference in developing quality softwares systems.

FORMAL METHODS IN XP: A CONCEPTUAL SOLUTION

Formal methods are set of practices for specification and verification but are not constrained with any specific software development methodology. Mostly published reports focusing on improving the formal techniques for different domain and application development and lacks a complete methodology that can be followed for developing object oriented systems more effectively. Here in this account of literature we are suggesting a conceptual solution for software development industry with the integration of formal techniques into the extreme programming development methodology. Through this, companies will be able to get benefited aspects of both the integrated domains to develop high quality software systems.

Figure 3 shows our proposed approach for development process of XP with the integration of formal methods. Here we have suggesting a conceptual solution.

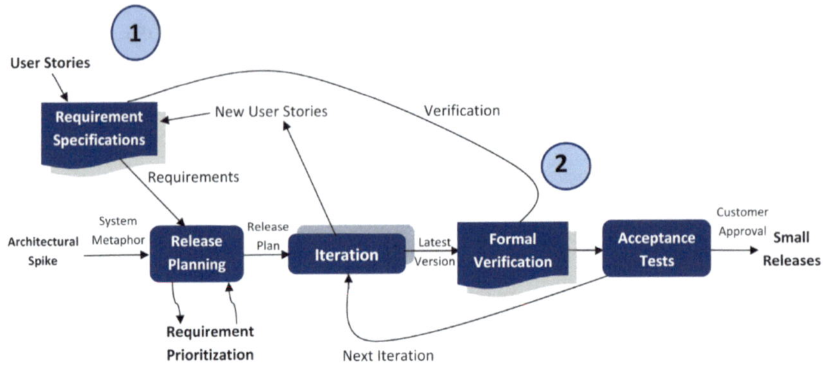

Figure 3. Proposed approach.

User Stories

User stories in XP serves as the basis for defining the functions of the system needs to perform, and to facilitate requirements managements described in an informal manner. In safety circle system use of formal specification techniques consider as the primary intention so the generated requirements can be error free hence using user stories in xp available in informal way needs to be describing through the formal specification techniques to make them more accurate and precise. And serve as the input for the release planning including system metaphor.

Requirement Specification

Formal specification is the description of a program's properties defined in alogic oriented mathematical language. Formal specification is not directly executable but capable of representing higher level of abstraction than a conventional programming language of the system. Here the focus of the formal specification tasks is generating abstract models from user stories and extracting requirements specification for development as well as for validation before the implementation in forthcoming development phases. Figure 4 explains the process of proposed approach for requirement specification.

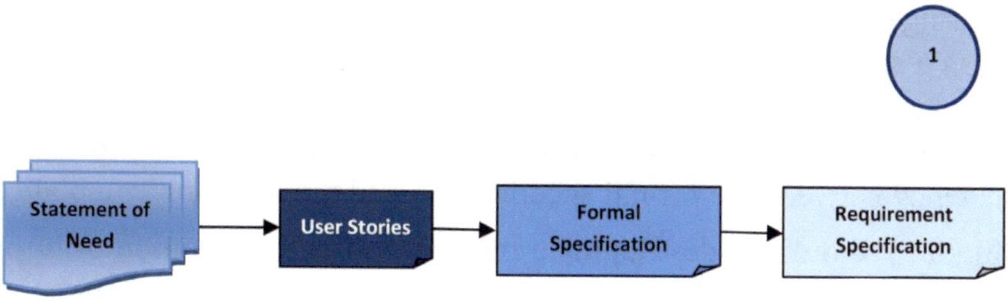

Figure 4. Formal specification in proposed approach.

Release Planning

In our proposed approach, requirements will be extracted from the described formal specification in the earlier phase and then the requirement prioritization will be done through the spike. Once the requirement specification are generated, will be forwarded to release planning phase in which on the basis of each requirement programmers estimate their resources and efforts as per customer needs. At the end of the release planning phase a release plan will be developed for the forth coming iteration. Figure 5 shows the inputs and outputs for the release planning phase.

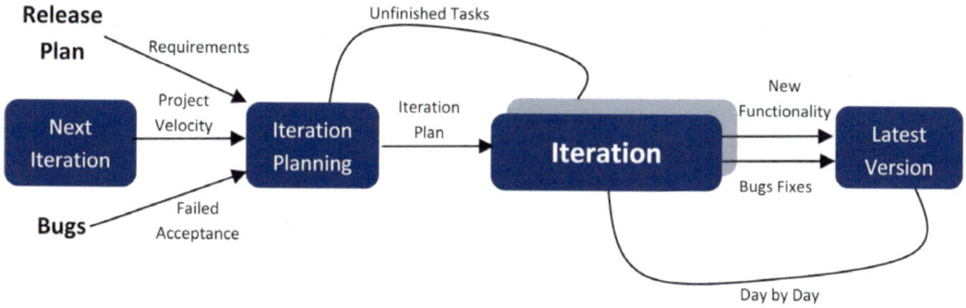

Figure 5. Planning game.

Iteration

During release plan, iteration plan has been forwarded for each iteration phase of 2-4 weeks as per plan. This phase followed through the developing system's functionality incrementally with increasing complexity of the system model. Refactoring and pair programming are the core activities of development iteration in XP process representing described in Figure 6.

During iteration daily stand up meetings and close customer collaboration serve as a source to measure development progress are essential in XP. In addition pair programming and continuous integration supports frequent and automated integration of tests and ensure knowledge sharing for system consistency between the formal specification and the final implementation.

Figure 7 shows the development process with the integration of formal verification phase. In XP, TDD developers are required to write automated unit tests before the code is implemented this can be done by developing formal specifications defines at the earlier stage. And formal verification can be performed easily from the requirement specifications using automated code driven tests. This can be cost and time effective activity.

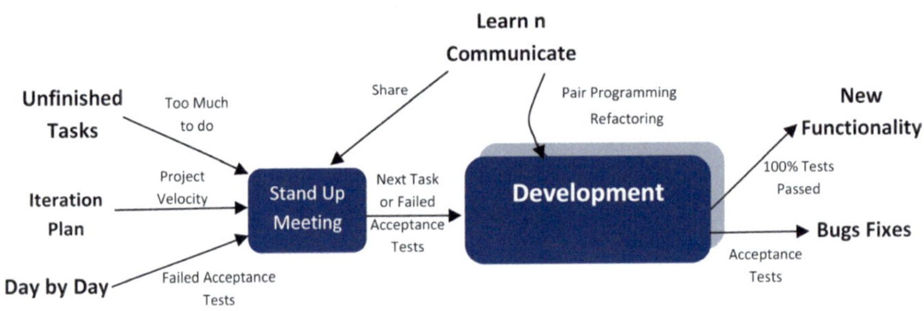

Figure 6. Iteration.

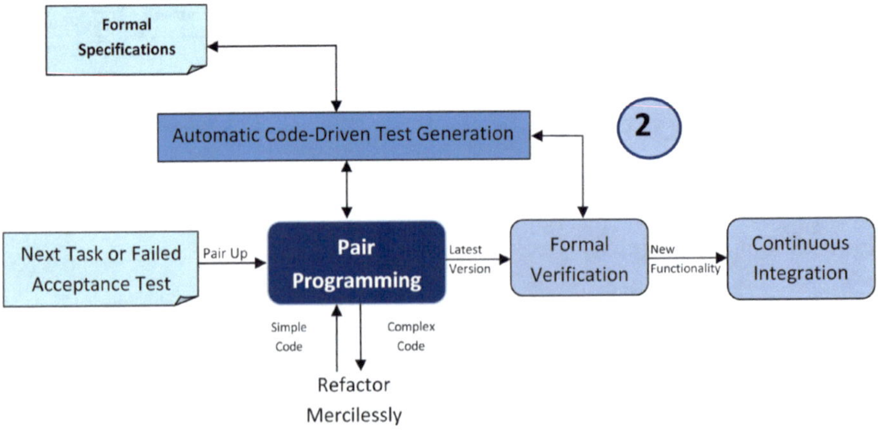

Figure 7. Formal verification.

Continuous Integration

Another very effective practice for producing high quality software in extreme programming is continuous integration in which teams keep the system fully integrated after every coded functionality with passed acceptance test. Once the code has been verified from the formal verification techniques, new unit coded functionality integrated into the system to increase the system quality and efficiency that reduces the system integration issues as well.

EVALUATION OF PROPOSED SOLUTION

To get practical support for our proposed methodology we have conducted a control experiment. To conduct the experiment we selected two groups of undergrad students having good understanding of XP with enough programming skills. Each group was comprised of five members; group-II has added knowledge of VDM and Z specifications as well. We have given a project titled police reporting system to both the groups, Group-I used the traditional XP, while Group-II followed proposed methodology for the system development. Groups were under continuous monitoring to get results with respect to time of system development phase, error rate and product quality. System details are eliminated here just for the sack of prise content and focusing only on the results.

Figure 8 represent the duration in days with the SDLC phases, because XP is iterative methodology and focuses more on development and implementation in contrast formal XP takes more time in planning and designing. Here we have presented cumulative time in days for each phase and implementation phase include development, testing and integration. Use of formal XP took initially longer time but reduces overall development time as compare to traditional XP that lead towards the higher productivity as result shows in Figure 9.

Following Figure 10 present the number of unit functionalities developed in each iteration.

Product quality evaluated on the basis of number of passed acceptance tests after each iteration in Figure 11 shows each iteration results.

Error rate evaluated during each unit development phase Figure 12.

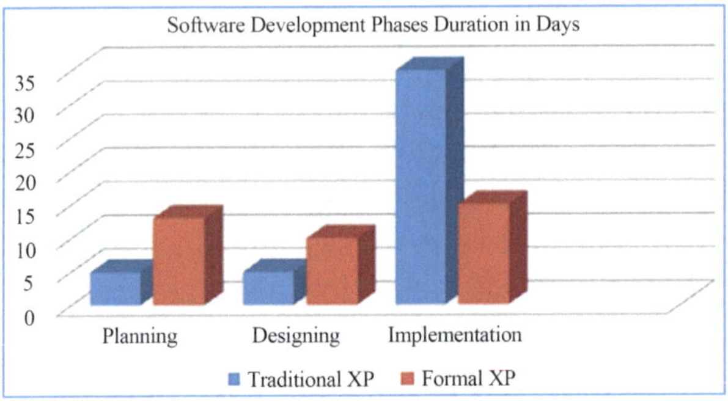

Figure 8. Cumulative number of days for each SDLC phase.

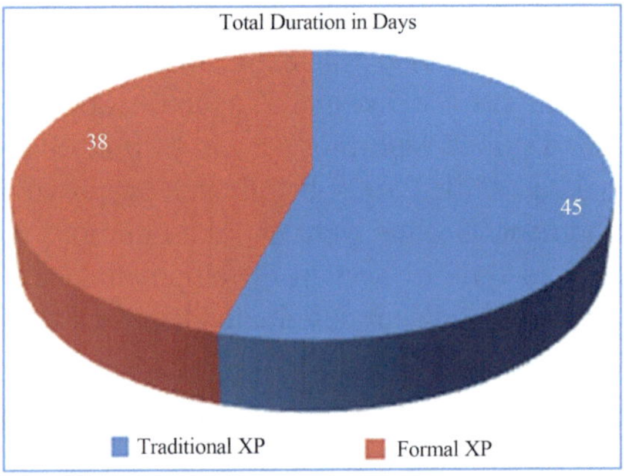

Figure 9. Project duration in days.

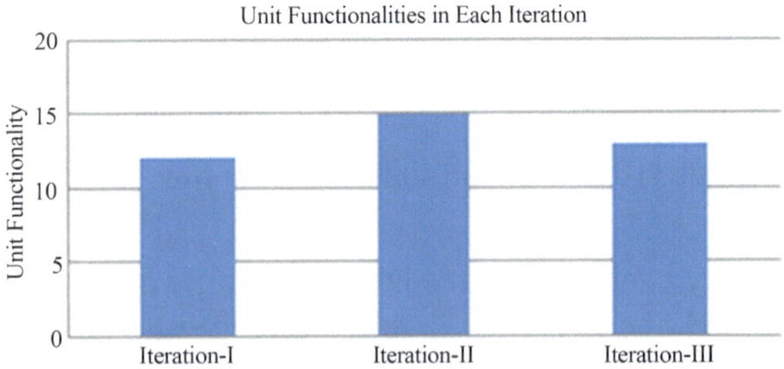

Figure 10. Number of unit functionalities developed during each iteration.

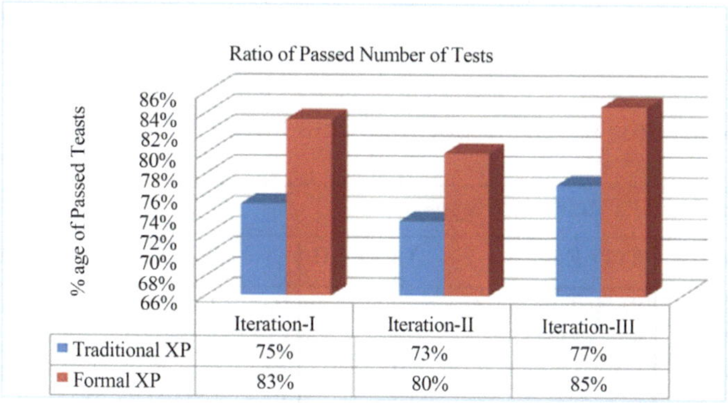

Figure 11. Number of passed acceptance tests after each iteration.

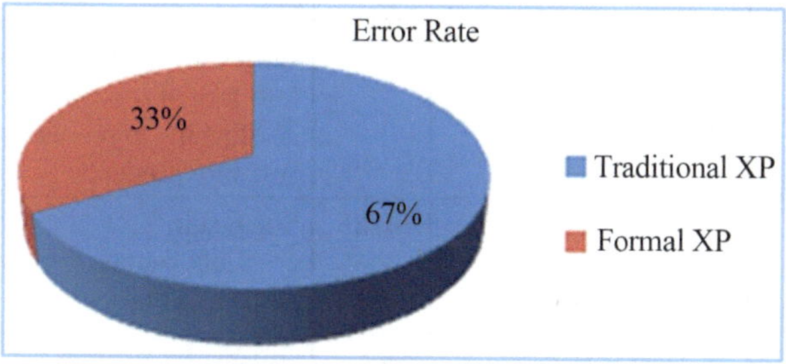

Figure 12. Error rate during development process.

DISCUSSION AND CONCLUSIONS

The work presented here is with the objective of devising a complete development method for the application of formal specification and verification within an agile system development approach i.e. extreme programming.

Many literary and industrial based evidences show the effectiveness of XP in the traditional software development and the literary studies also report some evidences of the successful integration of XP practices in different domains for the system specification and verification, but it lacks a complete development process. In our proposed approach, we have suggested a complete process development for the extreme programming with the formal specification, formal verification techniques and the limited level validation process which supports our notion that formal XP can lead to a higher quality product with reduced error rate and improved time efficiency. Table 2 represents the literature support for the proposed work.

Table 2. Present the use of XP practices with FM to improve software quality.

Conceptual Model's Validation Support				
STUDY ID		**TITLE**	**YEAR**	**SUPPORTING CENCEPT**
STD-1	[30]	Formal Agility. How much of each?	2003	Studied XP practices from the prism of FM to show that how some XP practices can admit the integration of Formal Methods.
STD-2	[31]	Using a formal method to model software design in XP projects	2005	Successfully introduces X-Machine in XP for a succinct and accurate software system
STD-3	[32]	Applying XP Ideas Formally: The Story Card and Extreme X-Machines	2003	Present an approach of using XP story cards and transform those into formal specifications through X-Machine to produce high quality software products.
STD-4	[3]	Scrum Goes Formal: Agile Methods for Safety-Critical Systems	2012	Suggest that XP practices can successfully support the formal method and techniques
STD-5	[33]	Agile Specification Driven Development	2004	Present an approach of using TDD practice for specification driven development that leads towards quality software development.
STD-6	[34]	On the Use of XP in the Development of Safety-Oriented Hypermedia Systems	2003	Uses XP practices in the development of safety-oriented hypermedia systems with formal methods for exhaustive testing
STD-7	[35]	Formal Methods and Extreme Programming?	2003	Evaluated how formal methods overcome the lack of upfront specification and design practices in XP
STD-8	[36]	20 Years of Teaching and 7 Years of Research: Research When You Teach	2008	results from multiple experiments found that there was a measurable quality premium in using XP and uses extreme x-machines for producing high quality products
STD-9	[5]	Formal versus agile: Survival of the fittest?	2009	Suggest that XP practices can get benefit from formal methods
STD-10	[37]	Formal Extreme (and Extremely Formal) Programming	2003	Analyse how Formal Methods (FM) can interact with agile process XP, and suggest that XP practices can improved using FM. can

Application of formal methods is believed as it improves system reliability at the cost of lower productivity whereas XP focuses on more productivity, So, in principle, using process development activities of FM and XP can improve its efficiency like pair programming, daily code development and integration, the simple design or metaphor and iterative development process. On the other hand, one criticism to XP is that it lacks formal or even semi-formal practices. So here in this paper we have tried to devise a XP process utilizing the formal method techniques and the result shows that the appropriate combination results in a more efficient and higher quality development method because each can be able to minimize others' issues.

Informal specification can have ambiguity and irrelevant details and self-contradictory and incomplete abstractions which cannot be handled easily in traditional XP. By defining the requirement specification through the process of formal specification, these issues can be effectively minimized.

The role of manager in XP is to synchronize and manages the work of all team members, with the application of the formal specification and verification. It is required that all managers, trackers and coaches have the implementation knowledge of formal models and their synchronization in the software development process. To make this possible, developer's focus should be on the improvement of the formal specification technique which is easier to be read and understood by the people who don't have the strong mathematical background like the graphical notations used in SOFL or the more familiar C-like syntax for VDM.

The process of formal verification in our proposed approach can be successfully used in minimizing the manual unit tests and regression testing process in traditional XP and reduces the programmer's efforts of continuous testing with efficient time utilization. As suggested in the solution, formal requirement specifications at first step can be easily transformed into automated code driven test generation which leads towards the error free code generation of requirements. There are also many tools available for the system verification developed through formal specifications.

The method suggested in this paper can provide effective guidelines for companies looking for an effective development methodology for formal methods and applying formal specification and/or verification techniques for software development.

LIMITATIONS AND FUTURE WORK

Shagufta Shafiq, Nasir Mehmood Minhas Here we have presented a theoretical model with a very limited evaluation process. But for the industrial applications, it should be verified from the industry. In future, we will try to develop complete specification process that includes how the user stories will be transformed into requirement specifications. In addition to the evaluation of the proposed conceptual solution, several things are needed in order to ensure higher acceptance of formal methods with industry and industrial practices.

REFERENCES

1. Boca, P., Bowen, J.P. and Siddiqi, J.I. (2010) Formal Methods: State of the Art and New Directions. Springer-Verlag London Limited, Berlin. http://dx.doi.org/10.1007/978-1-84882-736-3
2. Woodcock, J., Larsen, P.G., Bicarregui, J. and Fitzgerald, J. (2009) Formal Methods: Practice and Experience. ACM Computing Surveys, 41, 1-36. http://dx.doi.org/10.1145/1592434.1592436
3. Wolff, S. (2012) Scrum Goes Formal: Agile Methods for Safety-Critical System. 2012 Formal Methods in Software Engineering: Rigorous and Agile Approaches (FormSERA), Zurich, 2 June 2012, 23-29. http://dx.doi.org/10.1109/MC.2009.284
4. Schwaber, K. (2004) Agile Project Management with Scrum. Prentice Hall, Upper Saddle River.
5. Black, S., Boca, P.P., Bowen, J.P., Gorman, J. and Hinchey, M. (2009) Formal versus Agile: Survival of the Fittest? IEEE Computer, 42, 37-45.
6. Larsen, P.G., Fitzgerald, J. and Wolff, S. (2010) Are Formal Methods Ready for Agility? A Reality Check. 2nd International Workshop on Formal Methods and Agile Methods, Pisa, 17 September 2010, 13 Pages.
7. Johnson, S.C. and Butler, R.W. (2001) Formal Methods. CRC Press LLC, Boca Raton.
8. Grunerand, S. and Rumpe, B. (2010) GI-Edition. Lecture Notes in Informatics. 2nd International Workshop on Formal Methods and Agile Methods, Vol. 179, 13-25.
9. Beck, K. (1999) Extreme Programming Explained. Addison-Wesley, Boston.
10. Beck, K. (2003) Test-Driven Development. Addison-Wesley, Boston.
11. (2013) Extreme Programming: A Gentle Introduction. http://www.extremeprogramming.org/.
12. Wood, W.A. and Kleb, W.L. (2003 Exploring XP for Scientific Research. IEEE Software, 20, 30-36.
13. Dijkstra, E.W. (1972) Notes on Structured Programming, Structured Programming. In: Dahl, O.-J., Hoare, C.A.R. and Dijkstra, E.W., Eds., Structured Programming, Academic Press, London, 1-82.

14. Dijkstra, E.W. (1968) A Constructive Approach to the Problem of Program Correctness. BIT Numerical Mathematics, 8, 174-186. http://dx.doi.org/10.1007/BF01933419
15. Kemmerer, R.A. (1990) Integrating Formal Methods into the Development Process. IEEE Software, 7, 37-50. http://dx.doi.org/10.1109/52.57891
16. Eleftherakis, G. and Cowling, A.J. (2003) An Agile Formal Development Methodology. Proceedings of 1st South-East European Workshop on Formal Methods, SEEFM'03, Thessaloniki, 20 November 2003, 36-47.
17. Ostroff, J.S., Makalsky, D. and Paige, R.F. (2004) Agile Specification-Driven Development. Lecture Notes in Computer Science, 3092, 104-112.
18. Broy, M. and Slotosch, O. (1998) Enriching the Software Development Process by Formal Methods. Lecture Notes in Computer Science, 1641, 44-61.
19. Liu, S. and Sun, Y. (1995) Structured Methodology + Object-Oriented Methodology + Formal Methods: Methodology of SOFL. Proceedings of First IEEE International Conference on Engineering of Complex Computer Systems, Ft. Landerdale, 6-10 November 1995, 137-144.
20. Liu, S. (2009) An Approach to Applying SOFL for Agile Process and Its Application in Developing a Test Support Tool. Innovations in Systems and Software Engineering, 6, 137-143. http://dx.doi.org/10.1007/s11334-009-0114-3
21. Suhaib, S.M., Mathaikutty, D.A., Shukla, S.K. and Berner, D. (2005) XFM: An Incremental Methodology for Developing Formal Models. ACM Transactions on Design Automation of Electronic Systems, 10, 589-609. http://dx.doi.org/10.1145/1109118.1109120
22. Schwaber, K. and Beedle, M. (2002) Agile Software Development with Scrum. Prentice-Hall, Upper Saddle River.
23. Karlström, D. (2002) Introducing Extreme Programming—An Experience Report. Proceedings 3rd International Conference on Extreme Programming and Agile Processes in Software Engineering, Alghero.
24. Holström, H., Fixgerald, B., Agerfalk, P.J. and Conchuir, E.O. (2006) Agile Practices Reduce Distance in Global Software Development.

Information and Systems Management, 23, 7-18. http://dx.doi.org/10.1201/1078.10580530/46108.23.3.20060601/93703.2

25. ISO TC 22 SC3 WG16 Functional Safety, Convenor Ch. Jung. Introduction in ISO WD26262 (EUROFORM-Seminar, April 2007).
26. Drobka, J., Noftzd, D. and Raghu, R. (2004) Piloting XP on Four Mission Critical Projects. IEEE Software, 21, 70-75. http://dx.doi.org/10.1109/MS.2004.47
27. Wils, A., Baelen, S., Holvoet, T. and De Vlamincs, K. (2006) Agility in the Avionics Software World. 7th International Conference, XP 2006, Oulu, 17-22 June 2006, 123-132.
28. Boehm, B. and Turner, R. (2003) Balancing Agility and Discipline. Addison Wesley, Boston.
29. Pikkarainen, M. and Mäntyniemi, A. (2006) An Approach for Using CMMI in Agile Software Development Assessments: Experiences of Three Case Studies. 6th International SPICE Conference, Luxembourg, 4-5 May 2006, 1-11.
30. Herranz, Á. and Moreno-Navarro, J.J. (2003) Formal Agility, How Much of Each? Taller de Metodologías Ágiles en el Desar-Rollo del Software, VIII Jornadas de Ingeniería del Software Bases de Datos (JISBD 2003), Grupo ISSI, 47-51.
31. Thomson, C. and Holcombe, M. (2005) Using a Formal Method to Model Software Design in XP Projects. Annals of Mathematics, Computing and Tele-Informatics, 1, 44-53.
32. Thomson, C. and Holcombe, W. (2003) Applying XP Ideas Formally: The Story Card and Extreme X-Machines. 1st South-East European Workshop on Formal Methods, Thessaloniki, 21-23 November 2003, 57-71.
33. Ostroff, J.S., Makalsky, D. and Paige, R.F. (2004) Agile Specification-Driven Development. Lecture Notes in Computer Science, 3092, 104-112.
34. Canos, J., Jaen, J., Carsi, J. and Penades, M. (2003) On the Use of XP in the Development of Safety-Oriented Hypermedia Systems. Proceedings of XP 2003, Genova, 25-29 May 2003, 201-203.
35. Baumeister, H. (2002) Formal Methods and Extreme Programming. Proceedings of Workshop on Evolutionary Formal Software Development, in Conjunction with FME, Copenhagen, 189-193, 1-2.

36. Holcombe, M. and Thomson, C. (2007) 20 Years of Teaching and 7 Years of Research: Research When You Teach. Proceedings of the 3rd South-East European Workshop on Formal Methods, Thessaloniki, 30 November-1 December 2007, 1-13.

37. Herranz, A. and Moreno-Navarro, J.J. (2003) Formal Extreme (and Extremely Formal) Programming. In: Marchesi, M. and Succi, G., Eds., 4th International Conference on Extreme Programming and Agile Processes in Software Engineering, XP 2003, LNCS, No. 2675, Genova, 88-96.

Chapter 2

FORMAL METHODS FOR COMMERCIAL APPLICATIONS ISSUES VS. SOLUTIONS

Saiqa Bibi, Saira Mazhar, Nasir Mehmood Minhas, and Irfan Ahmed

UIIT-PMAS Arid Agriculture University, Rawalpindi, Pakistan

ABSTRACT

It was advocated that in 21st century, most of software will be developed with benefits of formal methods. The benefits include faults found in earlier stage of software development, automating, checking the certain properties and minimizing rework. In spite of their recognition in academic world and these claimed advantages, formal methods are still not widely used by commercial software industry. The purpose of this research is to promote formal methods for commercial software industry. In this paper we have identified issues in use of formal methods for commercial applications and devised strategies to overcome these difficulties which will provide motivations to use formal methods for commercial applications.

Citation: Bibi, S. , Mazhar, S. , Minhas, N. and Ahmed, I. (2014), "Formal Methods for Commercial Applications Issues vs. Solutions". Journal of Software Engineering and Applications, 7, 679-685. doi: 10.4236/jsea.2014.78062.

Keywords: Formal Methods, Commercial Applications, Issues of Formal Methods

INTRODUCTION

Formal languages are the languages in which syntax and semantics are properly defined by using mathematical notations. Formal languages are of mathematical nature so; they raise the assurance on the system by reducing uncertainty in the specification of system [1] . Commercial application softwares are designed for vending to serve a commercial need or on the demand of customer. These applications are larger in size.

Formal methods are actually precise techniques; tools support is provided for development of software as well as hardware systems. Mathematical techniques of formal methods enable developer to examine and prove models at any stage of the software development life-cycle: gathering requirements, specification, architecture, design, implementation, testing, maintenance, and development [2] .

The purpose behind the promotion of formal approaches was the detonation in software entanglement that began in the 1960s. Around then, software systems were rapidly getting to be more complex, however advance in devices and systems for improvement completed does not keep pace. Accordingly, there was a clear need for new techniques that might permit engineers to understand this complication. Formal methods made this practical by giving a mathematical framework for investigating projects [3] .

Formal methods are used in software requirement specification: preparing an accurate report of what the software needs to do, and avoiding conditions on how it is to be attained [2] .

Use of formal methods at the stage of formal specification can create very useful documentation [4] . A specification is a practical agreement between vendor and customer to offer them equally with a general acceptance of the software requirement. Absolute system specifications are essential because a design and implementation of system originate their quality in detail from the requirements specification. Modern research is representing the obvious advantages of formal and mathematical techniques for software requirements detain and design. Methods used for such mathematical technique to software requirements capture and design approach are jointly called formal methods for specification of software [5] .

At the implementation stage, the use of formal methods is utilized for checking code of software. Each system particular match totally states an accuracy hypothesis that, if most conditions are fulfilled, the project will attain the impact depicted by documentation. Confirmation of code is the endeavour to demonstrate this theorem, or in any event to figure out the reason of theorem ignorance to hold. The inductive statement strategy for project confirmation was imagined by Floyd and Hoare and involves explaining the system with scientific declarations, which are associations that are between the variables of system and the beginning principles; each one time control, achieves a specific focus of the system. Coding can additionally be produced immediately from models provided by formal methods [3] .

For many years, it was advocated that applying formal methods in software development would help industry congregate its goals of produce an enhanced software process and increase quality of software. The benefits that have been cited include finding defects in earlier stage of software development, automating checking of certain properties, and minimizing rework; despite these claimed benefits and usability in each phase of software development i.e. requirements specification, software architecture, software design, implementation, maintenance, testing, and evolution. In spite of its claimed benefits formal methods are still not widely used by commercial software companies. In this paper we have described the challenges in the use of formal methods for commercial applications industry, devised strategies to overcome these difficulties which will provide motivations to use formal methods for commercial applications. We have divided our work into two sections. In the first section we have identified barriers of formal methods for commercial applications and devised strategies to overcome these barriers and in the second section we draw motivations for the use of formal methods in commercial applications.

FORMAL METHODS: ISSUES VS. SOLUTIONS

Formal methods are still not seen widely in use for commercial applications due to a number of issues:

Issue-1: Lack of Skilled Persons with Mathematical Background

Formal methods for commercial applications development are not often commonly used or it does not well understood by many of the software engineers because implementation of formal methods demands the unambiguous concepts of discrete mathematics [6] . Formal verification needs mathematical skills not only due to the complex interactions between program subcomponents, but also for the deficiencies in current verification interfaces. These skill barriers economically resist the verification process by avoiding the selection of less skilled persons [7] .

Solution: Tutorials & Trainings Help to Build Mathematical Knowledge

To build knowledge for formal methods in software development organizations on their own, a high quality tutorials and self-learning materials can help. Self-training materials allow independent developers to become familiar more easily with formal methods on their behalf [3] .

Issue-2: Expensive

Business managers have faith that formal methods can enhance the software quality, but formal methods are not widely used because these methods are considered costly and unfeasible [8] .

The most considerable doubt in formal methods, mostly from a perspective of management, is that these methods are expensive because implementing successful formal methods in an organization also need to purchase the tools for supporting these methods, training of engineers and designers, and effort and time to incorporate formal methods in the existing software development process, with other expenditure [5] .

Solution: Ramp-Up Cost for Formal Methods Pay off over Many Projects

The ramp-up cost plays significant role for the implementation of formal methods. The vast number of software development tools and methodologies place their focus on long term cost savings. There is considerable support in the literature that formal methods provide real benefits in this area, increasing system reliability, and thereby decreasing long-term support and maintenance costs, while simultaneously maintaining or even decreasing initial development costs. So while formal methods may not be appropriate or cost effective for one-time use on a particular project, the evidence suggests that the initial investment can pay off over many projects [3] .

Issue-3: The Inadequate Tool Support

In United States a fact for formal methods indicates lack of tool support as a barrier. They also highlighted it the key reason for the lack of appreciation of real world considerations to be the part of formal method community and therefore they are still not used in commercial industry [9] [10] .

Solution: Formal Methods Supported by Variety of Tools

There are many tools available that provide support to formal methods such as Finite State Machines, VDM, Z, and OBJ. These tools are used to increase the productivity and accuracy in all the phases of a system. These tools possess different type of characteristics and used in industry according to the nature or requirement of the systems [11] .

In early 80s, tools for Computer-Aided Software Engineering and Computer-Aided Structured Programming were seen as the mean of increasing programmer's productivity and for reducing programming "bugs". Now the Tool support can see as a source of increasing productivity and accuracy for formal developments [12] .

It is our expectation that in near future more focus will be paid to Integrated Formal Development Support (IFDS) Environments that will be helpful in support of many of the phases of formal development. These toolkits will provide Integrated Programming Support Environments that will be support in configuration management and version control and facilitate all of the process activities and large scale developments more harmoniously.

Issue-4: Increase in Development Cycle

Although many established advantages of formal methods, these are badly accepted by industrial professionals. Many causes have been submitted for this situation one of them is that they increase the software development cycle [13] .

Solution: Early Error Detection Helps in Reduce Development Time

It reduces the development time by applying testing techniques in earlier phases of the lifecycle [14] . In the seventies, was reporting that over half the software development time was devoted to testing activities. Formal methods offer new possibilities for verification i.e. model checking. It enables us in more effective identification of software defects which allows reducing verification time [7] . The use of a formal methods or model removes ambiguities in specification and thus reduces the chance of errors being introduced during software development. Thus this reduces the testing effort and time [15] .

FORMAL METHODS: MOTIVATIONS FOR COMMERCIAL APPLICATIONS

Formal Method Maximize Automation with Automated Tools

Automated tools allow producing models for verification promptly and in a convenient way directly from the design of models [16] . Modern progress in analysis tools of formal methods have made it realistic to verify significant properties formally to provide guarantee that design faults are identified and approved correctly in early stage of software development lifecycle [17] . A survey was presented in 2010 for effect of formal methods in software industry. The satisfaction level with automated tools is greater than 80% shown in Figure 1 [8] .

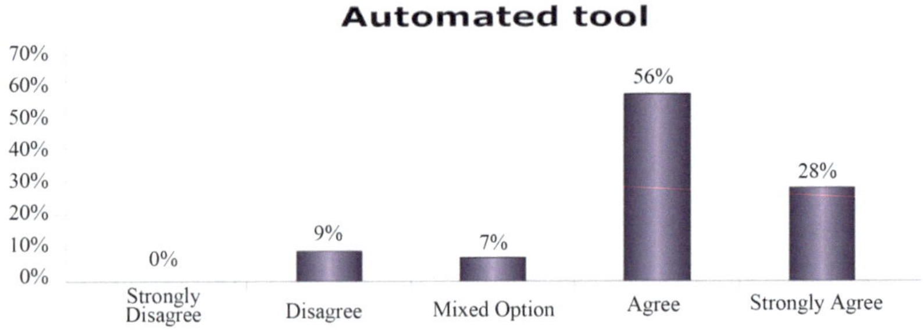

Figure 1. Satisfaction level with automated tools [8] .

Automatic Verification Improvement

By using the formal verification approach better verification quality can be achieved with 70 percent less time and effort [18] as compare to other approaches. Only 30% effort is required by using formal techniques. Formal verification results in 2007-2010 survey are shown in the Table1

Table 1. Formal verification vs. simulation [18] .

Subtasks	Simulation	Formal verification
Preparation	Simulation script is generated by register tool	Properties are generated by register tool
Execution	3 days of simulation time	1.5 days for automatic set-up of 31 register block set-up and exhaustive verification of 12,600 properties
Analysis effort	60,000 entries to be analysed	No additional effort
Quality of analysis	Not-exhaustive, semi-automatic, error prone	Exhaustive, automatic, fail-safe
Total effort	3 days compute time + 2 days manual effort	1.5 days compute time (70% less than simulation)

Formal Methods Reduce Cost

From formal specification, we can thoroughly gain effective test cases directly from the requirement. Test cases generation is a cost effective way [19] . Effects of Formal Methods on cost are presented in a survey in 2010 shown in Figure 2.

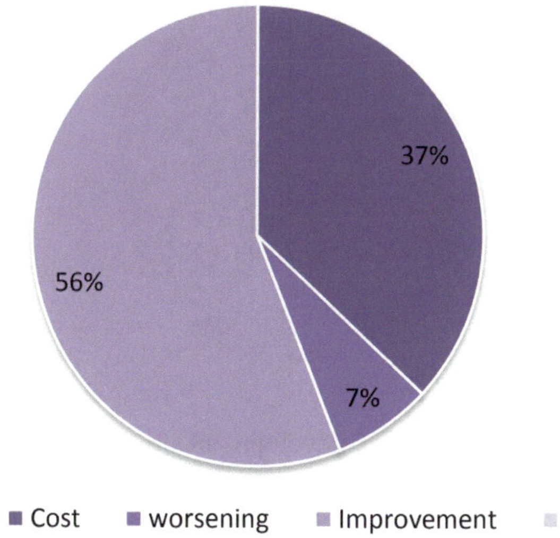

Figure 2. Formal approaches' effect on cost [8] .

Formal Methods Reduces Defects at Early Stage

Formal specification produces accurate requirements and designs so that it reduces the chances of unintentional fault injections. Correctness of software system is also proved by formal verifications. Axiomatic correctness is one of verification methods [20] . Formal description forces the writer to ask all sorts of questions that would be delayed until phase of coding. This helps to decrease the errors that may occur in coding phase [16] . The results presented in 2011 indicate that applying ASD as a formal technique for developing controls software could results in fewer defects [21] . 63% defects are reduced by using formal techniques. The Table 2 given below shows that the defects are reduced where Formal techniques are applied:

Figure 3 also present the effectiveness of formal methods as compare to traditional approach. Formal specification, formal verification techniques can lead to a higher quality product with reduced error rate and improved time efficiency [22] .

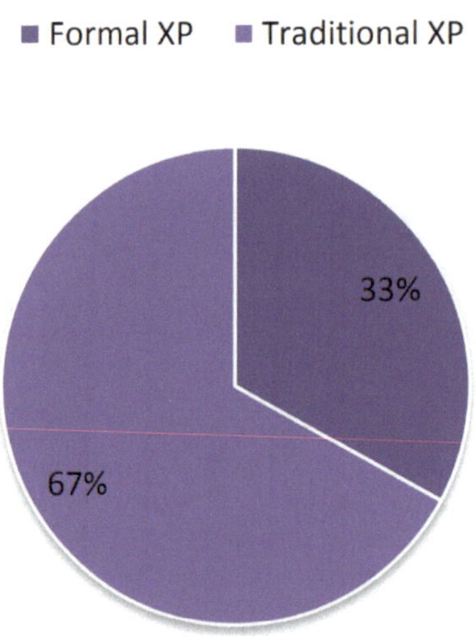

Figure 3. Less error rate during development process using formal approach [22] .

Table 2. Defects are reduced with formal approaches [21] .

ASD used	Unit	Lines of code				Defects			
		Manual LOC	ASD LOC	Total LOC	ASD%	Manual defects	ASD defects	Total defects	Defects/ KLOC
No	Acquisition	6140	0	6140	00.00%	33	0	33	5375
No	BEC	7007	0	7007	00.00%	44	0	44	6279
No	EPX	7138	0	7138	00.00%	7	0	7	0.981
No	FEAdapter	13190	0	13190	00.00%	18	0	18	1.365
YES	FEClient	15462	12153	27615	44.01%	9	2	11	0.398
YES	Orchestration	3970	8892	12862	69.13%	3	4	7	0.544
No	QA	23303	0	23303	00.00%	90	0	90	3.862
No	Status Area	8969	0	8969	00.00%	52	0	52	5.798
No	TSM	6681	0	6681	00.00%	7	0	7	1.048
No	UIGuidance	20458	0	20458	00.00%	23	0	23	1.124
No	Viewing	19684	0	19684	00.00%	294	0	294	14.936
YES	XRayIP	14270	2188	16458	13.29%	27	0	27	1.641

Formal Methods Improves Quality

A survey presented in 2010 for effect of formal methods in software industry. It is presented that use of formal techniques improves quality of software in industry, 92% cases reported that quality is increased against the other approaches, and there is no single case that reported a decline in software quality. Figure 4 shows the effect of formal methods on quality of software [8] .

Overall effect of formal methods in software industry is shown in Figure 5. By applying formal methods in commercial software industry batter results can be achieved as compare to other approached as shown by survey results. Researchers are hopeful about the flourishing use of formal approaches for commercial software industry in future.

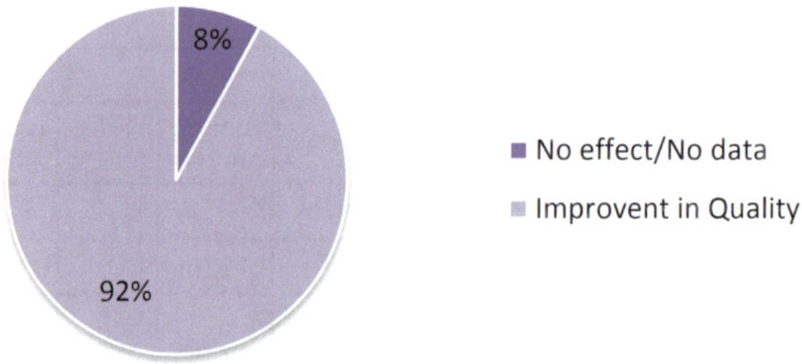

Figure 4. Formal approaches' effect on quality [8] .

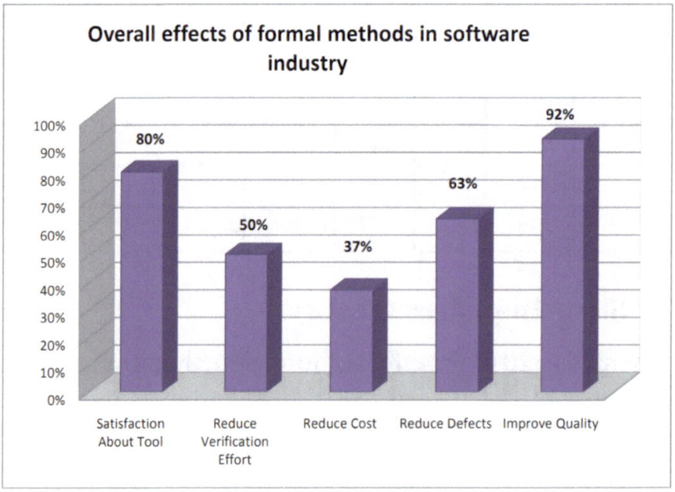

Figure 5. Overall effects of formal methods in software industry.

CONCLUSION

For many years, it was advocated that applying formal methods in software development would help industry congregate its goals of producing an enhanced software process and increasing quality of software. De-Saiqa Bibi, Saira Mazhar, Nasir Mehmood Minhas, Irfan Ahmed spite claimed benefits and usability in each phase of software development, formal methods are still not widely used by commercial software companies. Formal methods have not been widely used in industry due to a number of barriers. We have identified barriers of formal methods for commercial applications and then provide their solution. Formal methods offer several advantages i.e. maximize automation with automated tools, automatic verification improvement cost saving, defect reduction and quality improvement. These benefits are the stimulus to use formal methods in commercial software industry. By applying formal methods in commercial software industry batter results can be achieved as compared to other approaches shown by survey results. The purpose of this research is to promote formal methods for commercial application software in industry.

REFERENCES

1. Sammi, R., Rubab, I. and Qureshi, M.A. (2010) Formal Specification Languages for Real-Time Systems. 2010 International Symposium in Information Technology (ITSim), Kuala Lumpur, 15-17 June 2010, 1642-1647. http://dx.doi.org/10.1109/ITSIM.2010.5561643
2. Woodcock, J.I.M. and Bicarregui, J. (2009) Formal Methods: Practice and Experience Engineering College of Aarhus. ACM Computing Surveys, 41, 1-40.
3. Geer, P.A. (2011) Formal Methods in Practice: Analysis and Application of Formal Modeling to Information System.
4. Bowen, J.P. and Hinchey, M.G. (2006) Ten Commandments of Formal Methods... Ten Years Later. Computer, 39, 40-48.
5. Sommerville, L. (2009) Chapter 27 Formal Specification.
6. Stidolph, D.C. and Whitehead, J. (2003) Managerial Issues for the Consideration and Use of Formal Methods. Lecture Notes in Computer Science, 2805, 170-186.
7. Schiller, T.W. and Ernst, M.D. (2012) Reducing the Barriers to Writing Verified Specifications. ACM SIGPLAN Notices, 47, 95-112.
8. Fulara, J. and Jakubczyk, K. (2010) Practically Applicable Formal Methods. Lecture Notes in Computer Science, 5901, 407-418.
9. Stidolph, D.C. (2003) When Should Formal Methods Be Used?
10. Jhala, R. and Majumdar, R. (2009) Software Model Checking. ACM Computing Surveys, 41, 1-54. http://dx.doi.org/10.1145/1592434.1592438
11. Kefalas, P., Eleftherakis, G. and Sotiriadou, A. (2003) Developing Tools for Formal Methods.
12. Bowen, J.P. and Hinchey, M.G. (1994) Seven More Myths of Formal Methods?: Dispelling Industrial Prejudices. Lecture Notes in Computer Science, 873, 105-117.
13. Knight, J.C., Dejong, C.L., Gibble, M.S. and Nakano, L.G. (1998) Why Are Formal Methods Not Used More Widely?
14. Hierons, R.M., Bogdanov, K., Bowen, J.P., Cleaveland, R., Derrick, J., Dick, J., et al. (2002) Using Formal Specifications to Support Testing. ACM Computing Surveys, 41, Article No. 9.

15. Singh, M. (2013) Formal Methods: A Complementary Support for Testing. International Journal of Advanced Research in Computer Science and Software Engineering, 3, 320-322.
16. Cofer, D., Whalen, M. and Miller, S. (2008) Model-Based Development. 1-8.
17. Whalen, M., Cofer, D., Miller, S., Krogh, B.H. and Storm, W. (2008) Integration of Formal Analysis into a Model-Based Software Development Process. Lecture Notes in Computer Science, 4916, 68-84.
18. Knablein, B.J. and Sahm, H. (2010) Contributed Article Automated Formal Method Verifies Highly-Configurable HW /SW Interface. 1-7.
19. Batra, M., Malik, A. and Dave, M. (2013) Formal Methods?: Benefits, Challenges and Future Direction. Journal of Global Research in Computer Science, 4, 2-6.
20. Alves, M.C.B., Dantas, C.C. and Silva, R.B. (2007) A Topological Formal Treatment for Scenario-Based Software Specification of Concurrent Real-Time Systems. 1-7.
21. Groote, J.F., Osaiweran, A.A.H. and Wesselius, J.H. (2011) Benefits of Applying Formal Methods to Industrial Control Software. 1-10.
22. Shafiq, S. and Minhas, N.M. (2014) Integrating Formal Methods in XP—A Conceptual Solution. Journal of Software Engineering and Applications, 7, 299-310.

Chapter

WHY FORMAL METHODS ARE CONSIDERED FOR SAFETY CRITICAL SYSTEMS?

3

Monika Singh[1], Ashok Kumar Sharma[1], and Ruhi Saxena[2]

[1]Faculty of Engineering & Technology (FET), Mody University of Science & Technology, Sikar, India
[2]Computer Science & Engineering, Thapar University, Patiala, India

ABSTRACT

Formal methods are the mathematically techniques and tools which are used at early stages of software development lifecycle processes. The utter need of using formal methods in safety critical system leads to accuracy, consistency and correctness in proposed system. In safety critical real time application, requirements should be unambiguous and very accurate which can be achieved by using mathematical theorems. There is utter need to focus on the requirement phase which is the most critical phase of SDLC. This paper focuses on the use of Z notation for incorporating the accuracy,

Citation: Singh, M. , Sharma, A. and Saxena, R. (2015), "Why Formal Methods Are Considered for Safety Critical Systems?", Journal of Software Engineering and Applications, 8, 531-538. doi: 10.4236/jsea.2015.810050.

consistency, and eliminates ambiguity in safety critical system: Road Traffic Management System as a case study. The syntax, semantics, type checking and domain checking are further verified by using Z/EVES: a Z notation type checker tool.

Keywords: Formal Methods, Safety Critical System, Z Notation, Z/EVES, Syntax & Type Checking, Domain Checking

INTRODUCTION

Formal specification languages are mathematically based on languages which are adequately used for construction of accurate, consistent and unambiguous systems and software. As formal methods are equipped with tool, which can be used for both the prospective i.e. describing a system and later on for analyzing their functionalities. The major obstacles behind formal methods to be used in practices frequently are the time spent on specification [1] [2] . Nevertheless, formal methods do not guarantee correctness, but their use emphasize to increase the understanding of a system by divulging errors or facets of incompleteness that may be expensive to correct them at any later point of time. However, formal methods play a critical role in safety critical system as they focus on refinement of requirements in the early stage of development which consequently increase the system's accuracy and consistency. Various formal languages are used for this purpose like VDM, B-Methods, Petri Net, and Z notation etc. Z notation is a model based on formal specification language which uses the set theory and first order predicates [3] .

A lot of work has been done in this area of formal analysis of UML diagrams with formal approaches [4] -[8] . In article 8, UML based framework is presented to develop web applications. [5] represents the verification properties by HOL theorem prover. A formalization approach is developed for UML class diagrams in [6] . The paper [7] advocates how the formal methods can be used for safety properties of real time critical application such as railways. [8] explains an integrated approach of Z notation and Pertinet for analysis of safety critical properties.

In this article, Z notation is used for formal analysis of safety critical system i.e. Road Traffic Management System which is further verified by using the Z/EVES tool.

PROPOSED APPROACH & METHODOLOGY

In the first part of this section, the proposed approach is discussed. Then the tool and methodology used are discussed in section.

Proposed Approach

Figure 1 defines the proposed approach for designing the safety critical system using the formal methods.

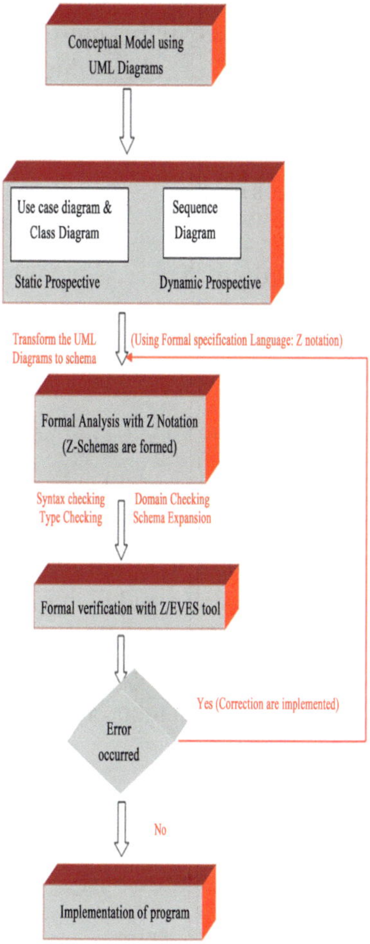

Figure 1. The proposed approach for formal analysis of safety critical application.

Z/EVES

This tool is used for verifying the specification written in Z notation language. This verification includes syntax, semantics, type checking, and domain checking of the given system's specification. Z/EVES present two type of interface: graphical user interface and the command line interface [3] [9] . In this paper, we used the graphical user interface for verifying and composing the specification which were written in Z notation language. Moreover, Z/EVES propose two mode of operations i.e. "Eager" and "Lazy". In our article we use the "Eager" mode since in this mode a paragraph is checked if and only if all the previous ones are checked which is highly recommended for safety critical real time application. By using Z/EVES, following can be done:

- syntax and type checking;
- schema expansion;
- precondition calculation;
- domain checking;
- general theorem proving.

UML

Unified Modeling language is in fact the blue prints for the system to be developed. It provides a better way to understands the requirements of the propose system. UML consists of nine diagrams which are used for capturing the both aspects of the system i.e. static and dynamic [10] -[12] . This paper aims at the static behaviour by composing the use case diagram of RTMS system which is further verified by using Z/EVES type checker tool. The conceptual model of Road Traffic Management System (RTMS) is given in Figure 2.

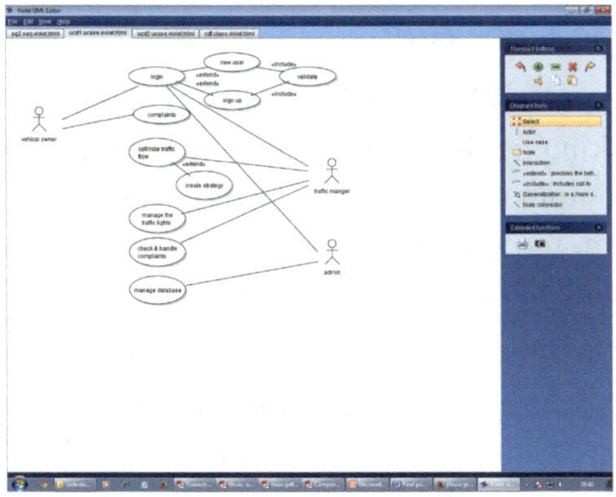

Figure 2. Use case diagrams of vehicle owner.

FORMALIZATION OF USE CASE DIAGRAM USING Z/EVES

Z schema is the notion for structuring the specification including the pre, post condition and the list of invariant & variables. Z schema has two parts i.e. declaration part and predicate part. The Z schema has both declaration as well as predicate part that is shown in Figure 3.

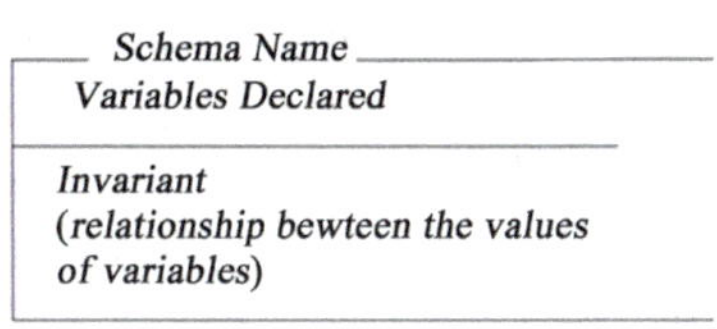

Figure 3. State space of schema.

The above part of central line consists of variables declaration and the below part of line describes the relationship the variable's various values. This paper emphasis on three main characteristics of formal analysis of safety critical system which are:

1. Syntax & Type checking;
2. Schema Expansion; and
3. Domain checking.

1) Syntax & Type Checking

The syntax and type checking facility is provided by the Z/EVES tool. The syntax & type checking facility enables that the syntax used in Z specification is correct which is automatically done by Z/EVES tool. In case of road traffic management system, the schema of Vehicle Owner is considered for syntax & type checking which is consists of two variables:

- Vowner is the set of names with RTMS registered.
- Regist Vowner is the function which when implemented on a particular Vehicle Owner name, provides the unique registration number associated with the person.

In Figure 4, the schema for Vehicle Owner with basic data type is given: [Name, Seqchar].

In Vehicle Owner schema, a partial function named "Regist Vowner" is defined which maps the corresponding vehicle owner with a registration number i.e.

Regist Vowner: Name→ Seqchar

Moreover, "Regist Vowner" is a one-to-one function which maps Vehicle Owner name with registration number. Since it is a one-to-one function, therefore every Vehicle Owner has a unique registration number and consequently, would be no ambiguity. The schema of Vehicle Owner is further verified by Z/EVES tool for syntax & type checking in Figure 5. The left most columns' value "Y" shows that the schema is implemented using correct syntax. If there would be any syntax error, it shows "N" instead of "Y" in syntax column [9] .

Vehicle Owner

Vowner: P Name
Age: PZ
Phno: Seqchar
regist Vowner: Name → Seqchar

Vowner = dom regist Vowner

Figure 4. Vehicle Owner schema with invariants.

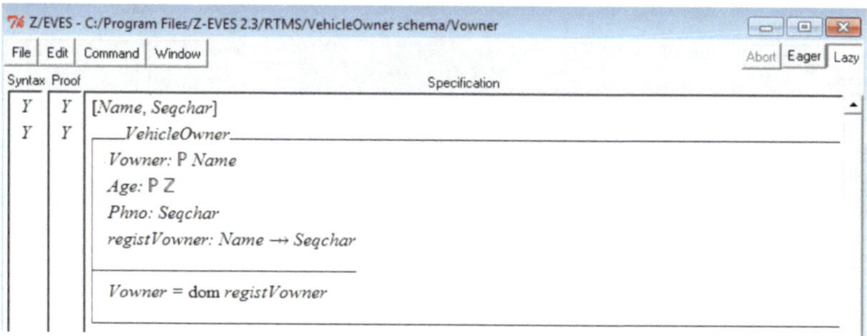

Figure 5. Syntax checking of Vehicle Owner schema by Z/EVES.

2) Schema Expansion

The schema expansion facility enables to extend the functionality of system and helps in understanding the complex schema structure in detail. Initially, the list of registered vehicle owner in RTMS is empty which is depicted by the "Init Vehicle Owner" schema in Figure 6.

Since the lower part of the schema explain the relation between the variables, the function Regist Vowner is assigned a value "φ", and means initially there is no registered vehicle owner in RTMS. Figure 7 shows the Z/EVES result of "Init Vehicle Owner".

Now, the Vehicle Owner may perform a list of tasks like: Login. If the Vehicle Owner is Login first time, he/she has to register him/her; otherwise he/she will sign in. In Figure 8, the schemas of Login operation is implemented.

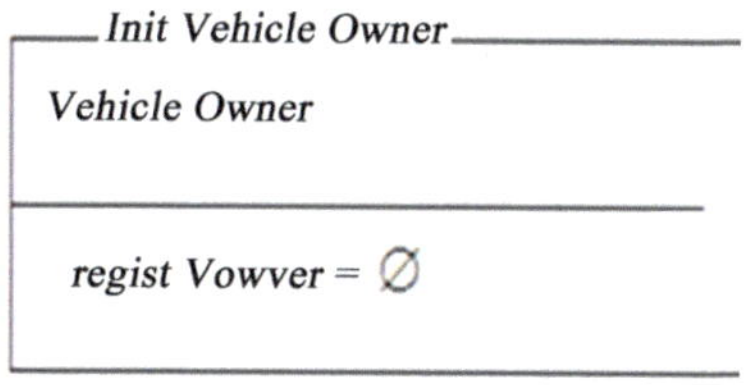

Figure 6. Initial state space of schema Vehicle Owner.

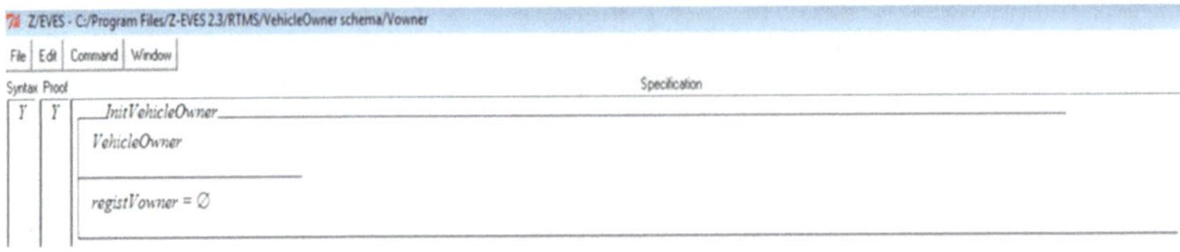

Figure 7. Initial Vehicle Owner schema.

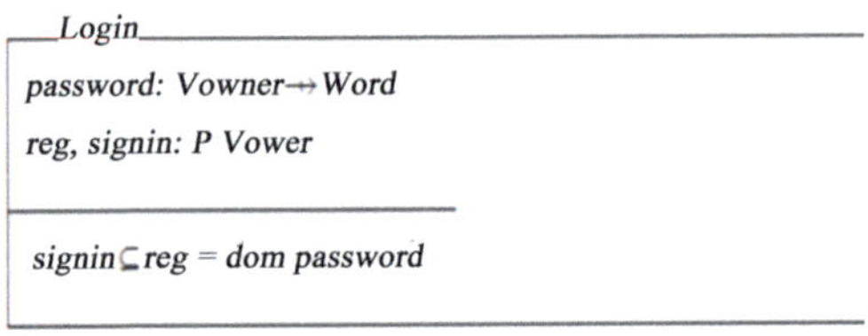

Figure 8. State space of schema Login.

In this schema:

Password: Vowner→Word

"Password" is a function which associates a username to password. Nevertheless, it is a one-to-one function which in turn provides accuracy and correctness to system. Now Signin set and registered set both is the member of power set of Vehicle Owner which is mathematically shown by using set theory as following.

Signin, Reg: P Vowner

Also the Signin set is a subset of registered set and the registered set having the values which are there in domain of "password" function i.e.

Signin ⊆ Reg = Dom Password

Initially, Login schema is empty which is here explained by assigning a value "φ" to both the set whether it's a registers one or a new one i.e.

Reg = φ; Signin = φ

This is called schema expansion which is one of the key features of Z/EVES tool i.e. from "Init Login" schema to "Login" schema.

In Figure 9, the schema expansion is shown and verified by Z/EVES as follow.

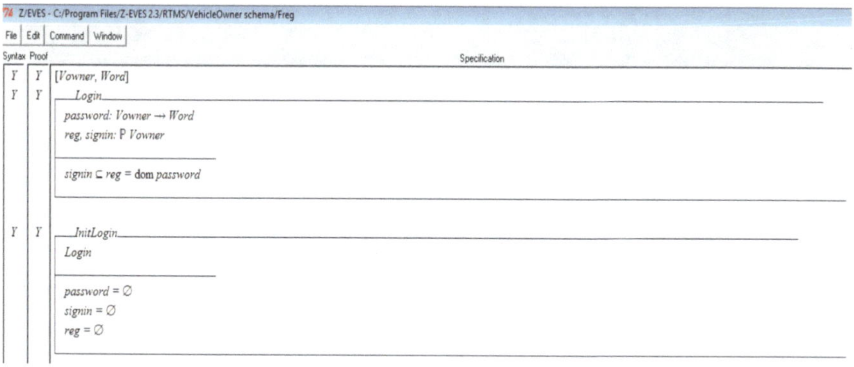

Figure 9. Z/EVES Schema expansion of Initial Login to Login schema.

3) Domain Checking

Domain checking feature of Z/EVES tool enables us to write the statements which are meaningful and in finding the domain errors. However, it has been found that as compared to syntax & type checking, domain checking is more crucial because where syntax and type checking is done automatically, one needs to work together with theorem prover to accomplish the domain checking. We also observed that proof "by reduce" in the proof window of the tool was sufficient for our formal specifications for domain checking. Now if you are already registered, you will opt for the sigin option. By investigating Figure 10, the value for syntax column is "Y", means no error, but the value in proof column is "N". This is related to domain checking. The proof can be initiated by selecting the theorem in the Specification window, right clicking, and selecting "Show proof" which is shown in Figure 9.

The proof can be done by various mean in Z/EVES by choosing "Action Point" by Reduction, Cases, Quantifiers, Normal Norms and Equality. In our case, we use the option "prove by reduction". Figure 11 describes the proof by reduce action point in case of "Signin" schema.

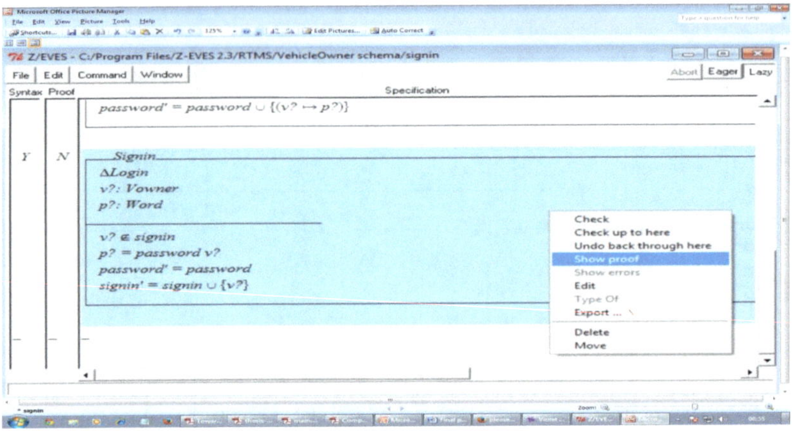

Figure 10. Domain checking with Z/EVES.

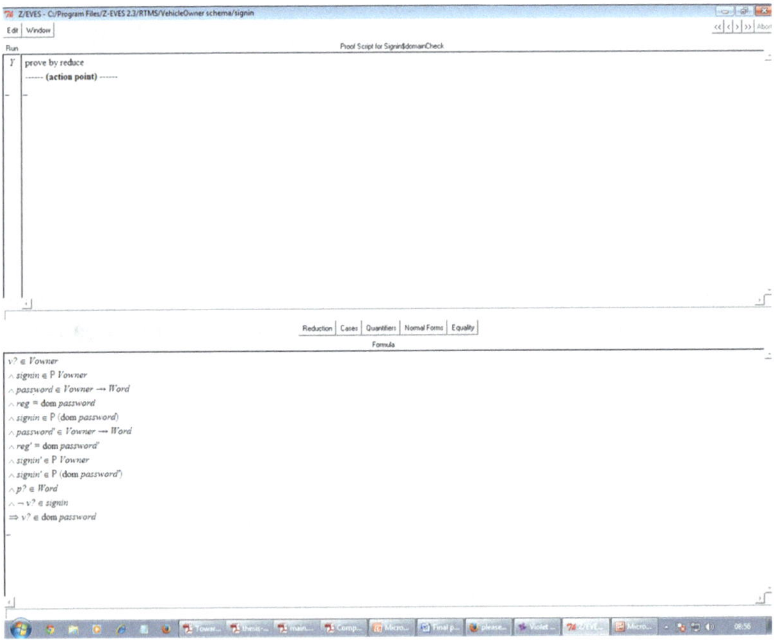

Figure 11. Proof script by using action point "proof by reduce" for "Signin" schema.

RESULT ANALYSIS

Any proposed model is incomplete without tool support. Nevertheless, use of formal language adequately increases the accuracy and completeness but, the use of computer tool indeed increases the level of confidence significantly for the system to be developed by fingering out the potential errors in syntax and semantics of formal narration. Table 1 depicts the result of formal analysis of proposed schemas of road traffic management system using Z/EVES. The attributes in the table are name of the schema followed by syntax & type checking, domain checking, proof and reduction. The second row in table, having status Y for all columns indicating that the schema named "Vehicle Owner" is correct with respect to syntax & type check errors, domain check and having correct proof by performing reduction on the set of predicates for making specification meaningful. The Y^1 symbol shows that the action point in proof window is chosen as "prove by reduce".

Table 1. Result analysis by Z/EVES.

Schema Name	**Syntax & Type Checking**	**Domain Checking**	**Schema Expansion**	**Proof by Reduction**
Vehicle Owner	Y	Y	Y	Y^1
Login	Y	Y	Y	Y^1
Signin	Y	Y	Y	Y^1

CONCLUSION

The use of formal methods in safety critical application increases quality in terms of accuracy, consistency, and in completeness. This paper describes the use of Z notation, a formal methods for Vehicle Owner, an actor of Road Traffic Management System; which will be further verified by Z/EVES, a typechecker tool for Z notation specification. In Future, the schema of Traffic Police, Admin, and Traffic Manager will be implemented and verified by Z/EVES theorem prover.

ACKNOWLEDGEMENTS

Authors are thankful to faculty of Engineering & Technology (FET), Mody University of Science & Technology for providing the facility to carry out the research work.

REFERENCES

1. Woodcock, J.C.P. (1989) Structuring Specifications in Z. IEE/BCS Software Engineering Journal, 4, 51-66. http://dx.doi.org/10.1049/sej.1989.0007
2. Hall, A. (2002) Correctness by Construction: Integrating Formality into a Commercial Development Process. Proceedings of International Symposium of Formal Methods Europe, 2391, 139-157. http://dx.doi.org/10.1007/3-540-45614-7_13
3. Spivey, J.M. (1989) The Z Notation: A Reference Manual. Prentice-Hall, Englewood Cliffs.
4. Hamdy, K.E., Elsoud, M.A. and El-Halawany, A.M. (2011) UML-Based Web Engineering Framework for Modeling Web Application. Journal of Software Engineering, 5, 49-63. http://dx.doi.org/10.3923/jse.2011.49.63
5. Hasan, O. and Tahar, S. (2007) Verification of Probabilistic Properties in the HOL Theorem Prover. Proceedings of the Integrated Formal Methods, 4591, 333-352. http://dx.doi.org/10.1007/978-3-540-73210-5_18
6. He, X. (2000) Formalizing UML Class Diagrams: A Hierarchical Predicate Transition Net Approach. Proceedings of 24th Annual International Computer Software and Applications Conference, Taipei, 25-28 October 2000, 217-222.
7. Zafar, N.A., Khan, S.A. and Araki, K. (2012) Towards the Safety Properties of Moving Block Railway Interlocking System. International Journal of Innovative Computing, Information and Control (ICIC International), 5677-5690.
8. Heiner, M. and Heisel, M. (1999) Modeling Safety Critical Systems with Z and Petri-Nets. Proceedings of International Conference on Computer Safety, Reliability and Security, London, 26-28 October 1999, 361-374. http://dx.doi.org/10.1007/3-540-48249-0_31
9. The Z/EVES 2.0 User's Guide: Mark Saaltink. October 1999 ORA Canada.
10. Mostafa, A.M., Manal, A.I., Hatem, E.B. and Saad, E.M. (2007) Toward a Formalization of UML2.0 Meta-Model Using Z Specifications. Proceedings of 8th ACIS International Conference

on Software Engineering, Artificial Intelligence, Networking and Parallel/Distributed Computing, 3, 694-701. http://dx.doi.org/10.1109/SNPD.2007.508

11. Jacobson, R.I. and Booch, G. (2006) The Unified Modeling Language Reference Manual. 2nd Edition.
12. Selic, B. and Rumbaugh, J. (1998) UML for Modeling Complex Real-Time Systems. Technical Report, Object Time.

Chapter

AN INTEGRATION OF UML SEQUENCE DIAGRAM WITH FORMAL SPECIFICATION METHODS– A FORMAL SOLUTION BASED ON Z

4

Nasir Mehmood Minhas, Asad Masood Qazi, Sidra Shahzadi, and Shumaila Ghafoor

University Institute of Information Technology, PMAS-University Institute of Information Technology, Rawalpindi, Pakistan

ABSTRACT

UML Diagrams are considered as a main component in requirement engineering process and these become an industry standard in many organizations. UML diagrams are useful to show an interaction, behavior and structure of the system. Similarly, in requirement engineering, formal specification methods are also being used in crucial systems where precise

Citation: Minhas, N. , Qazi, A. , Shahzadi, S. and Ghafoor, S. (2015), "An Integration of UML Sequence Diagram with Formal Specification Methods-A Formal Solution Based on Z". Journal of Software Engineering and Applications, 8, 372-383. doi: 10.4236/jsea.2015.88037.

information is required. It is necessary to integrate System Models with such formal methods to overcome the requirements errors i.e. contradiction, ambiguities, vagueness, incompleteness and mixed values of abstraction. Our objective is to integrate the Formal Specification Language (Z) with UML Sequence diagram, as sequence diagram is an interaction diagram which shows the interaction and proper sequence of components (Methods, procedures etc.) of the system. In this paper, we focus on components of UML Sequence diagram and then implement these components in formal specification language Z. And the results of this research papers are complete integrated components of Sequence diagram with Z schemas, which are verified by using tools and model based testing technique of Formal Specifications. Results can be more improved by integrating remaining components of Sequence and other UML diagrams into Formal Specification Language.

Keywords: Formal Specifications, Software Requirement Specifications, Formal Notations

INTRODUCTION

Formal Methods are based on mathematical techniques, which can be used in any phase of Project life cycle, especially in an initial stage. When requirements are gathered from clients, project team has to know about the system. There are many techniques of formal methods, like Model based Languages, Process Oriented and Algebraic Specifications.

In Software Engineering, formal specifications and UML Diagrams are very useful to understand the requirements and specifications of the system. Formal specifications and UML are used since many years in Software Engineering, and UML diagrams are considered as a standard tool in many organizations. There is a complete method in Software Engineering named as “Clean Room Software Engineering” [1] basically based on formal specifications. The idea behind the Clean Room SE is “Do it Right, at first Time”. It is composed of gathering requirements, and then transforms them into statistical methods, so there will no need of unit testing.

UML diagrams are important to understand the complexity of system. UML describes the behavior and structure of a program. Also, they describe the interaction of components with the system. These include Use Case, Class, Activity and many other diagrams. UML diagrams are easy to understand by the users, developers and domain experts whereas the formal methods are difficult to understand by the users, domain experts and developers as well.

Before go forward, we also need to know that where the specification part lies actually in the Requirement engineering process. A brief detail of requirement engineering process is given below:

Software requirement engineering involves requirements elicitation, requirements specification, requirements validation and requirements management [2] [3] . Requirements elicitation involves the ways of gathering the requirements which include many traditional, cognitive, model based techniques etc.

Whereas, the requirements Specification (where analysis and negotiation of requirements are performed), requirements of users are specify to make them understandable and meaningful for developers. Specifications can be formal as well as non-formal [4] . Formal techniques include the set of tools and techniques based on mathematical models whereas informal techniques are based on modeling the requirements in diagrams or making architecture of system. There are many techniques in both types of specification. Like in formal techniques of specifications, we have different formal specification languages like Z, VDM etc. and in in-formal or non-formal techniques, we have UML diagrams which include use-cases, sequence diagrams, collaboration and interaction diagrams etc.

In Requirements validation, the completeness of the requirements being checked which means either gathered requirements are correct, complete or not. The main objective to analyze the validation and verification of RE process to identify and resolve the problems and highly risk factors of software in early stages to make strengthen the development cycle [5] . Finally, in Requirements management phase, issues and conflicts of users are resolved. According to Andriole [6] , the requirement management is a political game. It is basically applied in such cases where we have to control the expectations of stakeholders from software, and put the requirements rather than in well-meaning by customers but meaning full by developers, so they can examine that, they actually full fill the user's requirements.

Authors of [7] include the Requirement change management under the Requirement Engineering process. RCM is a term which is used as the history or previous development of the similar software product (s). On the basis of historical development, we investigate the need of RCM or not.

Unique Features of Sequence Diagrams

There are some unique features of Sequence diagrams and also reasons for choosing sequence diagrams for this research purpose, which are:

- Sequence Diagrams are used to show the priorities of steps/ modules of system, Lower step denotes to the later.
- Reverse Engineering of UML sequence diagrams are used to support reverse engineering in software development process [8] .
- It shows a dynamic behavior of system and considered as good system architecture design approach [9] .
- Sequence diagrams include the Life line of the objects, and it can be easily integrated because of Time dimension [10] .
- We can use messages to make it understandable by all stakeholders.
- We can also use loops, alternatives, break, parallelism between complex components of system and many more [10] .

Our Idea is to work on integration of UML Sequence Diagram's attributes with formal specification methods like Z notations to bridge a gap between both (formal and informal) methods. So, once we make a sketch of any system in to Sequence diagram to show its sequence of steps, requirement priorities, time bar information and others, then we will transform these attributes into Z schemas. So it will be an easy for developers, to develop system if they had requirements in proper mathematical forms. Also, there was a research gap of properly integration of Sequence diagrams with formal specifications.

RELATED WORK

The Integration of formal and in formal methods of specification is not a new area, it is being used by many software industries as well as there is a complete area of research in Software engineering to tackle the limitations of both techniques and transform them into an intermediately solution. Many studies are there which are helpful to integrate the Z notations with Scrum, requirement elicitation and many other areas.

In [11] , authors represents a conceptual solution to formalize the Class Diagrams, in which they formalize the class diagrams through steps including representing classes in Z, representing associations in Z representing aggregation in Z and then represent generalization of classes in Z.

Similarly, in [12] researchers represent a conceptual solution to integrate XP methodology with Z. They work on user stories phase, where user stories will be verified through Formal Verification techniques. It can also be observed from [13] in which integration of the Z notations into Use case Diagram, because use case diagrams are very common in software development companies, as these are easy to understand by all stakeholders, so they apply Z on them, to bridge a gap between formal and informal techniques. It has been observed from [14] , which is study on an Integration of formal methods into agile methodology that formal methods can lead towards a better software development solution. In [15] , they apply formalization in the requirement specification phase of Requirement Engineering Process. They describe an analyzing phase, in which they will focus on such a specification which is being analyzed early.

Concept of making Z schemas of UML class and sequence diagrams on the basis of some semantic rules can also be found in the [16] [17] . Firstly, a Video on Demand case study is been taken in this study, then authors draw its class structure diagram to shows the hierarchy of the classes, and then Z schemas are defined. Secondly, sequence diagram is generated on same case study, furthermore its objects are defined in a complete way using Z schemas.

Tony Spiteri [18] takes a case study, transform it into UML diagrams, and then implement the specifications into formal method languages, then apply optimization methods to minimize the computing resources; total time and total cost. in [19] -[21] z notations based schemas are applied in some real life examples and case studies, furthermore, the authors also uses z/ eves tool for formal model check as well as z schemas verification. In [11] , very important concept can be found related to this study, in which sequence diagram is analyzed through the states of the system and their relationship according to the message using state transitions graphs.

EXPECTATIONS FROM SYSTEM SPECIFICATIONS

In any system development, we gather the requirements from users and then we try to understand "What" should be done, but formal specification methods also specify "Why" should be done. For moving from analysis to implementation we have to identify these (and many else) variables from [22] -[24] .

Domain Knowledge

Understanding about the system as well as its context, and it should be known by all stakeholders. For example for Library management system, there should be complete understanding about the library environment and ordinary procedures followed by Library. Similarly for Airline reservation system, there should be a complete knowledge about its possible components like scheduling, ticketing boarding etc.

User's Requirements

These are the requirements which are not the requirements of system. Actually, these requirements are defined by the client or user to make it efficient or easy to use, like cost, ease of use etc. are the examples of user's requirements.

System Requirements

These requirements are purely related to the system, which must be included in the system. For example a flight reservation system must include the flight place with time and date.

System Specification

To specify the gathered requirements, software engineers' uses many ways to transform the story based requirements into a meaningful form for developers. These specifications are not basically dependent to any design, these can be in form of abstract prototypes, formulas procedures or else. In this part, specifications are sent to the developers to implement them, and testers to test them and to the users to verify them.

Design Structure

In design structure, we have to focus on "How" part. For example how the functionality is allocated to the system component, how the system's components will communicate each other etc.

Problem Refinement

Formal notation methods like VDM and Z helps the software engineers to refine the problem. In user's stories which problem look more complex, and involve complex mathematical structure, we refine it by using formal methods, so by specify the relationship between components problem becomes in more refined shape.

PROPOSED SOLUTION

Formal specifications use mathematical notation and provide state requirements and mechanism for the verification of system correctness. Z specification provides a mathematical technique to model relation via predicate calculus that has states and relational functions [9] . Our Research will basically focus on implementation of UML Sequence diagram into Z. So, to implement our problem we proposed some sequence of steps which is given in Figure 1.

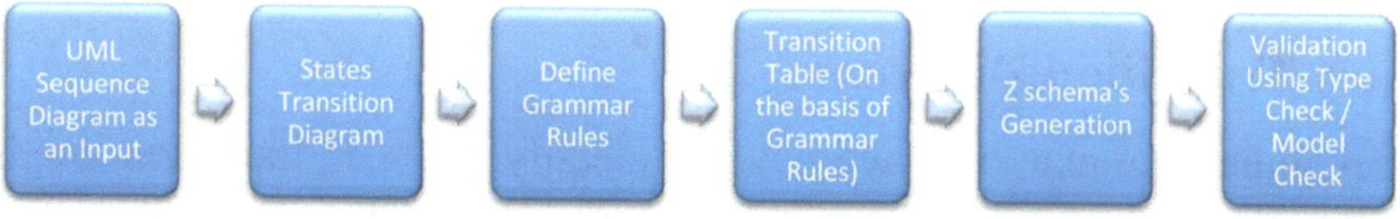

Figure 1. Proposed solution steps to integrate UML sequence architecture with formal specification methods.

FORMALIZATION OF FLIGHT RESERVATION SYSTEM

UML Sequence Diagram as an Input

A flight reservation system may contain the records of flights, which includes the place, date and time of flight, airline name, number of seating capacity or number of tickets, list of users etc. Now on the basis of this given data, following operations can be performed like Creation of Reservation, Cancel the reservation, Sign In/Sign Up etc. UML sequence diagram for Flight reservation system can be seen in Figure 2.

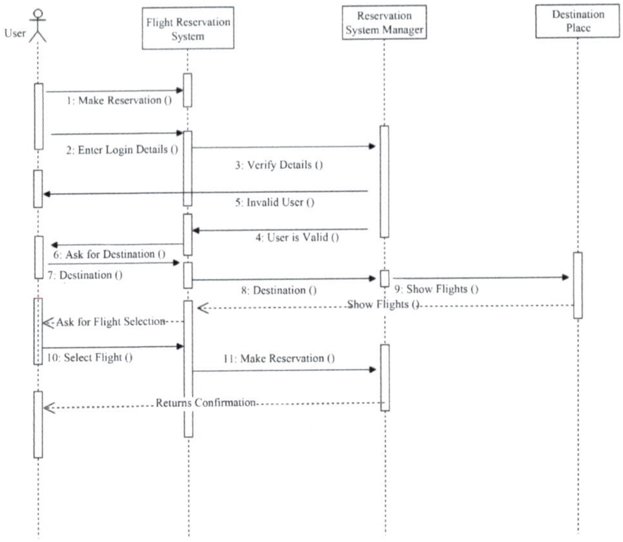

Figure 2. UML sequence architecture of flight reservation system.

States-Transition Diagram

To set some grammar rules we have to identify the state transition graph or diagram. So, on the basis of these grammar rules we can formalize our sequence diagram's components. The State UML diagram is given in Figure 3.

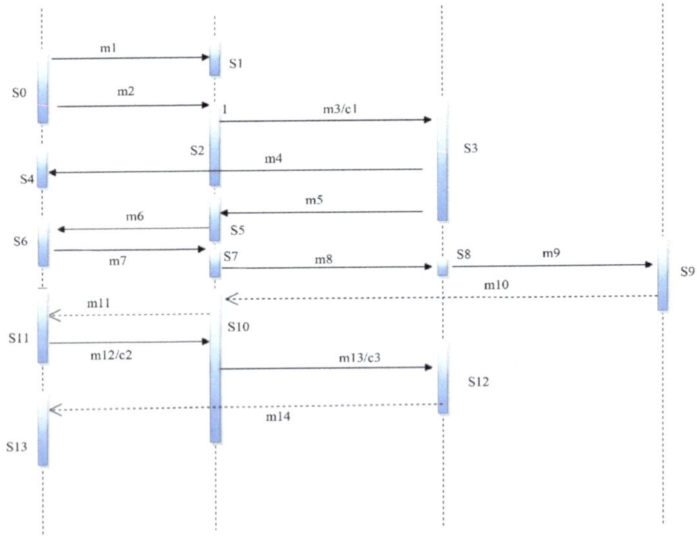

Figure 3. State Transition graph/diagram on the basis of UML sequence diagram.

Define Grammar Rules

In transformation procedure from state to grammar development is given in Table 1. In transformation there are certain states regarding each object and messages execute from one state another state, a production rule is created, for the execution of message if there is no condition null condition is supposed. We can elaborate this concept with the help of example for this consider row 1 where m1 (message 1: make reservation) execute from state S0 to S1 here no condition is imposed for execution so, there is null condition. We can determine final states from given table where S2, S10 are final states and rest of all are failure of operation.

Table 1. States-transition table with their termination conditions.

Sr.#	STATES/MESSAGE	OUTPUT
1	S0, m1, S1, null	S0 ⇒ m1S1, null
2	S0, m2, S2, null	S0 ⇒ m2S2, null
3	S2, m3, S3, c1	S2 ⇒ m3S3, c1
4	S3, m4, S4, null	S3 ⇒ m4S4, null
5	S3, m5, S5, null	S3 ⇒ m5S5, null
6	S5, m6, S6, null	S5 ⇒ m6S6, null
7	S6, m7, S7, null	S6 ⇒ m7S7, null
8	S7, m8, S8, null	S7 ⇒ m8S8, null
9	S8, m9, S9, null	S8 ⇒ m9S9, null
10	S9, m10, S10, null	S9 ⇒ m10S10, null
11	S9, m10, S10, null	S9 ⇒ m10S10, null
12	S10, m11, S11, null	S10 ⇒ m11S11, null
13	S10, m12, S11, c2	S10 ⇒ m12S11, c2
14	S10, m13, S12, c3	S10 ⇒ m13S12, c3
15	S12, m14, S13, null	S12 ⇒ m14S13, null

Using Table 1: After constructing rules regarding each message now, here for the termination of the process the null production are added, represented by derivation tree for parsing of a scenario.

Rule (r1): S0 ⇒ m1S1, null|m2S2, null,

Rule (r2): S1⇒∈,

Rule (r3): S2 ⇒ m3S3, c1,

Rule (r4): S3 ⇒ m4S4, null|m5S5, null,

Rule (r5): S4⇒∈,

Rule (r6): S5 ⇒ m6S6, null,

Rule (r7): S6 ⇒ m7S7, null,

Rule (r8): S7 ⇒ m8S8, null,

Rule (r9): S8 ⇒ m9S9, null,

Rule (r10): S9 ⇒ m10S10, null|m10S10, null,

Rule (r11): S10 ⇒ m11S11, null|m12S11, c2.

To check validation we can derive by above diagram which we have constructed in grammar rules, here is only the validation for S0, we can check for all states like this way:

According to r1 S0 ⇒ m2S2, now if we apply r3 we get m3S

S0 ⇒m2S2

(By applying r3 on S2 we get m3S3)

⇒m2m3S3

(By applying r4 on S3 we get m5S5)

⇒m2m3m5S5

(By applying r6 on S5 we get m6S6)

⇒m2m3m5m6S6

(By applying r7 on S6 we get m7S7)

⇒m2m3m5m6m7S7

(By applying r8 on S7 we get m8S8)

⇒m2m3m5m6m7m8S8

(By applying r9 on S8 we get m9S9)

⇒m2m3m5m6m7m8m9S9

(By applying r10 on S9 we get m10S10)

⇒m2m3m5m6m7m8m9m10S10

(By applying r11 on S10 we get m11S11)

⇒m2m3m5m6m7m8m9m11S11

State-Transition Table

Rules for Constructing Z Schemas:

Rule 1:

S0 4∧ S1 € states, if current (state) = i then new (state) = i+1, condition c = null for the execution, message = m1 will move from S0 to S1. Creation and termination time of object (passenger) is between start and end time of S0, S1, and same in the case of message m2.

Rule 2:

S2 ∧ S3 € states, if current (state) = i then new (state) = i+1, condition c = c1 for the execution, message = m3 will move from S2 toS3 regarding objects flight reservation system and reservation system manager. Termination and creation time of these objects (o1, o2) must not be greater than the time require for the start and end of states S2, S3.

Rule 3:

S3 ∧ S4 € states, if current (state) = i then new (state) = i+1, condition c = null for the execution, message = m4 will move from S3 toS4 regarding objects flight reservation system and reservation system manager. Termination and creation time of these objects (o2 to o1) must not be greater than the time require for the start and end of states S3, S4.

Rule 4:

S8 ∧ S9 € states, message m9 execute by fulfilling condition c9. Termination and creation time of these objects (o3 to o4) must not be greater than the time require for the start and end of states.

Z Schemas Generation

In schema generation, we are using Z as formal specification language. The schema's mentioned below are the schemas of the sequence diagram of flight reservation system, which are based on the set of grammar rules, which we define earlier. Schemas are defined at the Appendix.

TESTING AND VERIFICATION

We have taken a small case study to work on this particular area, so we can test and validate our schemas and model efficiently. Our Z schemas are written in Z word tool, in which there is an option for type checking Z schemas using tool fuzz. Our grammar rules are semantic based solutions, which can be clearly seen in our state transition diagram. For model check we have used the same tool. Our resultant schemas are error free, but the results can also be improved through using other tools and techniques like Z/eves, CZT and many others.

The procedure of our testing was based on Z word tool which uses fuzz tool for type checking. The tool can be downloaded from Internet, and after installing we can use following procedure as described in Figure 4 for Type Check.

By executing schemas, we achieve correctness of our schemas as described in Figure 5.

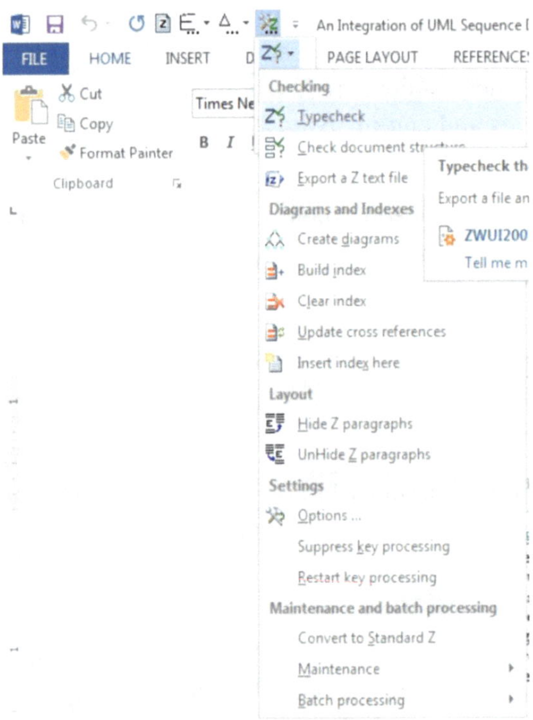

Figure 4. Type check specification.

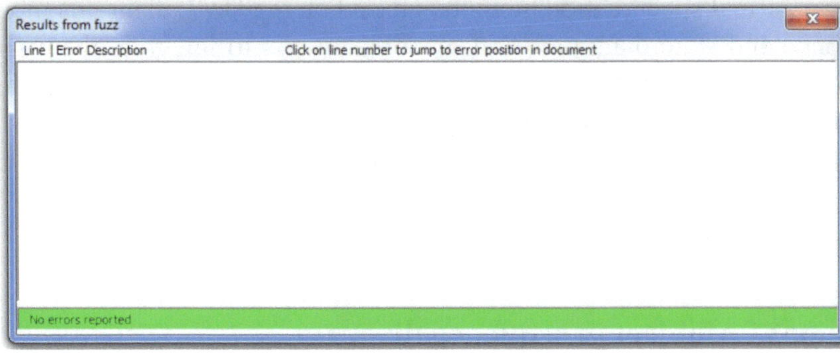

Figure 5. On Execution of Schemas the correctness is shown.

LIMITATIONS AND FUTURE WORK

The case study, which we take as a reference is a simple study, and does not cover all the features of Z, also this integration make the system more complex for understandable to normal stake holders like Users. Although a sequence diagram is decomposed into parts, which means modules, sub modules, their relations etc. are extracted but overall cost of system in terms of time and money can be increased, that is why, formal specifications were not cordial welcomed by software industry. But now in this study and related previous (referenced) studies a gap between formal and informal methods of requirement engineering and specifications are bridged.

Furthermore, there are many other informal techniques which are needed to be formalized like many development models, requirement elicitation techniques which are typically based on user stories etc. Also we can improve the results by using other formal and mathematical techniques and algorithms to optimize the results and decrease the overall cost of system.

CONCLUSIONS

In this paper, we have focused on the integration of UML sequence diagram into using Z specification language. For this we take a system "the Flight Reservation System" by following all procedure as described in our methodology, we formalize our system into Z specification as well as we try to accommodate maximum features of UML diagrams into our proposed solution by applying some grammar rules, which are used in our semantic based solution.

Our formal specification method is based on the UML diagrams include sequence and state diagrams, and our objective is to integrate them using Z schemas notations. But it was not an easy task to include all the features and applications in one paper or one solution. But overall Z schemas are analyzed and tested using fuzz as a type checking.

APPENDIX

Schemas for object in sequence diagram

```
_Passenger ____________________
| pname:PName
| pstart,pend:Time
| states:seq state
| attributes:□Atributes
| function: Atributes□Atributes
|______________________________
| states ≠ <>
| n-states≥1
| ⇒(∃ s0,s1:state | s0∧ s1∈active stats
| .states0= s0∧states(n-states)=s1
| ⇒pstart ≤ s0.stime ∧ s1.etime ≤ oend)
| ∀i:ℕn-states ≥ 1∧I ∈ 1.........n-states-1
| According to rule 1
| .∃s0,s1:states
| .states i= s0∧staes (i+1)=s1⇒s0.etime ≤ s1.stime
| ∀input , output:Atributes | (input,output) ∈functions
| .input ∈ atributes ∧ output ∈ atributes
|______________________________
```

Schemas for Messages in Sequence Diagram

Condition: = NULL |TRUE| FALSE

```
_Enter-Id______________________
| Starttime ,end time:ℕ
| condition: Condition
| from s0 to s1 :states
|______________________________
| starttime<end time
| s0.stime ≤ starttime ∧ end time ≤ s1 . etime
| according to rule 1
|______________________________
```

Schemas for Sequence Diagram

Flight Reservation system

objects : □*Object*
messages:□ *Messages*

$\forall o1, o2 : object \mid o1 \wedge o2 \in objects$
$\exists\ s1, s2 : state \mid s1 \wedge s2 \in ran\ \ o1.state \wedge o2\ .\ states$
$\exists m: Message \mid m \in messages\ .m\ from = s1 \wedge m.\ to = s2$
$\forall m : Message \mid m \in messages$
According to rule 2
$\exists\ o1\ , o2 : object \mid o1\ \wedge o2 \in objects$
$\exists\ s1\ , s2 : state \mid s1 \in ran..\ o1\ .staes\ \wedge\ s2\ \in ran\ o2.\ Staes$
Where $s1 = m.\ from\ \wedge\ s2\ m.\ to$

Operations in Reservation System

Show Flight

Ξ *Flights*
Place? : Place
Date! : Date
Time !: Time
Starttime ,end time:$\mathbb{N}$
condition: Condition
from s8 to s9 :states

$Place? \in\ \ Known$
$Date! = Flights\ (\ Date?)\ \ \wedge$
$Time! = Flights\ (Time?)$
starttime<end time
$s8.stime \leq starttime\ \wedge\ end\ time \leq s9\ .\ etime$

Login System

Ξ *Users Initialization*
Usr ?: USERS
Res ?: RESOURCES
Result! REPORT
Starttime ,end time :$\mathbb{N}$
condition :C1
from s0 to s1 :states

$Usr\ ? \in Users$
$Res\ ? \in Resources$
Result! = Welcome
Start –time <end time
$S0.stime \leq start\ time\ \wedge\ end\ time \leq s1.time$

_*Sign up users*________________________

Δ *Users Initialization*

Us ? : User

Crdls ?? : $\mathbb{P}$*CEDENTIAL*

U ? $\in$ *users*

U ? $\in$ *registered users*

Registered_users′ = Registered_Users $\cup$

{ (*U*? ↔ *crdls*

REFERENCES

1. Selby, R.W., Basili, V.R. and Baker, F.T. (2006) Cleanroom Software Development: An Empirical Evaluation. IEEE Transactions on Software Engineering, SE-13, 1027-1037.
2. Chikh, A. (2011) A Knowledge Management Framework in Software Requirements Engineering Based on SECI Model. Journal of Software Engineering and Applications, 4, 718-728. http://www.SciRP.org/journal/jsea http://dx.doi.org/10.4236/jsea.2011.412084
3. Flores, F., Mora, M., álvarez, F., et al. (2010) Towards a Systematic Service Oriented Requirement Engineering Process (S-SoRE). Proceedings of the International Conference, CENTERIS 2010, Viana do Castelo, 20-22 October 2010, 111-120. http://dx.doi.org/10.1007/978-3-642-16402-6_12
4. Batra, M., Malik, A. and Dave, M. (2013) Formal Methods: Benefits, Challenges and Future Direction. Journal of Global Research in Computer Science, 4.
5. Boehm, B.W. (1984) Verifying and Validating Software Requirements and Design Specifications. IEEE Software Journal, 1, 75-88.
6. Andriole, S. and Safeguard Sci. Inc. (1998) The Politics of Requirements Management. IEEE Software Journal, 15, 82-84. http://dx.doi.org/10.1109/52.730850
7. Flores, F., Mora, M., álvarez, F., O'Connor, R. and Macias, J. (2008) Handbook of Research on Modern Systems Analysis and Design Technologies and Applications. In: Global, I.G.I., Ed., Chapter VI: Requirements Engineering: A Review of Processes and Techniques, Minnesota State University; Mankato, 96-111.
8. Rountev, A. and Connell, B.H. (2005) Object Naming Analysis for Reverse-Engineered Sequence Diagrams. Proceedings of the International Conference on Software Engineering, St. Louis, 15-21 May 2005, 254-263.
9. Zafar, N.A. and Alhumaidan, F. (2013) Scenarios Verification in Sequence Diagram. The Journal of American Science, 9, 287-293. http://www.jofamericanscience.org
10. UML Basics: The Sequence Diagram. http://www.ibm.com/developerworks/rational/library/3101.html
11. Shroff, M. and France, R.B. (1997) Towards a Formalization of UML Class Structures in Z. The 21st Annual International Computer Software

and Applications Conference, 1997 (COMPSAC' 97), Washington DC, 11-15 August 1997, 646-651. http://dx.doi.org/10.1109/cmpsac.1997.625087

12. Sgafiq, S. and Minhas, N.M. (2014) Integrating Formal Methods in XP—A Conceptual Solution. Journal of Software Engineering and Applications, 7, 299-310. http://dx.doi.org/10.4236/jsea.2014.74029
13. Sengupta, S. and Bhattacharya, S. (2006) Formalization of UML Use Case Diagram—A Z Notation Based Approach.
14. Black, S., Boca, P.P., Bowen, J.P., Gorman, J. and Hinchey, M. (2009) Formal versus Agile: Survival of the Fittest? IEEE Computer, 42, 37-45. http://dx.doi.org/10.1109/MC.2009.284
15. Fernández-y-Fernández, C.A. and José, M.J. (2012) Towards an Integration of Formal Specification in the áncora Methodology.
16. Spivey, J.M. (1998) The Z Notation: A Reference Manual. Prentice Hall International, Oxford.
17. El Miloudi, K., El Armani, Y. and Attouhami, A. (2013) Using Z Formal Specification for Ensuring Consistency in Multi View Modeling. Journal of Theoretical and Applied Information Technology, 57, 407-411.
18. Staines, T.S. (2007) Supporting UML Sequence Diagrams with a Processor Net Approach. Journal of Software, 2, 64-73. http://dx.doi.org/10.4304/jsw.2.2.64-73
19. Alhumaidan, F. and Zafar, N.A. (2013) Automated Semantics Treatment of Sequence Diagram Defining Grammar Rules. http://worldcomp-proceedings.com/proc/p2013/FCS7057.pdf
20. Zafar, N.A. (2006) Modeling and Formal Specification of Automated Train Control System Using Z Notation. IEEE Multi-Topic Conference (INMIC'06), Islamabad, 23-24 December 2006, 438-443. http://dx.doi.org/10.1109/inmic.2006.358207
21. Zafar, N.A., Khan, S.A. and Araki, K. (2012) Towards the Safety Properties of Moving Block Railway Interlocking System. International Journal of Innovative Computing, Information & Control, 8, 5677-5690.
22. Heitmeyer, C.L., Jeffords, R.D. and Labaw, B.G. (1996) Automated Consistency Checking of Requirements Specifications. ACM Transactions on Software Engineering and Methodology, 5, 231-261. http://dx.doi.org/10.1145/234426.234431

23. Hall, A. (1996) Using Formal Methods to Develop an ATC Information System. IEEE Software, 13, 66-76. http://dx.doi.org/10.1109/52.506463

24. Bano, M. and Zwoghi, D. (2013) User's Involvement in Requirement Engineering and System Success. IEEE 3rd International Workshop on Empirical Requirement Engineering, Rio de Janeiro, 15 July 2013, 24-31.

SECTION 2: PROGRAMMING LANGUAGES SEMANTICS

Chapter

DECLARATIVE PROGRAMMING WITH TEMPORAL CONSTRAINTS, IN THE LANGUAGE CG

5

Lorina Negreanu

POLITEHNICA University of Bucharest, Splaiul Independentei 303, 060042 Bucharest, Romania

ABSTRACT

Specifying and interpreting temporal constraints are key elements of knowledge representation and reasoning, with applications in temporal databases, agent programming, and ambient intelligence. We present and formally characterize the language CG, which tackles this issue. In CG, users are able to develop time-dependent programs, in a flexible and straightforward manner. Such programs can, in turn, be coupled with evolving environments,

Citation: Lorina Negreanu, "Declarative Programming with Temporal Constraints, in the Language CG", The Scientific World Journal, volume 2015, article ID 540854, https://doi.org/10.1155/2015/540854.

thus empowering users to control the environment's evolution. CG relies on a structure for storing temporal information, together with a dedicated query mechanism. Hence, we explore the computational complexity of our query satisfaction problem. We discuss previous implementation attempts of CG and introduce a novel prototype which relies on logic programming. Finally, we address the issue of consistency and correctness of CG program execution, using the Event-B modeling approach.

INTRODUCTION

Specifying and reasoning about phenomena that evolve in time are essential traits of any intelligent system. Their key components are usually identified as (i) representing the temporal behaviour of a system and (ii) extracting information which is otherwise implicit in the system representation. In the traditional line of research, temporal representation and reasoning are deployed for (program) verification [1]. Thus, the entire system behaviour is encoded by some form of labelled transition graph (Kripke Structure), and temporal logic is used for expressing specific properties of the underlying system. Finally, model checking [2] is employed for verifying whether the property is entailed by the system at hand. Unlike the traditional approach, we focus on capturing nonnecessarily deterministic evolutions of a system. Thus, instead of characterizing all possible behaviours, by unfolding, for example, a transition system and examining all paths, we look at a single "*evolution path*." We consider that our approach has interesting advantages with respect to the traditional line of research based on temporal logics such as LTL, CTL (for a more detailed motivation see, e.g., [3]). We are less interested in eventuality (e.g., fairness constraints) or maintenance (e.g., safety constraints) of properties, which are typical for model checking [4–7] and for deductive reasoning [8–11] in temporal logics. Instead, we would like to identify temporal relations in the occurrence of properties, in the spirit of Allen's Interval Algebra [12]. As an example, consider identifying "*those individuals which were married at least twice*." This amounts to finding those properties "*married*" which occur one *after* the other and which enrol the same individual.

Our framework consists of (i) a *representation* of an evolution path of a system, one which is specifically tailored for capturing temporal relations between the system properties, (ii) a *temporal language* which we employ for expressing complex temporal constraints between properties, as shown in the above example, and (iii) a *rule-based programming language*, CG, which allows the programmer to specify time-dependent programming. Rule-based programming languages operate on a working memory of factual information: they check rule applicability against the working memory and subsequently modify the latter, by effectively applying the rules. CG follows the same principle; only here the working memory has a *temporal structure*, which is precisely (i). Specifying when a rule is applicable is done using (ii). Applying the rule means executing actions which are aimed at coercing the system evolution according to the programmer's intentions.

CG can be highly effective in specifying intelligent device behaviour in intelligent houses, as illustrated in [13–16]. Also, CG has been employed for temporal data mining [3]; finally, (i) and (ii) were also used as a means for representing game outcomes of multiagent systems [17].

The aim of this paper is to (a) build an all-encompassing view of our approach, (b) present our already established main theoretical results, (c) introduce a novel implementation based on Prolog, and, finally, (d) examine aspects pertaining to *correctness* of our approach. (a) has already been discussed in different variants, in [3, 13–17]; (b) has been the subject of [3, 14, 17]. (c) and (d) however are new contributions which, to our knowledge, have not been considered yet.

The rest of the paper is structured as follows. In Section 2, we introduce the main primitives of our modeling approach. In Section 3, we review the temporal language $L_{\mathscr{H}}$ and its computational properties. In Section 4 we illustrate the rule-based language CG and in Section 5 we examine aspects pertaining to its correctness. In Section 6 we illustrate a lightweight implementation for CG and finally, in Section 7, we conclude.

MODELING EVOLVING APPLICATIONS

Our approach relies on describing the state of the modelled domain as a set of relationships between the actors of the domain, relationships called *qualities* in what follows: quality relation instances of the form $R(i_1,...,i_n)$ where R designates the property at hand, n is the arity of R, and $i_1,...,i_n$ are the individuals enrolled in the relationship. For instance, the quality *Married(John,Alice)* designates a binary relationship between two individuals, while *On(ac)* designates a property of the device *ac* (air conditioner).

A state, as seen in the conventional approach, is unpacked into a set of qualities, which portrays the status of the domain over a finite time interval, given that no changes are present during the interval at hand. A state transition corresponds to a change in the domain: the commencement of new qualities or the termination of existing ones. Such a change is triggered by *actions*. An action is also an instance of the form $(i_1,...,i_n)$, which designates an instantaneous event of type a, which enrols individuals $i_1,...,i_n$. For instance, *Marries(John,Alice)* is an action which changes the status of John and Alice: having been initially single, they now become married. Similarly, *TurnOn(ac)* is an action which changes the status of the air conditioner.

State unpacking is illustrated in Figure 1. Above, a conventional transition system is used to describe the evolution of a domain: John and Alice are initially single, they become married, and Alice awaits a child that also comes later on. Below, we use a quality-oriented description: the focus shifts from states labelled with certain properties to qualities introduced and terminated by actions. In the former approach, the lifespan of properties is implicit: one must examine the sequence of states on which the property continuously holds. For example, *Married(John,Alice)* holds from s_2 to s_4. In our approach, the lifespan of qualities is represented explicitly, by their initiating and terminating actions. For instance, *Married(John,Alice)* holds from the moment a_3 was executed until a_6 was executed. We assume a_3 designates the marriage action while a_6 is a special action belonging to the current moment. We have not labelled actions to avoid cluttering the figure.

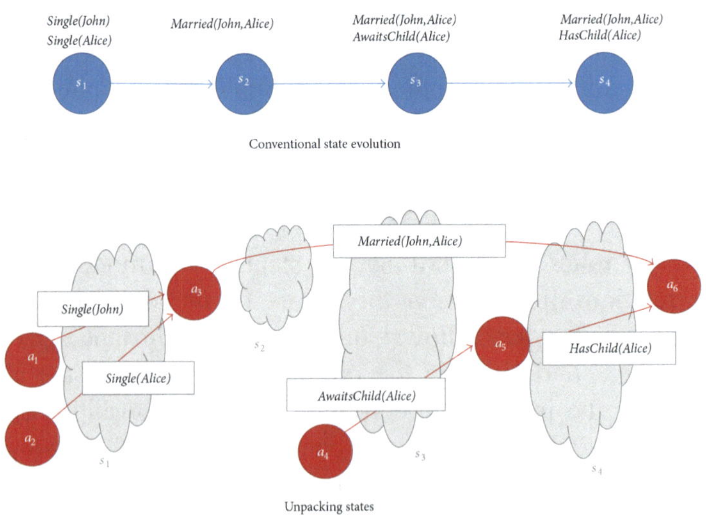

Figure 1. Unpacking states.

It is easy to see that the two description styles are equivalent. Nevertheless, we argue that our quality-oriented description suits better applications where the timing is important and, moreover, where the temporal relationship between qualities is an essential issue. Also, by avoiding unnecessary relabellings of sequences of states, we obtain a more compact representation which speeds up processing and saves space.

Domain Representation

In what follows, we distinguish between an ontological representation of a domain itself and a temporal one. The former is *temporally flat* and provides the taxonomy which characterizes the domain. The latter is, in essence, a temporal structure which instantiates the taxonomy, as we will further show.

Individuals

The actors of a described domain are *individuals*. They are atomic, unique, and identifiable by themselves. They are used to represent entities from the domain (John and Alice or the air conditioner, in the above examples), as well as primitive values of use in the language (e.g., *20 degrees*, the time-stamp *18:50:00*, etc.) or even the environment seen as an entity in itself. Seen from the programming perspective, individuals behave much like atoms in the language of Prolog: they are string literals without an explicit type.

Actions

An action corresponds to an instantaneous stimulus applied to one or more individuals. Actions are represented as relation instances $(i_1,...,i_n)$ where a designates the action type, n is the arity of a, and $i_1,...,i_n$ are the individuals that the action enrols.

Qualities

A quality designates a time-dependent property (i) of individual i or an n-ary relationship $R(i_1,...,i_n)$, between individuals $i_1,...,i_n$.

Time

Individuals, actions, and qualities are merely taxonomical entities. In what follows, we add temporal dimension to each one. First, we consider individuals as perennial. Their existence is unaltered by the evolution of the domain. The temporal dimension of an action is an *action node*. A group of action nodes uniquely identify a moment of time when they occur, provided that their occurrence is simultaneous. We call a collection of such action nodes a *hypernode*. The temporal dimension of a quality $q = (i_1, ..., i_n)$ is a *quality edge* (a, b) which spans action nodes a and b. a models the event which has initiated the enrolment of $(i_1, ..., _n)$ in R, while b models the event responsible for its termination. The lifespan of q is given by the temporal moments when a and b occur, respectively.

These temporal components are glued together in a structure called temporal graph (short t-graph).

Definition 1 (temporal graph). A temporal graph is an oriented graph $\mathscr{H} = (A, E)$, where A designates the set of action nodes and E that of quality edges, together with a partition H over A. One denotes the elements $h_i \in H$ as hypernodes. One assumes elements of A and E have a unique label of the form $(i_1, ..., i_n)$, which one denotes by $\mathscr{L}(a)$ for $a \in A$ and $\mathscr{L}(a, b)$ for $(a, b) \in E$, respectively. For a more rigorous treatment, one refers the reader to [3].

The domain evolution described in Figure 1 is captured by the t-graph from Figure 2 (we have omitted the representation of the quality *AwaitsChild(Alice)*, due to limited space). We have represented action labels in blue. Also, in order to make the figure more legible, we have only labelled those actions subject to our discussion.

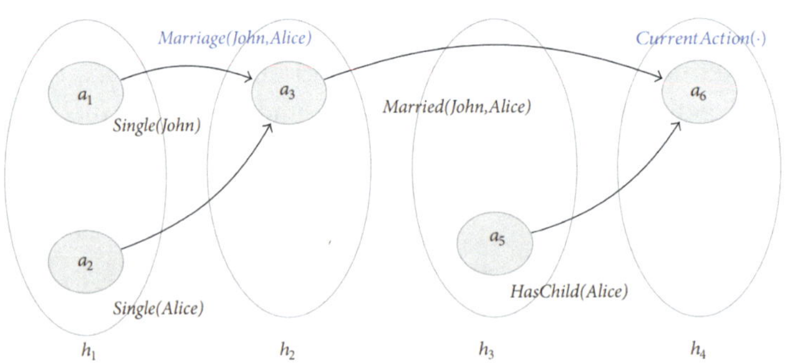

Figure 2. The temporal graph $\mathscr{H}_1$ describing John and Alice's evolution.

Definition 2 (temporal ordering, precedence). A hypernode h *immediately precedes* another (h') in a t-graph, if and only if there exists a quality edge (a, b) such that $a \in$ h and $b \in$ h'. Immediate precedence is a partial ordering of hypernodes, as illustrated in Figure 2. For instance, h_1 immediately precedes h_2 and h_3 immediately precedes h_4; however the same cannot be said about h_2 and h_3. Although represented in sequence in Figure 2, h_2 and h_3 need not occur in this particular order. Thus, it might be the case that Alice has a child prior to the marriage to John or that the child comes after the marriage. Such information is absent from $\mathscr{H}_1$ and neither conclusion could be made. However, in Figure 3, the ambiguity is lifted by the presence of the quality (a_3, a_5), labelled *AwaitChild(John,Alice)*.

We denote by $\succ$ the transitive closure of the immediate precedence relationship, described previously. If h $\succ$ h', we say h precedes h'.

An action node $a \in$ h (immediately) precedes $a' \in$ h' if and only if h $\succ$ h'.

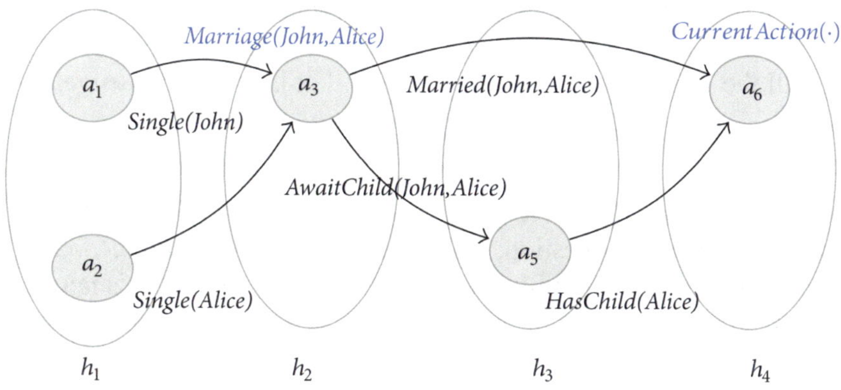

Figure 3. The temporal graph $\mathscr{H}'_1$ describing a more precise evolution of John and Alice.

Let $q = (a, b)$ and $q' = (a', b')$ be two quality edges. q occurs *before* (after) q' if and only if b precedes a' (b' precedes a); q occurs *just before* (just after) q' if and only if a precedes a' and a' precedes b (a' precedes a and a precedes b'); q overlaps with q' if and only if a, a' and b, b' are simultaneous, respectively; q meets q' if and only if b, a' are simultaneous or coincide; q contains q' if and only if a precedes a' and b precedes b'. The relationships between quality edges are inspired from Allen's Interval Algebra [12].

For instance, in Figure 2, (a_1, a_3) meets with (a_3, a_6). Similarly, in Figure 3, (a_1, a_3) is before (a_5, a_6). The same does not hold in Figure 2.

ASKING TEMPORAL QUESTIONS: QUERIES

The Language $L_{\mathcal{H}}$

Temporal graphs store time-dependent information. They act as a temporal knowledge base for an ever changing domain. In what follows, we present a means for interrogating the knowledge base, the language $L_{\mathcal{H}}$.

Consider a possible query such as *Married(John,Alice)*. Intuitively, the question intended here is whether John is married to Alice. Judged with respect to time, the question becomes as follows: "*Is it the case that John was married to Alice, at any point in the evolution of the domain?*" The answer to such a query formulated with respect to a temporal graph $\mathcal{H}$ will return all the quality edges which satisfy it, that is, all quality edges (a, b) from $\mathcal{H}$ such that $\mathcal{L}(a,b) = Married(John, Alice)$.

Next, consider the query *Married(X,Alice)*. Here, *X* is a variable. We use the Prolog-style convention and denote variables by capitals. The query encodes the following question: "*Was Alice ever married to someone?*" The answer will produce a possibly empty set of *records*. If the set is nonempty, then each *record* is a (different) witness that the answer to the above question is *yes*. In our case, each record will contain a substitution $X =$ (e.g., $X = John$) as well as the quality edge (a, b) such that $\mathcal{L}(a,b) = Married(i, Alice)$.

Further on, consider the query *Married(X,Y) after Married(Y,John)*. The query will identify all marriages of some individual *X* to *Y* which precede those of *Y* to John. In this case, each record will store the individual values for *X* and *Y*, together with the quality edges labelled accordingly.

Also, we have

(a) $\|Single(X)\|_{\mathcal{H}_1} = \{\{(a_1, a_3)\}, \{(a_2, a_3)\}\}$;

(b) $\|Single(X) before\ hasChild(X)\|_{\mathcal{H}'_1} = \{\{(a_2, a_3), (a_5, a_6)\}\}$;

(c) $\|Single(X) before\ hasChild(X)\|_{\mathcal{H}_1} = \emptyset$;

(d) $\|Single(X) before\ hasChild(X) \wedge$
$Single(X) meets\ Married(X, John)\|_{\mathcal{H}'_1} =$
$\{\{(a_2, a_3), (a_3, a_6), (a_5, a_6)\}\}$.

The evaluation of query (a) (in $\mathcal{H}_1$) will produce two records, each containing the quality edge which satisfies the label. The evaluation of query (b) (in $\mathcal{H}'_1$) will produce one record of two qualities, one for each label which occurs in the formula. The evaluation of query (c) will produce no record, while that of query (d) will produce one record of three qualities. Each record implicitly contains a mapping function between each satisfied label and its corresponding quality. Since such a function is not vital for the discussion of our approach in this paper, we have chosen to omit it.

Definition 3 (the language $L_{\mathcal{H}}$). Let $\mathbb{V}_{ars}$ designate a set of variables. A term, denoted by t, is either a variable or an individual. The syntax of $L_{\mathcal{H}}$ is recursively defined as follows:

$$\begin{aligned} \varphi ::= \ & R(t_1, \ldots, t_n) \mid \\ & R(t_1, \ldots, t_n) \propto \varphi_1 \mid \\ & \neg\varphi_1 \mid \\ & \bigwedge_i R(t_1, \ldots, t_n) \propto \varphi_i, \end{aligned} \tag{1}$$

where $\propto$ designates any temporal precedence relations between quality edges specified in Definition 2, R is some quality type of arity n, and $t_1,\ldots,t_n$ are terms.

We denote by $\|\varphi\|_{\mathcal{H}}$ the set of records which satisfy the query φ in the temporal graph $\mathcal{H}$. $\|\cdot\|_{\mathcal{H}}$ constitutes the semantics of $L_{\mathcal{H}}$, which we will not discuss in detail. Instead, we refer the reader to [3].

Negation and conjunction require some clarifications. The formula $\neg\varphi$, interpreted in a t-graph $\mathcal{H}$, should be interpreted as φ *is not true in* $\mathcal{H}$; hence $\|\varphi\|_{\mathcal{H}} = \emptyset$. Hence, a (sub)formula of the type $\neg\varphi$ will not generate a record, when satisfied.

Conjunction is used to express multiple temporal constraints over the same quality. For instance, the formula *Single(X) before hasChild(X)* $\wedge$ *Single(X) meets Married(X,John)* expresses two constraints on the quality

Single(X), which must be simultaneously satisfied by each quality edge from the record of *Single(X)*.

$L_{\mathscr{H}}$ Complexity

Proposition 4 (see [3, 18]). *Let* $\mathscr{H}$ *be a t-graph and* φ *a formula of* $L_{\mathscr{H}}$. *Checking* $\|\varphi\|_{\mathscr{H}} \neq \emptyset$ *is NP-complete.*

Sketch. We prove hardness only. For membership, see [3, 18]. As a reduction, we use the conjunctive query problem [19]: given a structure S and a sentence of the form

$$\varphi_c = \exists x_1 \cdots \exists x_n C_1 \wedge \cdots \wedge C_n \tag{2}$$

where each C_i is an atomic formula containing no free variables, the problem asks if S makes the formula true. From φ_c we build an $L_{\mathscr{H}}$ formula as follows: for each C_i we build the $L_{\mathscr{H}}$ formula (gadget): C_i *overlaps* $Fi(e)$. φ is the conjunction of such gadgets. Next, from the structure S, we build a t-graph $\mathscr{H}$: (i) we create a quality edge $q = (a, b)$ labelled *Fix(e)*; (ii) for each relation instance $C_i(\overline{c})$ in S, we build the quality edge $q_i = (a_i, b_i)$ labelled $C_i(\overline{c})$, such that q_i overlaps with q.

(⇒). Assume S makes φ_c true. In particular, S makes $\exists x_1 \cdots \exists x_n. C_i$ true, for each C_i. Hence, there exists a quality edge in $\mathscr{H}$ which satisfies the query C_i *overlaps* $Fi(e)$, for each C_i. Thus, $\|\varphi\|_{\mathscr{H}}$ is nonempty.

(⇐). Assume $\|\varphi\|_{\mathscr{H}}$ is nonempty and let r be some record of $\|\varphi\|_{\mathscr{H}}$. r must contain, for each $L_{\mathscr{H}}$ subformula C_i *overlaps* $Fi(e)$, a quality edge (a_i, b_i) which satisfies it; hence one labelled $C_i(\overline{c_i})$, which is also a relation instance of S. Therefore, for all $C_i, \overline{c_i}$, are evidence that the conjunctive query φ_c is true in S.

The computational complexity of the query satisfaction problem may seem discouraging at first sight. However, the source of complexity can be found in the maximal arity of the underlying qualities. For instance, given a formula Q(X,Y,Z), substitution would require building n^3 possible labels Q(i,j,k), where n is the total number of individuals. For formulae where the arity is an unbounded m, the possible labels become exponential: n^m. However, in practice, it is less likely that queries will be formulated with qualities of arity larger than 4. Thus, under this assumption, the computational complexity of satisfying queries becomes manageable.

TEMPORAL INFERENCE: CG

Updating Temporal Graphs

As illustrated up to this point, the language $L_{\mathscr{H}}$ is a means for investigating the evolution of a domain described as a temporal graph. The latter acts as a structured log and offers no means of interfering with the domain's current and future evolution. In this section we make a step forward and describe a means of achieving this. We introduce yet another language, which we call CG, which can be used to make changes to a domain. Unlike $L_{\mathscr{H}}$, CG is not a logical/temporal language, but a *programming language*, operating on a knowledge base which constitutes a temporal graph. The basic programming unit of CG is the *rule*. A rule consists of (i) a set of preconditions, (ii) an action, and (iii) a set of effects. An example is given below (in what follows, we abandon our "John and Alice" example theme for a more practical one, related to the field of application of CG, namely, that of agent programming):

```
rule r1:
On(X) as q
OperatesUnderCG(X)
=> turnOff(X) as a =>
terminate q in a, create Off(X) from a.
```

Each precondition is given as an $L_{\mathscr{H}}$-formula, where qualities can be named for later use (e.g., On(X) as q). The action (ii) specifies a stimulus under which the rule-at-hand is activated. In our example, turnOff is the respective action. Once a rule is activated, each precondition must be evaluated, in order to establish whether the rule should be applied or not. Let $\varphi_1 \cdots \varphi_n$ be the preconditions of a rule. By evaluating φ_1 we obtain the set $\|\varphi_1\|_{\mathscr{H}}$, which contains a list of records. Each such record r will contain the qualities which have satisfied φ_1, together with a substitution for each variable occurring in φ_1. The evaluation of the next precondition (φ_2) will be achieved with respect to the substitution in each r. When evaluating the last precondition with respect to all previous records, we will obtain complete substitutions of all variables occurring in the preconditions. Each such substitution, together with the matched qualities, is an *activation record.*

If at least one activation record exists for a rule, we say it is applicable, which means the effects (iii) can be enforced on the temporal graph. The effects of a rule consist in adding new qualities to the temporal graph or terminating existing ones. Both initiation and termination are relative to existing qualities from the temporal graph and to the current moment. For instance, terminate q in a will have the effect of terminating the matched quality q, in the action node corresponding to a. Similarly, create Off(x) from a will create a new quality edge labelled Off(x), which spans a and a special *current action*, belonging to the current moment of time and which implicitly terminates all qualities which are known to hold at the current moment.

Rules such as r1 model ontological knowledge. They maintain temporal graphs by making implicit information explicit. In the previous example, the occurrence of a signal turnOff(a) will produce the disconnection of the device a, provided that it is controlled by the application (OperatesUnderCG(a) is true). The rule explicitly states this, by adding the Off(a) quality, starting from the exact time when turnOff(a) is executed. Such rules are *reactive to actions only*.

We also allow programmers to execute their own actions and thus steer the evolution of the domain in the desired way. For instance, the very simple rule

```
rule r2:
On(X),AirConditioner(X),Open(win)
=> turnOff(X)
```

will turn off all air conditioners, whenever the window is opened.

In what follows we provide a grammar for the language CG:

```
<rule>::= rule <id>: <pl> [=> <act>]
=> <el>
<pl>:: =φ [,<pl>]
<act>,<qual>::= R(t_1,...,t_n)
<el>::= <eff> [,<el>]
<eff>::= <act>
          | create <qual> from <id>
          | terminate <id> in <id>.
```

We have denoted by pl and el *precondition list* and *effect list*. Act and qual designate the *action* and *quality* tokens, while eff designates an effect.

id is a program identifier used to denominate rules, matched qualities, and/or actions.

CHECKING THE CORRECTNESS OF CG PROGRAMS

Rule-based programs are usually validated by submitting some sample results to human experts. While it can be helpful it obviously does not provide enough coverage. A formal specification provides an independent standard of accuracy that can be used to check the program output. Our goal is to develop a formal specification for CG programs. We have avoided defining new action logics based on $L_{\mathscr{H}}$—in the spirit of PDI (propositional dynamic logics) [20] or situation calculus [21]—for specifying rules, their preconditions, and effects and opted for existing methods. To this end we will use Event-B specification method [22] and the Rodin platform [23].

A rule-based program has two components: a database of rules and a rule interpreter. The correctness of the rule-based program involves the correctness of the database and the correctness of the interpreter. A correct database of rules is a database where the rules do not contradict. A correct rule interpreter infers all the pertinent conclusions entailed by its facts and rules and does not infer any conclusions that are not justified by them. In order to fit into the Event-B modeling framework, in this section, we have opted for viewing preconditions, actions, and effects as *facts*, thus ignoring the differences between them. This abstraction does not affect the generality of our results and merely serves to make our model more legible.

Event-B Modeling of Facts and Rules

We model the facts by the abstract set FACT. Rules associate a set of facts—the premises—with another fact—the conclusion. If the premises all hold, the conclusion will also hold. In our model we represent the rule database as a relation from the set of facts to facts:

$$rules \in \mathscr{P}(\text{FACT}) \longleftrightarrow \text{FACT}. \tag{3}$$

New facts are generated by examining the whole set of facts, applying all the applicable rules, and adding all the new facts to the set. The whole process is repeated until no new facts appear. We model this process as an application of the function *infer*:

$$infer \in \mathscr{P}(\text{FACT}) \longleftrightarrow \mathscr{P}(\text{FACT}) \tag{4}$$

having as domain the initial set of facts and codomain the final set of facts. While the initial set of facts also appears in the final set, *infer* may add new facts:

$$infer = \lambda facts \cdot facts \in \mathscr{P}(\text{FACT}) \mid facts$$
$$\cup\, rules\left[\mathscr{P}\left(facts\right)\right]. \tag{5}$$

The expression $rules[\mathscr{P}(facts)]$ is the set of all the conclusions of all the rules whose premises match some combination of the initial set of facts, where $\mathscr{P}(facts)$ is the set of all combinations of the initial facts. The expression rules $[\mathscr{P}(facts)]$ use the relational image brackets [·] in order to get the set of conclusions of all the pairs in the relation rules whose premises appear in the set $\mathscr{P}(facts)$.

The function *infer* should be applied until no more conclusions can be inferred. We model the repetitive application of *infer* by the function $closure$, the transitive closure of *infer*, defined by the following axioms:

$$closure\left(infer\right) \in \mathscr{P}(\text{FACT}) \longrightarrow \mathscr{P}(\text{FACT})$$

$$infer \subseteq closure\left(infer\right)$$

$$closure\left(infer\right); \quad infer \subseteq closure\left(infer\right)$$

$$\forall p \cdot infer \subseteq p \wedge p; \quad infer \subseteq p \Longrightarrow closure\left(infer\right) \subseteq p \tag{6}$$

that specify the characteristic properties of the irreflexive transitive closure. Given a relation r from a set S to itself, the irreflexive transitive closure of r, denoted by $closur(r)$, is also a relation from S to S. The characteristic properties of $closur(r)$ are as follows:

(i) relation r is included in $closure(r)$;
(ii) the forward composition of $closure(r)$ with r is included in $closure(r)$;
(iii) relation $closure(r)$ is the smallest relation dealing with (i) and (ii).

Rule Consistency

We assume that rules are consistent if, starting from a consistent set of facts, there is no way to infer inconsistent facts. In order to model the inconsistency of facts we use the set $inconsistent$ containing mutually exclusive categories, that is, sets of facts that we know are inconsistent:

$$inconsistent \in \mathcal{P}(\mathcal{P}(\text{FACT})). \tag{7}$$

A consistent set of facts contains no more than one element from each set of mutually exclusive categories:

$$consistent \in \mathcal{P}(\mathcal{P}(\text{FACT}))$$

$$\forall facts, mutually\_exclusive \cdot facts \in consistent \wedge$$

$$mutually\_exclusive \in inconsistent$$

$$\Longrightarrow card\left(mutually\_exclusive \cap facts\right) < 2, \tag{8}$$

where $facts$ ranges over all sets of consistent facts and $mutually\ exclusive$ ranges over all sets of inconsistent facts.

Specification of Rule-Based Programs

A rule-based program infers conclusions relevant to some goal based on some data. We model the inference process by the event $Inference$ (see Specification 1).

```
Inference
  ANY
    data
    goals
  WHERE
    grd1:   data ∈ 𝒫(FACT)
    grd2:   goals ∈ 𝒫(FACT)
  THEN
    act1:
    facts :| facts' ∈ 𝒫(FACT) ∧ facts ⊆ facts' ∧ facts' ⊆ (closure(infer))(facts ∪ data) ∧ conclusions ⊆ facts'
    act2:   conclusions = goals ∩ (closure(infer))(facts ∪ data)
  END
```

Specification 1. Specification of the inference process.

The variables $facts$ and $facts'$ represent the state of the system before and after the execution of the $Inference$ event. The nondeterministic assignment operator :| expresses that a modification is possible using a before-after-predicate (expressed in the above specification immediately after the first occurrence of :|, from the act1 action). The data are the new facts that are introduced, and conclusions are the facts that are inferred. The expression $(closur(infer))(facts \cup \text{data})$ denotes the set of valid facts that can be inferred from the initial set of facts and the new data; $facts \subseteq facts'$ models the assumption that the program never retracts any conclusions;$facts' \subseteq (closure(infer))(facts \cup \text{data})$ expresses that all new facts are valid inferences; conclusions $\subseteq facts'$ models the inference of valid goals.

The complete specification is shown in Specification 2.

```
CONTEXT
Rules_c0
SETS
  FACT
CONSTANTS
  rules
  infer
  closure
  consistent
  inconsistent
AXIOMS
  axm1:  rules ∈ ℙ(FACT) ↔ FACT
  axm2:  infer ∈ ℙ(FACT) → ℙ(FACT)
  axm3:  infer = (λfacts · facts ∈ ℙ(FACT) | facts ∪ rules[ℙ(facts)])
  axm4:  closure(infer) ∈ ℙ(FACT) → ℙ(FACT)
  axm5:  infer ⊆ closure(infer)
  axm6:  closure(infer); infer ⊆ closure(infer)
  axm7:  ∀p · infer ⊆ p ∧ p; infer ⊆ p ⟹ closure(infer) ⊆ p
  axm8:  consistent ∈ ℙ(ℙ(FACT))
  axm9:  inconsistent ∈ ℙ(ℙ(FACT))
  axm10:
  ∀facts, mutually_exclusive · facts ∈ consistent ∧ mutually_exclusive ∈ inconsistent → card(mutually_exclusive ∩ facts) < 2
  axm11:  ∀facts · facts ∈ consistent ⟹ (closure(infer))(facts) ∈ consistent
END
MACHINE
  Rules0
SEES
  Rules_c0
VARIABLES
  facts
  conclusions
INVARIANTS
  inv1:  facts ∈ ℙ(FACT)
  inv2:  conclusions ∈ ℙ(FACT)
EVENTS
  INITIALISATION:
    THEN
      act1:  facts = ∅
      act2:  conclusions = ∅
    END
  Inference:
    ANY
      data
      goals
    WHERE
      grd1:  data ∈ ℙ(FACT)
      grd2:  goals ∈ ℙ(FACT)
    THEN
      act1:  facts :| facts′ ∈ ℙ(FACT) ∧ facts ⊆ facts′ ∧ facts′ ⊆ (closure(infer))(facts ∪ data) ∧ conclusions ⊆ facts′
      act2:  conclusions = goals ∩ (closure(infer))(facts ∪ data)
    END
END
```

Specification 2. The complete specification.

Our model is an abstract one that can be further refined by refining the *Inference* event, by explicitly specifying control methods (e.g., backward/forward chaining), which explicitly specify how new facts are deduced.

Model Validation

The model has been specified and validated using Rodin, an Eclipse-based IDE for Event-B that provides support for refinement and mathematical proofs [23]. The model is validated by discharging proof obligations. The state of development is described in Table 1 with the required proof obligations.

Table 1

Name	Total	Auto	Manual	Reviewed	Undischarged
Rules	9	6	3	0	0
Rules_c0	6	4	2	0	0
Rules0	3	2	1	0	0

IMPLEMENTATION

Our previous implementation efforts were either (i) driven by the application context [15, 16, 24, 25] or (ii) attempted to follow closely the algorithm description, in order to highlight correctness [3, 14, 18]. There are lessons to be learned from either approach. For instance, approaches such as [16, 25] are highly dependent on the web service (WS) architecture which is vital for communicating with intelligent devices. Although proficient for the envisaged scenario, (i) lacks portability as well as scalability. While the WS approach has its well-known advantages [26], the author believes WS development can occasionally be hardened by the platform, IDE, and other application constraints. On the other hand, approaches such as [18] which is split between two implementations in two different languages (Haskell/Frege [27] and CLIPS [28]) and is not application-dependent may lack usability. We believe (ii) to be highly dependent on the implementations of the two languages (the former, Frege, a rather experimental language, is known to have unintuitive dissimilarities from Haskell).

Thus, we take a step back and opt for a new approach, one which preserves the application independence of (ii) but which is more user-friendly, easy to use, and reliant on a unique programming environment. The reader may have noticed some similarities between Prolog and CG. One common feature is the query (/clause) matching process which, essentially, is the same for both languages.

Actually, CG can be seen as a temporal layer on top Prolog: the flat knowledge base is replaced by a temporal one (a temporal graph). Each CG rule can be seen as a collection of Prolog clauses. Goal (re)satisfaction corresponds to rule execution for each found activation record.

In order to represent temporal graphs in Prolog, we use the following 5 metapredicates:

```
node(A).
edge(A,B).
in(H,A).
quality(A,B,q).
action(A,a).
```

The factual knowledge node(A) indicates that A is an action node, while action(A,a) assigns the label a to A. Similarly, edge(A,B). indicates that (A, B) is a quality edge, while quality(A,B,q) assigns the label q to (A, B). Both a and q are arbitrary Prolog predicates. Finally, in(H,A) indicates that H is the hypernode to which action node A belongs. Hypernodes are not prespecified by another predicates, since this would be superfluous. Thus, the set of hypernodes can be identified as the entities H satisfying the goal in(H,_).

The challenge when transforming a CG rule to a set of Prolog sequences is expressing temporal constraints between qualities. To achieve this, we compute the transitive closure of the appropriate direct precedence relation, introduced in Definition 2.

We illustrate this by a simple example, given by the following query:

```
q(X,Y) before r(X).
```

Such a query is transformed into a clause of the following form:

```
find(A,B):- quality(A,B,q(X,_)),
            quality(C,_,r(X)),
            precedes(B,C).
```

Thus, in order to identify the quality edge(s) satisfying the above query, one must find a quality edge labelled r(X) whose initiating action node C must be preceded by B. Precedence is computed as the transitive closure of the edge and simultaneous relations:

```
precedes(X,Y):- edge(X,Y), !.
precedes(X,Y):- simultaneous(X,Y), !.
precedes(X,Y):- edge(X,Z),
                precedes(Z,Y), !.
precedes(X,Y):- simultaneous(X,Z),
                precedes(Z,Y), !.
```

Note that we have used !(cuts), in order to avoid unnecessary explorations of the knowledge base, once precedence has been established. The simultaneous clause is defined as follows:

```
simultaneous(X,Y):- in(H,X),
                    in(H,Y),
                    not(X=Y), !.
```

And it is satisfied if the two action nodes X and Y belong to the same hypernode.

Specifying more complicated queries is achieved compositionally, following the scheme presented above. Executing the effects of a rule reduces to enriching the metarelations defined at the beginning of this section, in a *transactional* manner, in the spirit of [14]. This means, in short, that all effects resulting from rules which are applicable at the same moment of time are added to the knowledge base in a manner which is perceived as simultaneous by the programmer. This implies that the effects of one rule cannot invalidate another, if both rules have been applicable at the same moment.

CONCLUSION

Temporal graphs coupled with $L_{\mathcal{H}}$ in CG are a powerful method for performing temporal reasoning and for enforcing time-dependent behaviour within intelligent systems. One major advantage of CG is that time is not encoded explicitly in other program-dependent structures. Time is a language primitive in itself, and this design choice makes program development straightforward, even for the inexperienced programmer. Besides being easy to read, declarative programs can also be easy to verify, as illustrated in Section 5. The prototype described in Section 6 relies on the cost-expensive resolution of Prolog; however more efficient implementations are possible. We leave such an endeavour for future work.

ACKNOWLEDGMENT

The author wishes to acknowledge the help and support from Professor Cristian Giumale, who was the first to think about temporal graphs and whose guidance patroned the work presented in this paper.

REFERENCES

1. M. Y. Vardi, "From church and prior to PSL," in *25 Years of Model Checking*, vol. 5000 of *Lecture Notes in Computer Science*, pp. 150–171, Springer, Berlin, Germany, 2008.
2. E. M. Clarke Jr., O. Grumberg, and D. A. Peled, *Model Checking*, The MIT Press, 1999.
3. M. Popovici, "Using evolution graphs for describing topology-aware prediction models in large clusters," in *Computational Logic in Multi-Agent Systems*, M. Fisher, L. van der Torre, M. Dastani, and G. Governatori, Eds., vol. 7486 of *Lecture Notes in Computer Science*, pp. 94–109, Springer, Berlin, Germany, 2012.
4. E. M. Clarke, E. A. Emerson, and A. P. Sistla, "Automatic verification of finite state concurrent systems using temporal logic specifications: a practical approach," in *Proceedings of the Conference Record of the 10th Annual ACM Symposium on Principles of Programming Languages*, pp. 117–126, Austin, Tex, USA, January 1983.
5. A. P. Sistla, "On characterization of safety and liveness properties in temporal logic," in *Proceedings of the 4th Annual ACM Symposium on Principles of Distributed Computing*, pp. 39–48, Ontario, Canada, August 1985.
6. A. Biere, C. Artho, and V. Schuppan, "Liveness checking as safety checking," *Electronic Notes in Theoretical Computer Science*, vol. 66, no. 2, pp. 160–177, 2002.
7. O. Kupferman and M. Y. Vardi, "Model checking of safety properties," *Formal Methods in System Design*, vol. 19, no. 3, pp. 291–314, 2001.
8. L. Zhang, U. Hustadt, and C. Dixon, "A resolution calculus for the branching-time temporal logic CTL," *ACM Transactions on Computational Logic*, vol. 15, no. 1, article 10, 2014.
9. J. Gaintzarain and P. Lucio, "Logical foundations for more expressive declarative temporal logic programming languages," *ACM Transactions on Computational Logic*, vol. 14, no. 4, p. 28, 2013.
10. J. Gaintzarain, M. Hermo, P. Lucio, M. Navarro, and F. Orejas, "Invariant-free clausal temporal resolution," *Journal of Automated Reasoning*, vol. 50, no. 1, pp. 1–49, 2013.
11. M. Fisher, C. Dixon, and M. Peim, "Clausal temporal resolution," *ACM Transactions on Computational Logic*, vol. 2, no. 1, pp. 12–56, 2001.

12. J. F. Allen, "Planning as temporal reasoning," in *Principles of Knowledge Representation and Reasoning*, J. F. Allen, R. H. Fikes, and E. Sandewall, Eds., pp. 3–14, Morgan Kaufmann, San Mateo, Calif, USA, 1991.
13. C. Giumale, L. Negreanu, M. Muraru, M. Popovici, A. Agache, and C. Dobre, "Modeling with fluid qualities," in *Proceedings of the 18th International Conference on Control Systems and Computer Science (CSCS '11)*, 2011.
14. M. Popovici, M. Muraru, A. Agache, C. Giumale, L. Negreanu, and C. Dobre, "A modeling method and declarative language for temporal reasoni ng based on fluid qualities," in *Proceedings of the 19th International Conference on Conceptual Structures for Discovering Knowledge (ICCS '11)*, pp. 215–228, Springer, Berlin, Heidelberg, 2011.
15. C. Giumale, L. Negreanu, M. Muraru, and M. Popovici, "Modeling ontologies for time-dependent applications," in *Proceedings of the 12th International Symposium on Symbolic and Numeric Algorithms for Scientific Computing*, pp. 202–208, Timisoara, Romania, September 2010.
16. M. Popovici, M. Muraru, A. Agache, L. Negreanu, C. Giumale, and C. Dobre, "Integration of a declarative language based on fluid qualities in a service-oriented environment," in *Proceedings of the 14th IASTED International Conference on Artificial Intelligence and Soft Computing*, Crete, Greece, June 2011.
17. M. Popovici and L. Negreanu, "Strategic behaviour in multi-agent systems able to perform temporal reasoning," in *Intelligent Distributed Computing*, pp. 211–216, 2013.
18. M. Popovici, *A logical language for temporal knowledge representation and reasoning [Ph.D. thesis]*, 2012.
19. A. K. Chandra and P. M. Merlin, "Optimal implementation of conjunctive queries in relational data bases," in *Proceedings of the 9th Annual ACM Symposium on Theory of Computing (STOC '77)*, pp. 77–90, New York, NY, USA, 1977.
20. R. S. Streett, "Propositional dynamic logic of looping and converse," in *Proceedings of the 13th Annual ACM Symposium on Theory of Computing (STOC '81)*, pp. 375–383, New York, NY, USA, 1981.

21. G. Lakemeyer, "The situation calculus: a case for modal logic," *Journal of Logic, Language and Information*, vol. 19, no. 4, pp. 431–450, 2010.

22. J.-R. Abrial, *Modeling in Event-B: System and Software Engineering*, Cambridge University Press, New York, NY, USA, 1st edition, 2010.

23. Rodin platform, 2014, http://wiki.event-b.org/.

24. M. Popovici, C. Dobre, M. Muraru, and A. Agache, "Modeling of standards and the open world assumption," in *Proceedings of the Future Business Technology (FUBUTEC '11)*, pp. 5–17, April 2011.

25. M. Popovici, M. Muraru, A. Agache, L. Negreanu, C. Giumale, and C. Dobre, "An ontology-based dynamic service composition framework for intelligent houses," in *Proceedings of the 10th International Symposium on Autonomous Decentralized Systems (ISADS '11)*, pp. 177–184, Tokyo, Japan, March 2011.

26. M. P. Papazoglou and D. Georgakopoulos, "Introduction: service-oriented computing," *Communications of the ACM*, vol. 46, no. 10, pp. 24–28, 2003.

27. "The frege programming language," 2014, https://github.com/Frege/frege.

28. J. C. Giarratano and G. D. Riley, *Expert Systems: Principles and Programming*, Brooks/Cole, Pacific Grove, Calif, USA, 2005.

Chapter

6

LOLISA: FORMAL SYNTAX AND SEMANTICS FOR A SUBSET OF THE SOLIDITY PROGRAMMING LANGUAGE IN MATHEMATICAL TOOL COQ

Zheng Yang and Hang Lei

School of Information and Software Engineering, University of Electronic Science and Technology of China, No.4 Section 2 North Jianshe Road, Chengdu 610054, China

ABSTRACT

The security of blockchain smart contracts is one of the most emerging issues of the greatest interest for researchers. This article presents an intermediate specification language for the formal verification of Ethereum-based smart contract in Coq, denoted as Lolisa. The formal syntax and semantics of Lolisa contain a large subset of the Solidity programming language developed for the Ethereum blockchain platform. To enhance type safety, the formal syntax of Lolisa adopts a stronger static type system than Solidity.

Citation: Zheng Yang and Hang Lei, "Lolisa: Formal Syntax and Semantics for a Subset of the Solidity Programming Language in Mathematical Tool Coq", Mathematical Problems in Engineering, volume 2020, article ID 6191537, https://doi.org/10.1155/2020/6191537.

In addition, Lolisa includes a large subset of Solidity syntax components as well as general-purpose programming language features. Therefore, Solidity programs can be directly translated into Lolisa with line-by-line correspondence. Lolisa is inherently generalizable and can be extended to express other programming languages. Finally, the syntax and semantics of Lolisa have been encapsulated as an interpreter in mathematical tool Coq. Hence, smart contracts written in Lolisa can be symbolically executed and verified in Coq.

INTRODUCTION

The blockchain platform [1] is one of the emerging technologies developed to address a wide range of disparate problems, such as those associated with cryptocurrency [2] and distributed storage [3]. Presently, this technology has gained interest from the finance sector [4]. Ethereum is one of the most widely adopted blockchain systems. One of the most important features of Ethereum is that it implements a very flexible general-purpose Turing-complete programming language denoted as Solidity [5]. This allows for the development of arbitrary applications and scripts that can be executed in a virtual runtime environment denoted as the Ethereum Virtual Machine (EVM) to conduct blockchain transactions automatically. These applications and scripts (i.e., programs) are collectively denoted as smart contracts, which have been widely used in many critical fields, such as the medical [6] and financial fields. The growing use of smart contracts has led to an increased scrutiny of their security. Smart contracts can include particular properties (i.e., bugs) making them susceptible to deliberate attacks that can result in direct economic loss. Some of the largest attacks on smart contracts are well known, such as the attack on decentralized autonomous organization (DAO) and Parity wallet [7] contracts. In fact, many classes of subtle bugs, ranging from transaction-ordering dependencies to mishandled exceptions, exist in smart contracts [8].

The present article capitalizes upon our past work by defining the formal syntax and operational semantics for a large subset of the Solidity version 0.4. This subset is denoted herein as Lolisa and has the following features.

Consistency Lolisa formalizes most of the types, operators, and mechanisms of Solidity according to Solidity documentation. As such, programs written in Solidity can be translated into Lolisa, and vice versa, with a line-by-line correspondence without rebuilding or abstracting, which are operations that can negatively impact consistency.

Static Type System The formal syntax in Lolisa is defined using generalized algebraic datatypes (GADTs) [9], which impart static type annotation to all the values and expressions of Lolisa. In this way, Lolisa has a stronger static type system than Solidity for checking the construction of programs.

Executable and Provable In contrast to similar efforts focused on building formal syntax and semantics for high-level programming languages, the formal semantics of Lolisa are defined based on the GERM framework in conjunction with EVI. Therefore, it is theoretically possible for ethereum-based smart contracts written in Lolisa to be symbolically executed and have their properties simultaneously verified automatically in higher-order logic theorem-proving assistants directly when conducted in conjunction with a formal interpreter developed based on GERM framework.

Mechanized and Validated The syntax and semantics of Lolisa are mechanized using the Coq proof assistant [10]. We also develop a formal verified interpreter in Coq to validate whether Lolisa satisfies the above *Executable and Provable* feature and the meta-properties of the semantics. The details regarding the implementation of our formal interpreter have been presented in another paper [11].

The remainder of this paper is structured as follows. Section 2 introduces related work regarding the programming language formalization. Section 3 introduces the overall structure of the specification language framework and provides predefinitions of Lolisa syntax and semantics. Section 4 elaborates on the formal abstract syntax of Lolisa and compares this with the formal abstract syntax of Solidity. Section 5 presents the formal dynamic semantics of Lolisa, including the program execution semantics and the formal standard library for the built-in data structures and functions of EVM. Section 6 describes the integration of the Lolisa programming language and its semantics within the formal verified interpreter FEther. Section 7 discusses the contributions and limitations of our current work. Finally, Section 8 presents the conclusions of our work.

RELATED WORK

Software engineering techniques employing such static and dynamic analysis tools as Manticore [12] and Mythril [13] have not yet been proven to be effective at increasing the reliability of smart contracts.

KEVM [14] is a formal semantics for the EVM written using the K-framework, like the formalization conducted in Lem [15]. KEVM is executable, and therefore can run the validation test suite provided by the Ethereum foundation. The symbolic reasoning conducted for KEVM programs involves specifying properties in Reachability Logic and verifying them with a separate analysis tool. While these represent currently available mechanized formalizations of operational semantics, axiomatic semantics, and formal low-level programming verification tools for EVM and Solidity bytecode [16], they are not well-suited for high-level programming languages, such as Solidity. In response, the Ethereum community has placed open calls for formal verification proposals [17] as part of a concerted effort to develop formal verification strategies [18]. Fuzzing testing is an efficient and effective testing technique. Presently, numerous projects develop fuzzing in smart contracts to analyze vulnerabilities, such as ReGuard [19]. Securify [20] is a type of Ethereum-based smart contracts security analyzer based on static analysis. It verifies the behavior of target smart contracts based on the given security properties at the Ethereum virtual machine bytecode level. Securify provides a kind of domain-specific language which can write security properties according to the attack reports and the basic practices. MadMax [21] is a static program analysis framework that takes the Ethereum bytecode as analysis source code and automatically analyzes common vulnerabilities such as the integer and memory overflows vulnerabilities. Besides, it is the first tool that allows for loop specifications to be defined by a dynamic property. In this manner, this tool can avoid loop explosion during the verification process. Similarly to OYENTE, Ehtir [22] is also a type of rule-based static analyzer for the bytecode of Ethereum smart contracts. This tool can produce control flow graphs and includes the whole possible execution addresses. VeriSolid [23] is a formal verification framework which can be accessed through the web directly. Its foundational concept is FSolidM [24]. In brief, the VeriSolid presents a formal verification framework which provides an approach for semiautomatically developing the correct formal specifications of smart contracts. A new approach is presented in Abdellatif and Brousmiche [25] which can model the execution behaviors of target smart contracts based on a formal model checking language. This technique can be applied to verify the execution behavior and authority of target smart contracts by using model checking methods.

In other fields of computer science, a number of interesting studies have focused on developing mechanized formalizations of operational semantics for different high-level programming languages. The Park project [26] presents completely formalized denotational semantics and the corresponding syntax in the JavaScript language. The CompCert project [27] is another influential verification work for C and GCC that developed a formal semantics for a subset of C denoted as Clight. This work formed the basis for VST [28] and CompCertX [29]. In addition, a number of interesting formal verification studies have been conducted for operating systems based on the CompCert project. In addition, the operational semantics of JavaScipt also have been investigated [30], which is of particular importance to the present study because Solidity is a programming language like JavaScipt. However, few of the frameworks defined in these related works can be symbolically executed or analyzed in higher-order logic theorem-proving assistants directly.

FOUNDATIONAL CONCEPTS

The overall architecture of Lolisa is shown in Figure 1. Table 1 summarizes the helper functions used in the dynamic semantic definitions. Table 2 lists the state functions used to calculate commonly needed values from the current state of the program. All of these state that functions will be encountered in the following discussion. Components of specific states will be denoted using the appropriate Greek letter subscripted by the state of interest. As shown in Table 2, the context of the formal memory space is denoted as *M*, where σ is employed to denote a specific memory state; the context of the execution environment is represented as ε; and we assign Λ to denote a set of memory addresses, where the meta-variable α is employed to represent an arbitrary address. Similarly, we define the function return address Λ_{fun}. In addition, struct is an important data structure in Lolisa. Therefore, we adopt Σ to represent the Lolisa struct information context, and Θ is employed to represent the set of pointers of the struct types. Also, the following type of assignments may include variables, so our types will include references to variable-typing contexts, which we will denote as Γ, Γ_1, etc. Such contexts are finite mappings from variable names to types. Because programs may also contain references to the declared functions of a Solidity program, another mapping is needed from function identifiers to types. This mapping will be succinctly denoted as Φ, Φ_1, etc. Furthermore, we assign Ω as the native value set of the basic logic system. For brevity in the following discussion, we will assign $\mathscr{F}$ to represent the overall formal system combination of Σ, Γ, Θ, Ω, Φ, and Λ. Due to limitation of length, the details of Lolisa's formalization have been presented in our online report (https://arxiv.org/abs/1803.09885).

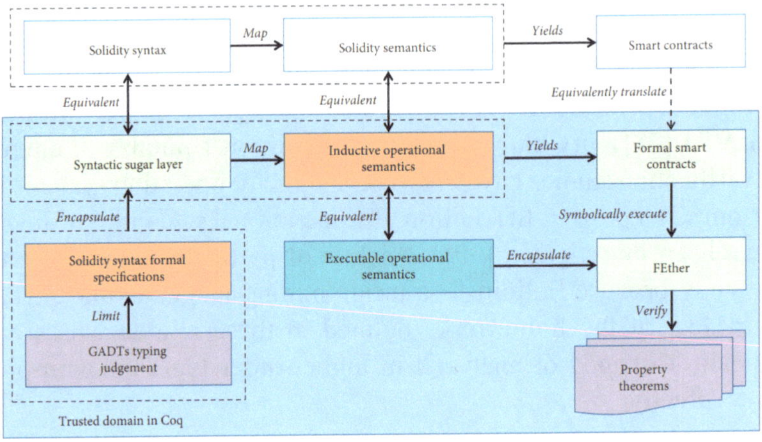

Figure 1. Overview of Lolisa's architecture.

Table 1. Helper functions.

Symbol	Definition	Symbol	Definition
map_{addr}	Searches the indexed address of a mapping type	map_{get}	Obtains the value in a mapping type term
$eval_{bop}$	Evaluates binary operation expressions	$eval_{uop}$	Evaluates unary operation expressions
$mems_{find}$	Searches the required struct member	env_{check}	Validates the current environment
set_{env}	Changes the current environment	$inherit_{chec}$	Validates the inheritance information
$init_{var}$	Initializes the variable address	$init_{re}$	Initializes the function return address

Table 2. State functions.

Symbol	Definition	Symbol	Definition
M	Memory space	ε	Environment information
Λ	Memory address set	Λ_{fun}	Function return address
Σ	Struct information	Θ	Struct pointer set
Γ	Context structure information	Ω	Native value set
Φ	Function information	$\mathscr{F}$	Overall formal system

FORMAL SYNTAX OF LOLISA

Types

The formal abstract syntax of Lolisa types is given in Figure 2. Supported types include arithmetic types (integers in various sizes and signedness), byte types, array types, mapping types, as well as function types and struct types. Although Solidity is a JavaScript-like language, it supports pointer reference. Therefore, Lolisa also includes pointer types (including pointers to functions) based on label address specification. Furthermore, these types of annotations and relevant components can be easily formalized by enumerating inductively in Coq or other higher-order logic theorem-proving assistants. Lolisa does not support any of the type qualifiers such as const, volatile, and restrict, and these qualifiers are simply erased during parsing.

Signedness:	$signedness ::= Signed \mid Unsigned$
Integer sizes:	$intsize ::= I8 \mid I16 \mid I32 \mid I64 \mid I128 \mid I256$
Byte sizes:	$bytesize ::= B4 \mid B8 \mid B16 \mid B32$
Array Index of map	$id_{map_{array}} ::= iAconst\_id(n) \mid iAvar\_id(\alpha)$
	$\mid iAstr\_id(\alpha, mems) \mid iArray\_id(\alpha, id_{map_{array}})$
Normal type$_{map}$:	$\tau_{nf_{map}} ::= Iundef \mid Iint(signedness, intsize)$
	$\mid Ibool \mid Istring \mid Ifloat \mid Ibytes(bytesize) \mid Istruct(\alpha)$
	$\mid Ivid(o\alpha) \mid Ipid(o\alpha) \mid Ifid(o\alpha) \mid Icid(o\alpha)$
Map Types Sig:	$\tau_{map} ::= Iarray(id_{map_{array}}, \tau_{map}) \mid \tau_{nf_{map}}$
	$Iaddress ::= Istruct(\_0xaddress) \mid IInt ::= Iint\ Signed\ I64.$
Array Index	$id_{array} ::= Aconst\_id(n) \mid Avar\_id(\alpha)$
	$\mid Astr\_id(\alpha, mems) \mid Amap\_id(\alpha, id_{map_{array}}) \mid Array\_id(\alpha, id_{array})$
Normal type:	$\tau_{nf} ::= Tundef \mid Tint(signedness, intsize)$
	$\mid Tbool \mid Tstring \mid Tfloat \mid Tbytes(bytesize) \mid Tstt$
	$\mid Tmodi \mid Tstruct(\alpha) \mid Tvid(o\alpha) \mid Tpid(o\alpha) \mid Tfid(o\alpha) \mid Tcid(o\alpha)$
Types Sig:	$\tau ::= Tmap(\tau_{map}, \tau) \mid Tarray(id_{array}, \tau) \mid \tau_{nf}$
	$Taddress ::= Tstruct(\_0xaddress) \mid TInt ::= Tint\ Signed\ I64$
	$Tuint ::= Tint\ Unsigned\ I64$

Figure 2. Abstract syntax of Lolisa types.

The types fill two roles in Lolisa. Firstly, they serve as type declarations of identifiers in statements and, secondly, they serve as signatures to specify the GADTs-style constructor of values and expressions for transmitting type information, which will be explained in the following sections. In Coq formalization, the term τ is declared as type according to rule 1, as follows:

$$\tau\text{: type.} \tag{1}$$

Note that many types are defined in Figure 2 as parameterized types recursively. In this way, a specific type is dependent on the specified parameters and can abstract and express many different Solidity types.

One of the most important data types of Solidity is mapping types. In Solidity documentation [4], mapping types are declared as mapping (KeyType⇒ValueType). Here, _KeyType can be nearly any type except for a mapping, a dynamically sized array, a contract, and a struct. As shown in Figure 2, _KeyType is defined as Tmap (τ_{map}, τ), where τ_{map} represents the _KeyType and τ represents the _ValueType. The best way to keep the terms in Lolisa well-typed and to ensure type safety is to maintain type isolation rather than adding corollary conditions. Therefore, we define a coordinate type $type_{map}$ for _KeyType employed in mapping. In particular, the address types in Lolisa are treated as a special struct type, so that _KeyType is allowed to be a struct type in Lolisa. In Coq formalization, $type_{map}$ shares the same constructor with that of type except for Tmap, and a term with type $type_{map}$ is recorded as τ_{map} according to rule 2, as follows:.

$$\tau_{\text{map}}\text{: type}_{\text{map}}. \tag{2}$$

In Solidity, array types, which are defined according to an array index id_{array} as Tarray (id_{array}, τ) in Coq, can be classified as fixed-size arrays and dynamic-size arrays. For fixed-size arrays, the size and index number are allowed to be declared by different data structures including constants, variables, struct, mapping, and field access values. These are respectively formalized as Array Index in Figure 2. Because the size of array types in Solidity can be dynamic, the dynamic-size array type in Lolisa is treated as a special mapping type of τ_{map} (Iint Signed I64).

As shown in Figure 3, (n)-dimensional mapping types, as well as array types, are widely defined in smart contracts. Due to the recursive inductive definition, Lolisa can express n-dimensional array types and n-dimensional mapping types easily, which is illustrated below by rules 3 and 4, respectively:

$$\text{Tarray}\left[\text{id}_0\text{Tarray}\left[\text{id}_1\text{Tarray}\left[\ldots\left[\text{Tarray id}_n\tau_{\text{final}}\right]\right]\right]\right], \tag{3}$$

$$\text{mapping}\left[\tau_{\text{map}_0}\Rightarrow\text{mapping}\left[\tau_{\text{map}_0}\Rightarrow\text{mapping}\left[\ldots\text{mapping}\left[\tau_{\text{map}_n}\Rightarrow\tau_{\text{final}}\right]\right]\right]\right]. \tag{4}$$

```
contract MappingExample{
    mapping(address => mapping(char => uint)) public balances;
        function update(uint amount) returns (address addr){
        ...;
        return msg.sender;
    }
}
```

Figure 3. A simple example of mapping types in Solidity.

We classify τ and τ_{map} into normal form types and nonnormal form types. The normal form types refer to types whose typing rules disallow recursive definition, whereas recursive definition is allowed for nonnormal form types. For example, the normal form of Tarray(id_{array}, Tbool) should be Tbool. In Figure 2, the normal types are defined separately as *Normal type*.

Expressions

Having formally specified all the possible forms of values that may be declared and manipulated in Solidity programs, we now discuss the expressions used in programs to encapsulate values. As introduced in Section 4.1, all expressions and their subexpressions are defined with GADTs, which are annotated by two types of signatures according to rule 5, as follows:

$$\text{expr: } \tau_0 \longrightarrow \tau_1 \longrightarrow \text{Type.} \tag{5}$$

Here, τ_0 refers to the current expression type and τ_1 refers to the normal form type after evaluation. For instance, we would define the type of an integer variable expression *e* as $\mathrm{expr}_{\mathrm{Tvid}_{o\alpha}}$. In this way, the formal syntax of expressions becomes clearer and abstract, and allows the type safety of Lolisa expressions to be maintained strictly. In addition, employing the combination of the two types of annotations facilitates the definition of a very large number of different expressions based on equivalent constructors. Of course, the use of τ_0 and τ_1 may be subject to different limitations depending on the situation.

Constant expressions are used to denote the native values of the basic formal system, which are transformed from the respective Lolisa values. Therefore, τ_0 and τ_1 should satisfy rule 6 given below:

$$\tau_0 = \tau_1 \Lambda \tau_0,$$

$$\tau_1 \in \tau_{nf}. \tag{6}$$

To satisfy the limitation TYPE-FORM, the array types and mapping types should be analyzed and simplified according to the type definitions given by Figure 2 into $\tau_{\text{final}} \in \tau_{\text{nf}}$, which can be formulated as $\Sigma\Theta \leftarrow \tau \longrightarrow \tau' \longrightarrow \cdots \longrightarrow \tau^n \Lambda \tau^n \in \tau_{\text{nf}}$. We denote this process as $\Downarrow_\tau$.

In addition, as mentioned previously, the type information of the value level is successfully transmitted into a constant expression. For example, a value v has type val τ_1, and the constant expression Econst has type $\forall(\tau$: type), val $\tau \longrightarrow$ expr $\Downarrow_\tau \Downarrow_\tau$. Therefore, τ in Econst (v) is determined by τ_1. For example, Econst(Vbool(b)) has type expr Tbool Tbool, where τ is specified by the Tbool of Vbool(b). The type information of the expression level can also be transmitted to the statement level in the same way, which will be described specifically in the next section.

For operator expressions, Lolisa supports nearly all binary and unary operators and we adopt op_{class}(operator) to simplify the formal abstract syntax. In Coq formalization, binary and unary operators are abstracted as an inductive type op that is also defined by GADTs, and specific operators serve as their constructors. In this way, operator expressions are made more clear and concise, and can be extended more easily than when employing a weaker static-type system. The binary and unary operators are annotated by two type signatures, as respectively given in rule 7, as follows:

$$\text{op: } \tau_0 \longrightarrow \tau_1 \longrightarrow \text{Type}. \tag{7}$$

Statements

Figure 4 defines the syntax of Lolisa statements. Here, nearly all the structured control statements of Solidity (i.e., conditional statements, loops, structure declarations, modifier definitions, contracts, returns, multivalue returns, and function calls) are supported, but Lolisa does not support unstructured statements such as goto and unstructured switches like the infamous "Duff's device". Besides, anonymous functions are forbidden in Lolisa because all functions must have a binding identifier to ensure that they are well formed. As previously discussed, the assignment $e_1 = e_2$ of a right-value (r-value) e_2 to a left-value (l-value) e_1, and modifier declarations, as well as function calls and structure declarations are treated as statements.

In addition, statements are also classified according to normal form and nonnormal form categories, where the normal form statement, given as stt_{nf}, represents a statement that halts after being evaluated. Actually, while Solidity is a Turing-complete language, smart contract programs written in Solidity have no existing halting problems because program execution is limited by gas, which we have defined in ε for Lolisa.

$$
\begin{aligned}
& strt_{mem} ::= str\_nil(addr) \mid str\_mem(\tau, name, strmem) \\
& stt_{nf} ::= Var(oacc, \varepsilon_{Tvar}) \mid Struct(a, strt_{mem}) \mid Assignv(\varepsilon_{addr}, \varepsilon) \\
& \quad \mid Return(\varepsilon) \mid Returns(\varepsilon_s) \mid Throw \mid Snil \mid Fstop \\
\textit{Statement:} \quad & s ::= Contract(\varepsilon_{Tpar}) \mid Modifier(\varepsilon_{Tpar}) \\
& \quad \mid Fun(oacc, oflag, opars\ \varepsilon_{Tfun}, modis, s) \\
& \quad \mid Funs(oacc, oflag, opars, \varepsilon_{Tfun}, ts, modis, s) \\
& \quad \mid Loop_{while}(\varepsilon_{Tbool}, s) \mid Loop_{for}(s_1, \varepsilon_{Tbool}, s_2, s_3) \\
& \quad \mid Fun_{call}(\tau, \tau', fitype, fipar) \mid Seq(s, s') \mid If(\varepsilon_{Tbool}, s, s') \mid stt_{nf}
\end{aligned}
$$

Figure 4. Abstract syntax of Lolisa statements.

As defined in Figure 4, we still inductively classify statement definitions into a normal form stt_{nf}, whose typing assignments must be conducted without recursive definition, and non-normal form statements. The normal form statements of Lolisa are defined as stt_{nf}. The remaining statements are nonnormal form statements.

Macro Definition of Formal Abstract Syntax

The Lolisa formal syntax is too complex to be adopted by general users. Lolisa syntax includes the same components as those employed in Solidity; however, it has stricter formal typing rules. Therefore, Lolisa syntax must include some additional components not supported in Solidity, such as type annotations and a monad-type option. Moreover, Lolisa syntax is formally defined in Coq formalization as inductive predicates. Thus, a Lolisa code looks much more complicated than the corresponding Solidity code, even though both the codes demonstrate line-by-line correspondence. An example of this difficulty is illustrated in the code segments shown in Figures 5 and 6. The formal Lolisa version of the conditional statement in the *pledge* function in Figure 6 is much more complicated than that in the original Solidity version in Figure 5.

```
function pledge(bytes32 _message) {
    if (msg.value ==0 || complete || refunded) throw;
    ...;
}
```

Figure 5. Conditional statement in Solidity.

```
Definition fun_pledge :=
(Fun (Some public) (Efun (Some pledge ) Tundef )
  ( pcons (Epar (Some _message) Tbyte32) pnil))
  ( nil )
  (
    (If ((Ebop forbOfBool
           (Ebop forbOfBool
            (Ebop feqbOfInt
              (Econst (Vfield Tuint (Fstruct _0xmsg msg) (@cons str_name (Nvar values) (@nil str_name)) None))
              (Econst (Vint (INT I64 Signed 0))))
            (Evar (Some complete) Tbool))
          (Evar (Some refunded) Tbool)
          )
        )
        (Throw;;;) (Snil;;;)
    );;;
    ...
);;;
.
```

Figure 6. Formal version of the conditional statement shown in Figure 5 in Lolisa.

The degree of complexity poses a challenge for general users to write Lolisa codes manually and develop a translator between Lolisa and Solidity or another language. This is a common issue in nearly all similar higher-level language formalization studies.

Fortunately, Coq and other higher-order theorem-proving assistants provide a special macro-mechanism. In Coq, this mechanism is referred to as the notation mechanism. Here, a notation is a symbolic abbreviation denoting a term or term pattern automatically parsed by Coq. For example, the symbols in Lolisa can be encapsulated as shown in Figure 7.

```
Notation "s '~>' ss" := (@cons str_name (Nvar s) ss)
  (at level 59, right associativity).
Notation " '\\' " := (@nil str_name)
  (at level 59, right associativity).
...
Notation " e1 '(||)' e2 " := (Ebop forbOfBool e1 e2)
  (at level 58, right associativity).
Notation " e1 '(<)' e2 " := (Ebop fltbOfInt e1 e2)
  (at level 52, right associativity).
Notation " e1 '(==)' e2 " := (Ebop feqbOfInt e1 e2)
  (at level 53, right associativity).
...
```

Figure 7. Macro definitions of Lolisa formal abstract syntax tree.

The new formal version of this example yields the notation in Figure 8, which demonstrates that the notation is nearly equivalent to the original Solidity syntax.

```
Definition fun_pledge :=
(Fun (Some public) (Efun (Some pledge ) Tundef )
  ( pcons (Epar (Some _message) Tbyte32) pnil))
  ( nil )
  (
    (If ((SCField Tuint Msg (values ~> \\) None) (==) (SCInt 0) (||) (SVBool complete) (||) (SVBool refunded))
        (Throw;;;) (Snil;;;)
    );;;
  ...
);;;
.
```

Figure 8. Simplified formal version of Figure 5 using syntactic abbreviations.

Through this mechanism, we can hide the fixed formal syntax components used in verification and thereby provide users with a simpler syntax. Moreover, this mechanism makes the equivalence between real-world languages and Lolisa far more intuitive and user friendly. In addition, this mechanism improves verification automation. Similar to converting Figures 5–8, we develop a translator, constructed by a lexical analyzer and a parser, to automatically convert the Solidity program to the macro definitions of the Lolisa abstract syntax tree. The translation process is given in Figure 9. The textual scripts of Ethereum smart contracts will be analyzed by the lexical analyzer of translator, which will generate the Solidity token stream. According to the syntactic sugar of Lolisa, the lexical analyzer will generate the respective Lolisa token stream. Next, the parser will take the Solidity token stream as parameters and generate the parse tree of smart contracts. Finally, the tokens of the parse tree will be replaced by the Lolisa token stream, and then the parser will rebuild the Lolisa parse tree and output the respective formal smart contracts rewritten by Lolisa. In this manner, the translation process can be guaranteed to be completed mechanically.

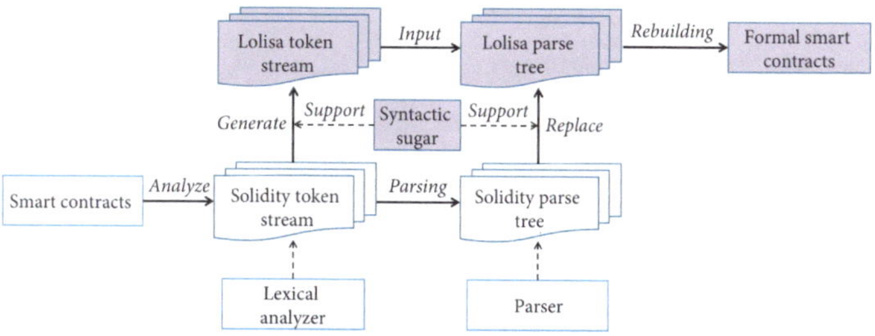

Figure 9. Translation process from smart contracts to its formal version.

FORMAL SEMANTICS

Evaluation of Expressions

The semantics of expression evaluation are the rules governing the evaluation of Lolisa expressions into the memory address values of the GERM framework, and this process includes two parts: the l-value position evaluation and the r-value position evaluation. In contrast, modifier expressions are a special case that cannot be evaluated according to these expression evaluation semantics, but their evaluation is conducted according to rule 8:

$$\vdash_{\Downarrow} \langle \sigma, \text{env}, \text{fenv}, E\text{modifier}(\,*\,)\rangle \Rightarrow \langle \sigma, \text{env}, \text{fenv}, \text{Error}\rangle. \tag{8}$$

Here, $\Downarrow_e$ represents the process of evaluating a modifier expression both in the l-value position and the r-value position. And the example semantics are summarized in Figure 10.

$$\frac{\begin{array}{c}\mathcal{E} \vdash env, fenv \qquad M \vdash \sigma, b_{infor} \\ \mathcal{F}, b_{infor} \vdash v = Varray(name\,[id_{array}]\,\tau) \\ \mathcal{E}, M, \mathcal{F} \vdash \mathcal{A}\big|\sigma, name, \Downarrow_{id_{array}}, \tau, address_{offset}\big| \xrightarrow{yields} Some\ addr\end{array}}{\mathcal{E}, M, \mathcal{F} \vdash \langle \sigma, env, fenv, Econst(v)\rangle \Rightarrow \langle \sigma, env, fenv, Some\ addr\rangle}\ (\text{EVAL-LEXP-ARR})$$

$$\frac{\begin{array}{c}\mathcal{E} \vdash env, fenv \qquad M \vdash \sigma, b_{infor} \\ \mathcal{F}, b_{infor} \vdash v = Vmap(name\,[id_{map}]\,(\tau_{map} \Rightarrow \tau)\ snd) \\ \mathcal{E}, M, \mathcal{F} \vdash \mathcal{A}\big|\sigma, name, \Downarrow_{id_{map}}, \tau, \tau_{map}, snd\big| \xrightarrow{yields} Some\ addr\end{array}}{\mathcal{E}, M, \mathcal{F} \vdash \langle \sigma, env, fenv, Econst(v)\rangle \Rightarrow \langle \sigma, env, fenv, Some\ addr\rangle}\ (\text{EVAL-EXP-MAP})$$

$$\frac{\begin{array}{c}M \vdash \sigma, b_{infor} \qquad \mathcal{F}, b_{infor} \vdash v \\ \mathcal{E}, M, \mathcal{F} \vdash \sigma \Downarrow_{v} \Rightarrow (\sigma, Some\ m_v)\end{array}}{\mathcal{E}, M, \mathcal{F} \vdash \langle \sigma, env, fenv, Econst(v)\rangle \Rightarrow \langle \sigma, env, fenv, Some\ m_v\rangle}\ (\text{EVAL-REXP-CONS})$$

$$\frac{\begin{array}{c}\mathcal{E} \vdash env, fenv \qquad M \vdash \sigma, b_{infor} \qquad \mathcal{F}, b_{infor} \vdash mems \triangleq \{v_0; v_1; \ldots; v_n\} \\ \mathcal{E}, M, \mathcal{F} \vdash \forall i \in \mathbb{N}. 0 \leq i \leq n,\ \sigma_i \Downarrow_{v_i} \Rightarrow (\sigma_i{}', out_i) \\ \forall i \in \mathbb{N}. 0 \leq i \leq n, out_i \neq Error \supset mems = Some\,[out_0.(m_{v_0}); out_1.(m_{v_1}); \ldots; out_n.(m_{v_n})]\end{array}}{\mathcal{E}, M, \mathcal{F} \vdash \langle \sigma, env, fenv, Estruct(head\,\{v_0; v_1; \ldots; v_n\})\rangle \Rightarrow \langle \sigma, env, fenv, Some\,\big(Str(head, mems, env, b_{infor})\big)\rangle}\ (\text{EVAL-REXP-STR})$$

$$\frac{\begin{array}{c}\mathcal{E} \vdash env, fenv \qquad M \vdash \sigma \\ \mathcal{E}, M, \mathcal{F} \vdash \sigma \Downarrow^{r}_{e_0}(\sigma, Some\ m_{v_0}) \wedge \sigma \Downarrow^{r}_{1}(\sigma, Some\ m_{v_1}) \qquad eval_{bop}(\sigma, op_2, m_{v_0}, m_{v_1}) \hookrightarrow re\end{array}}{\mathcal{E}, M, \mathcal{F} \vdash \langle \sigma, env, fenv, Ebop(op_2, e_0, e_1)\rangle \Rightarrow \langle \sigma, env, fenv, re \sqsubseteq \{Error, Some\ m_v'\}\rangle}\ (\text{EVAL-REXP-BOP})$$

$$\frac{\begin{array}{c}\mathcal{E} \vdash env, fenv \qquad M \vdash \sigma \\ \sigma \Downarrow^{r}_{e_0} \Rightarrow Some\ m_{v_0} \qquad eval_{uop}(\sigma, op_1, m_{v_0}) \hookrightarrow re\end{array}}{\mathcal{E}, M, \mathcal{F} \vdash \langle \sigma, env, fenv, Euop(op_1, e_0)\rangle \Rightarrow out\ \langle \sigma, env, re \sqsubseteq \{Error, Some\ m_v'\}\rangle}\ (\text{EVAL-REXP-UOP})$$

Figure 10. Formal operational semantics of Lolisa left and right expressions, including the array, mapping, constant, struct, and binary and unary operators.

Evaluating Expressions in the L-Value Position

In the following, we assign $\Downarrow_{e}^{l}$ to denote the evaluation of expressions in the l-value position to yield respective memory addresses. First, most expressions constructed by Econst obviously cannot be employed as the l-value because most of these represent a Lolisa constant value at the expression level directly. For brevity, we assign $\mathscr{A}]\sigma,\ *\ ]$ to denote the recursive processes of array and map employed for searching the indexed addresses. Note that struct and field are forbidden to specify expressions in the l-value position to ensure that Lolisa is well-formed and well-behaved. The only means allowed in Lolisa of altering the fields of structures are using Estruct to either change all fields or declaring a new field. Although this limitation may be not friendly for programmers or verifiers, it avoids potential risks.

In the previous section, we defined the semantics of array values. Accordingly, we can define the address searching process based on the semantics of arrays as rule 9, which takes name, $\Downarrow_{id_{array}}$, τ, and $address_{offset}$ as parameters. Similarly, we can define rule 10 below for mapping values:

$$\mathscr{A}]\sigma,\ \text{name},\ \Downarrow_{id_{array}},\ \tau,\ \text{address}_{\text{offset}}\Big], \tag{9}$$

$$\mathscr{A}]\sigma,\ \text{name},\ \Downarrow_{id_{map}},\ \tau,\ \tau_{\text{map}},\ \text{snd}\Big]. \tag{10}$$

Evaluating Expressions in the R-Value Position

In the following, we assign $\Downarrow_{e}^{r}$ to denote the evaluation of expressions in the r-value position to yield the respective memory addresses.

As shown in Figure 10, the rules EVAL-REXP-CONS define the evaluation of constant expressions. Here, we note that, because constant expressions store Lolisa values directly, the results can be obtained by applying $\Downarrow_{val}$ directly. In the expression level, the r-value position is specified with a struct type. This is also the only means of initializing or changing the value of a struct-type term. The rules EVAL-REXP-STR defines this process. Here, if the evaluation of Estruct fails, the process of evaluating a member's value yields an error message. Otherwise, the member's value set is obtained and the respective struct memory value is returned. Finally, the semantics of binary and unary operations are defined according to the rules EVAL-REXP-BOP and EVAL-REXP-UOP.

Due to the static type limitations in the formal abstract syntax definition based on GADTs, the expressions, subexpressions, and operations are all guaranteed to be well-formed, and the type dependence relations need not be checked using, e.g., informal assistant functions, as required by other formal semantics such as Clight. The functions eval_{bop} and eval_{uop} take the results of expression evaluations and required operations as arguments, and combine them together to generate new memory values.

Evaluation of Statements

In the following, we assign $\Downarrow_{stt}$ to denote the evaluation process of statements, and parts of the necessary operational semantics are summarized in Figure 11. Most evaluations employ the helper functions env_{check} and set_{gas}. The helper function env_{check} takes the current environment env and the super-environment fenv as arguments, and checks conditions such as gas limitations and the congruence of execution levels. Contract declarations are one of the most important statements of Solidity. In Lolisa, contract declaration involves two operations. First, the consistency of inheritance information is checked using the helper function inherit_{check}, which takes the inheritance relations in module context $\mathcal{C}$ and the source code as arguments. Second, the initial contract information, including all member identifiers, is written into a designated memory block. As defined in Figure 11, the formal semantics of contract declaration are defined as EVAL-STT-CON below.

$$\frac{\begin{array}{c} M \vdash \sigma, b_{infor} \quad \mathcal{F}, b_{infor} \vdash v \quad C \vdash inhertis_c, con_{infor} \quad \Lambda \vdash inherits \\ \mathcal{E} \vdash env, fenv \quad \Lambda \vdash Econ(Some\ \alpha) \quad \mathcal{E}, M, \mathcal{F} \vdash s \\ \mathcal{E}, M, \mathcal{F} \vdash env_{check}(env, fenv) \hookrightarrow true \wedge inherit_{check}(inhertis_c, inherits) \hookrightarrow true \wedge set_{gas}(s, env) \hookrightarrow Some\ env' \\ write_{dir}\big((\sigma, \alpha, Cid(\,(cid\ oaddr), infors, con_{infor}, env, b_{infor}))\big) \hookrightarrow \sigma' \end{array}}{\mathcal{E}, M, \mathcal{F} \vdash \langle \sigma, env, fenv, Contract\big((Econ(Some\ \alpha)), inherits, s\big) \rangle \Rightarrow \langle \sigma', env', fenv, normal \rangle} \ (\text{EVAL-STT-CON})$$

$$\frac{\begin{array}{c} \mathcal{E} \vdash env, fenv \quad M \vdash \sigma, b_{infor} \quad \Lambda \vdash str_\tau \quad \Sigma, \theta, \Lambda \vdash mems \\ \mathcal{E}, M, \mathcal{F} \vdash env_{check}(env, fenv) \hookrightarrow true \wedge set_{gas}(s, env) \hookrightarrow Some\ env' \\ write_{dir}\big(\sigma, str_\tau, Str_{type}(str_\tau, mems, env, b_{infor})\big) \hookrightarrow \sigma' \end{array}}{\mathcal{E}, M, \mathcal{F} \vdash \langle \sigma, env, fenv, Struct(str_{type}, mems) \rangle \Rightarrow \langle \sigma', env, fenv, stop \rangle} \ (\text{EVAL-STT- STRUCT})$$

$$\frac{\begin{array}{c} \mathcal{E} \vdash env, fenv \quad M \vdash \sigma, b_{infor} \quad \mathcal{E}, M, \mathcal{F} \vdash e \quad \mathcal{E}, M, \mathcal{F} \vdash inputs \\ \mathcal{E}, M, \mathcal{F}, b_{infor} \vdash Efun(Some\ \alpha) \\ env_{check}(env, fenv) \hookrightarrow true \ \wedge set_{gas}(s, env) \hookrightarrow Some\ env' \wedge read_{chck}(\sigma, env, b_{infor}, \alpha) \hookrightarrow Some\ m_v \\ set_{env}(env, 0, \alpha) \hookrightarrow env' \end{array}}{\mathcal{E}, M, \mathcal{F} \vdash \langle \sigma, env, fenv, Fun_{call}\big((Efun(Some\ \alpha)), inputs\big) \rangle \Rightarrow \langle \sigma \Downarrow_{m_v \cdot (stt)}, inputs, env', fenv, normal \rangle} \ (\text{EVAL-STT-FUN-CALL})$$

$$\frac{\begin{array}{c} \mathcal{E} \vdash env, fenv \quad \mathcal{E}, M, \mathcal{F} \vdash fpars \quad M \vdash \sigma, b_{infor} \quad \mathcal{E}, M, \mathcal{F} \vdash s \quad \mathcal{E}, M, \mathcal{F} \vdash inputs \quad M \vdash \sigma \\ \mathcal{F}, \Phi, b_{infor} \vdash Efun(Tundef, (Some\ a)) \\ env_{check}(env, fenv) \hookrightarrow true \wedge set_{gas}(Modifier(e, fpars, s), env) \hookrightarrow Some\ env' \\ (repeat\ init_{var}(\sigma, env, fenv, oacc, \tau, fpar)) \leftarrow fpars \equiv [fpar_0, fpar_1, \ldots, fpar_n] \hookrightarrow Some\ \sigma' \\ \text{①} length(inputs) \le length(fpars) \supset set_{par}(\sigma', fpars, inputs) \hookrightarrow Some\ \sigma'' \\ \text{②} init_{re}(\sigma'', \Lambda_{fun}, [Tundef]) \hookrightarrow Some\ \sigma''' \\ \text{③} write_{dir}(\,\sigma'', a, s) \hookrightarrow \sigma'''' \end{array}}{\mathcal{E}, M, \mathcal{F}, \Phi \vdash \langle \sigma, env, fenv, Modifier(e, fpars, s) \rangle \Rightarrow \langle \sigma'''' \Downarrow_s, env', fenv, normal \rangle} \ (\text{EVAL-STT-MODI})$$

Figure 11. Part of formal statement semantics of Lolisa, including environment and gas checking, contract, struct, modifier, and function call statements.

As rule EVAL-STT-STRUCT, the address is the new struct type identifier, and the struct-type information is written into the respective memory block directly.

In Lolisa, a function call statement is used to apply the function body indexed by the call statement. The process of applying an indexed function is defined by the rules EVAL-STT-FUN-CALL below.

Modifier declarations are a kind of special function declaration that requires three steps, and includes a single limitation. The parameter values are set by the set_{par} predicate. As defined by the rule EVAL-STT-MODI in Figure 11, the first step (denoted as ①) initializes and sets the parameters. The second step (denoted as ②) stores the modifier body into the respective memory block. The third step (denoted as ③) attempts to initialize the return address Λ_{fun}. Due to the multiple return values, $init_{re}$ takes a return type list as an argument. Particularly, the modifier body can only yield an initial memory state, and therefore cannot change memory states. The difference between modifier semantics and function semantics is that function semantics include checking the modifier limitations restricting the function. Specifically, taking EVAL-STT-FUN as an example, before invoking a function, the modifier restricting the function will be executed. If the result of a modifier evaluation is σ_{init}, it means that the limitations checking of the modifier fails and the function invocation will be thrown out. Otherwise, the function will be executed.

Development of Standard Library and Evaluation of Programs

As discussed previously, we have developed a small standard library in Lolisa that incorporates the built-in data structures and functions of EVM to facilitate execution and verification of Solidity programs rewritten in Lolisa using higher-order logic theorem-proving assistants. Here, we discuss the standard library in detail. Then, based on the syntax, semantics, and standard library formalization, we define the semantics governing the evaluation (i.e., execution) of programs written in Lolisa.

Development of the Standard Library and Evaluation of Programs

Note that we assume the built-in data structures and functions of EVM are correct. This is reasonable because, first, the present focus is on verification of high-level smart contract applications rather than the correctness of EVM. Second, Lolisa is sufficiently powerful to implement any data structure or function employed by EVM. Thus, we only need to implement the logic of these built-in EVM features using Lolisa based on the Solidity documentation [4] to ensure that these features are well formed. For example, an address is a special compound type in Solidity that has the *balance*, *send*, and *call* members. However, we can treat an address as a special struct type in Lolisa and define it using the Lolisa syntax, as shown in Figure 12. All other built-in data structures and functions of EVM are defined in a similar manner. Typically, *requires* is a special standard function that does not need a special address and, according to the Solidity documentation, is defined in Lolisa as rule 11:

$$\text{requires} \underset{=}{\text{def}} \lambda\left(\forall\, \tau_0\!: \text{type},\ e\!:\ \exp r_{\tau_0\ \text{Tbool}}\right).\ (\text{If}\ e\ (\text{Snil})\ (\text{Throw})). \tag{11}$$

$$\begin{aligned} Address \overset{\text{def}}{=} {} & Str_{type}\ \_0xaddress\ (str_{mem}\ TInt\ (Nvar\ addr) \\ & str_{mem}\ TInt\ (Nvar\ balance) \\ & str_{mem}\ (Tfid\ (Some\ \_Send))\ (Nvar\ send) \\ & str_{mem}\ TInt\ (Nvar\ gas)\ (str\_nil))))\ \_0xaddress\ 2\ occupy) \end{aligned}$$

Figure 12. Address type declaration in Solidity and its equivalent as a special struct type in Lolisa syntax.

Next, we pack these data structures and functions together as a standard library in Lolisa, which is executed prior to executing user programs. Thus, all built-in functions and data structures of EVM can be formalized in Lolisa, which allows the low-level behavior of EVM to be effectively simulated rather than building a formal EVM. Currently, this standard library is a small subset that only includes *msg*, *address*, *block*, *send*, *call*, and *requires*.

Program Evaluation

The semantics governing the execution of a Lolisa program (denoted as *P*(stt)) is defined by rules 12 and 13, where ∞ refers to infinite execution and *T* represents the set of termination conditions for finite execution.

$$\frac{\begin{array}{l}\varepsilon \vdash \text{env}, \text{fenv}\, M \vdash \sigma, b_{\text{inf or}} \mathscr{F} \vdash \text{opars}\, \varepsilon, M, \mathscr{F} \vdash P(\text{stt}) \varepsilon, M, \mathscr{F} \vdash \text{lib} \\ \text{env} = \text{set}_{\text{gas}}(\text{init}_{\text{env}}(P(\text{stt})))\text{fenv} = \text{init}_{\text{env}}(P(\text{stt})) \\ \sigma = \text{init}_{\text{mem}}(P(\text{stt}), \text{lib})\end{array}}{\varepsilon, M, \mathscr{F} \vdash \langle \sigma, \text{env}, \text{fenv}, \text{opars}, \Downarrow_{P(\text{stt})} \rangle \Rightarrow \langle \sigma', \text{env}, \text{fenv} \rangle}, \tag{12}$$

$$\frac{\begin{array}{l}\varepsilon \vdash \text{env}, \text{fenv}\, M \vdash \sigma, b_{\text{inf or}} \mathscr{F} \vdash \text{opars}\, \varepsilon, M, \mathscr{F} \vdash P(\text{stt}) \varepsilon, M, \mathscr{F} \vdash \text{lib} \\ \text{env} = \text{set}_{\text{gas}}(\text{init}_{\text{env}}(P(\text{stt})))\text{fenv} = \text{init}_{\text{env}}(P(\text{stt})) \\ \sigma = \text{init}_{\text{mem}}(P(\text{stt}), \text{lib})\end{array}}{\varepsilon, M, \mathscr{F} \vdash \langle \sigma, \text{env}, \text{fenv}, \text{opars}, \Downarrow_{P(\text{stt})} \rangle \Rightarrow \langle \sigma', \text{env}', \text{fenv} \rangle \wedge \text{env}'.(gas) \leq \text{fenv}.(\text{gasLimit}) \Rightarrow} \langle \sigma', \text{env}', \text{fenv} \rangle. \tag{13}$$

These rules represent two conditions of *P*(stt) execution. Under the first condition governed by rule 12, *P*(stt) terminates after a finite number of steps owing to a returned *stop*, *exit*, or *error*. Under the second condition governed by rule 13, *P*(stt) cannot terminate via its internal logic and would undergo an infinite number of steps. Therefore, *P*(stt) is deliberately stopped via the gas limitation checking mechanism. Here, *opars* represents a list of optional arguments. In addition, as discussed in Section 5.1, the initial environment env and super-environment fenv are equivalent, except for their gas values, which are initialized by the helper function init_{env}, and the initial gas value of env is set by set_{gas}. Finally, the initial memory state is set by init_{mem}, considering *P*(stt) and the standard library lib as arguments.

FORMAL VERIFICATION OF SMART CONTRACT USING FETHER

As introduced in Section 1, we have implemented a formal verified interpreter in Coq for Lolisa, denoted as FEther [11], which incorporates about 7000 lines of Coq code (not including proofs and comments). This interpreter is developed strictly following the formal syntax and semantics of Lolisa based on the GERM framework. To be specific, FEther is implemented by computational functions (considered as the mechanized computational semantics), which are equivalent to the natural semantics of Lolisa given in this paper. The implementation is conducted following the details presented in our previous study [11] using Gallina, which is the functional programming language provided by Coq. Accordingly, FEther can parse the syntax of Lolisa to symbolically execute formal programs written in Lolisa. While efforts are ongoing to prove the consistency between the semantics of FEther and Lolisa, FEther can be employed to prove the properties of real-world programs. This process is effective at exposing errors not only in the test suites that exemplify expected behaviors but also in normal smart contracts. Specifically, a simple case study is presented to demonstrate the symbolic execution and verification process based on Lolisa and FEther. Its source code is presented in Appendix A, and the respective formal version written in Lolisa is presented in Appendix B. Here, it is clear that the program will be thrown out if the message sender in the index mapping list and the current time now are less than *privilegeOpen* or are greater than *privilegeClose*. This is easily proven manually with the inductive predicate semantics defined previously. Meanwhile, we can verify this property by symbolically executing the program with the help of FEther in Coq directly, as shown in Figure 13. The formal intermediate memory states obtained during the execution and verification of this Lolisa program using FEther are shown in Figure 14. Then, we can compare the mechanized verification results and the manually obtained results to validate the semantics of Lolisa. In addition, the application of FEther based on Lolisa and the GERM framework also certifies that our proposed EVI theory is feasible.

```
Lemma no_in_time : forall pump pump_val m m' m0 m1 m2 m3 z blc gs (env : environment) (s : statement),
  let (_, _, cur, dn) := env in
  m = init_msg init_m z 19 blc gs IcoController msg ->
  m' = write_by_address m (Map priviledges
        (Some (iStr _0xaddress (Some (cons (iInt (Some (INT I64 Signed 19)) public occupy)
                (cons (iFid (fid (Some _0xsend)) (Some nil) public occupy)
                (cons (iInt (Some (INT I64 Signed blc)) public occupy)
                (cons (iInt (Some (INT I64 Signed gs)) public occupy)
                 nil)))))) public occupy, Bool (Some true) cur 2 public occupy))
                 Iaddress Tbool None cur dn public occupy) priviledges ->
  m0 = write_by_address m' (Int (Some (INT I64 Unsigned 0)) cur 2 public occupy) privilegeOpen ->
  m1 = write_by_address m0 (Int (Some (INT I64 Unsigned 3)) cur 2 public occupy) privilegeClose ->
  m2 = write_by_address m1 (Int (Some (INT I64 Unsigned 4)) cur 2 public occupy) now ->
  m3 = mem_address (mem_msg m2) ->
  pump > 100 ->
  pump_val > 100 ->
  test pump pump_val m3 (Some nil) env env example
   = Some init_m'
.
Proof.
  destruct env0; intros. unfold mem_msg, mem_address in *. initmem.
  rewrite H4; clear H4.
  do 7 (step; unfold_modify_s).
  step.
  next pump_val.
  push. unfold init_addr_str; cbn in *.
  do 5 (next pump_val).
  push.
  do 4 (step; unfold_modify_s).
  repeat step.
Qed.
```

Figure 13. Execution and verification of the Lolisa program in Appendix B using the formal interpreter FEther in Coq.

```
1 subgoal
pump, pump_val : nat
m, m', m0, m1, m2, m3 : memory
z, blc, gs : Z
l : list address
o : option address
a0 : address
d : dnum
s : statement
H5 : S pump > 100
H6 : pump_val > 100
______________________________________(1/1)
test pump pump_val
  {|
  m_init := initData;
  m_send := initData;
  m_send_re := initData;
  m_msg := Str_type _0xmsg
             (str_mem Taddress (Nvar sender)
                (str_mem Tint (Nvar values) str_nil)) _0xmsg 3 occupy;
  m_address := Str_type _0xaddress
                 (str_mem Tint (Nvar addr)
                    (str_mem Tint (Nvar balance)

                       (str_mem (Tfid (Some _Send)) (Nvar send)
                          (str_mem Tint (Nvar gas) str_nil)))) _0xaddress 3
                 occupy;
  m_throw := false;
  m_0x00000000 := initData;
  m_0x00000001 := initData;
  m_0x00000002 := initData;
  m_0x00000003 := initData;
  m 0x00000004 := initData;
  m_0x0000005f := initData;
  m_0x00000060 := initData;
  m_0x00000061 := initData;
  m_0x00000062 := initData;
  m_0x00000063 := initData |} (Some nil) (Evn l o a0 d) (Evn l o a0 d)
  (Var (Some public) (Evar (Some close) Tuint);;
   Var (Some public) (Evar (Some quota) Tuint);;
   Var (Some public) (Evar (Some rate) Tuint);;
   Var (Some public) (Evar (Some partiLimit) Tuint);;
   Var (Some public) (Evar (Some totalLimit) Tuint);;
   Var (Some public) (Evar (Some finalLimit) Tuint);;
   If

     (Econst (Vmap priviledges (Mstr_id Iaddress msg (sender ~>> \\\)) None))
     (Assignv (Evar (Some open) Tuint) (Evar (Some privilegeOpen) Tuint);;
      Assignv (Evar (Some close) Tuint) (Evar (Some privilegeClose) Tuint);;
      Assignv (Evar (Some quota) Tuint) (Evar (Some privilegeQuota) Tuint);;
      Assignv (Evar (Some rate) Tuint) RATE_PRIVILEGE;; nil)
     (Assignv (Evar (Some open) Tuint) (Evar (Some ordinaryOpen) Tuint);;
      Assignv (Evar (Some close) Tuint) (Evar (Some ordinaryClose) Tuint);;
      Assignv (Evar (Some quota) Tuint) (Evar (Some ordinaryQuota) Tuint);;
      Assignv (Evar (Some rate) Tuint) RATE_ORDINARY;; nil);;
   If
     (Evar (Some now) Tuint (<) Evar (Some open) Tuint
      (||) Evar (Some now) Tuint (>) Evar (Some close) Tuint) (Throw;; nil)
     (Snil;; nil);;
   If
     (Evar (Some subscription) Tuint (+) Evar (Some rate) Tuint
      (>) TOKEN_TARGET_AMOUNT) (Throw;; nil) (Snil;; nil);;
   If (Evar (Some index) Tuint (==) Econst (Vint (INT I64 Unsigned 0)))
     (Throw;; nil) (Snil;; nil);;
   If
     (Econst (Vmap deposits (Mvar_id Iuint index) None)
```

Figure 14. Formal memory states during the execution and verification of the Lolisa program in Appendix B using FEther in Coq.

DISCUSSION

Contributions

First, Lolisa formalizes most of the types, operators, and mechanisms of Solidity, and it includes most of the Solidity syntax. In addition, a standard library was built based on Lolisa to represent the built-in data structures and functions of EVM, such as msg, block, and send. As such, programs written in Solidity can be translated into Lolisa, and vice versa, with a line-by-line correspondence without rebuilding or abstracting, which are operations that can negatively impact consistency.

Second, the formal syntax in Lolisa is defined using generalized algebraic datatypes, which impart static type annotation to all the values and expressions of Lolisa. In this way, Lolisa has a stronger static type system than Solidity for checking the construction of programs. As such, it is impossible to construct ill-typed terms in Lolisa, which also assists in discovering ill-typed terms in Solidity source code. Moreover, the formal syntax ensures that all expressions and values in Lolisa are deterministic.

Finally, the syntax and semantics of Lolisa are mechanized using the Coq proof assistant. Besides, a formal verified interpreter FEther is developed in Coq to validate whether Lolisa satisfies the above Executable and *Provable* feature and the meta-properties of the semantics. In contrast to similar efforts focused on building formal syntax and semantics for high-level programming languages, the formal semantics of Lolisa are defined based on the FSPVM-E framework. As such, it is possible for programs written in Lolisa to be symbolically executed and have their properties simultaneously verified automatically in Coq proof assistant directly as program execution in the real world when conducted in conjunction with FEther.

Limitations

Although the novel features in the current version of Lolisa specification language confer a number of advantages, some limitations remain.

First, because the Lolisa is large subset of Solidity, some of Solidity characteristics, such as inline assembly, have been omitted in Lolisa. Hence, some complicated Ethereum smart contracts are not supported by the current version of Lolisa current. These characteristics will be supported in the updated version of Lolisa.

Second, the Lolisa is formalized at the Solidity source-code level. Although it will analyze vulnerabilities before the compiling process, it cannot guarantee the correctness of the corresponding bytecode when the compiler is untrusted. One possible solution is developing a low-level version of Lolisa, which executes the bytecode generated by the compiler, then proving the equivalence between Solidity execution results and the respective execution results of the bytecode.

Finally, although the current version of Lolisa can be verified in FEther symbolically, this process is not yet fully automated. In occasional situations, programmers must analyze the current proof goal and choose suitable verification tactics. Fortunately, this goal can be achieved by optimizing the design of the tactic evaluation strategies.

CONCLUSION AND FUTURE WORK

In this paper, we defined the formal syntax and semantics for a large subset of Solidity, which we denoted as Lolisa. The formal syntax of Lolisa is strongly typed according to GADTs. The syntax of Lolisa includes nearly all the syntax in Solidity, and the two languages are therefore equivalent with each other. As such, Solidity programs can be translated to Lolisa line-by-line without rebuilding or abstracting, which are operations that are too complex to be conducted by general programmers, and may introduce inconsistencies. Moreover, we have mechanized Lolisa in Coq completely, and have developed a formal interpreter FEther in mathematical tool Coq based on Lolisa, which was employed to validate the semantics of Lolisa. By basing the formal semantics of Lolisa on our FSPVM-E framework [31], programs written in Lolisa can be symbolically and automatically executed in Coq, and thereby verify the corresponding Solidity programs simultaneously. As a result of the present work, we can now directly verify smart contracts written in Solidity using Lolisa.

The source files containing the formalization of Lolisa abstract syntax tree are accessible at https://gitee.com/UESTC_EOS_FV/LolisaAST/tree/master/SPEC

Presently, we are working toward verifying the correctness of FEther, and developing a proof of the equivalence between computable semantics and inductive semantics. Subsequently, we will implement our proposed preliminary scheme based on the notation mechanism of Coq to extend Lolisa along two important avenues.

Our ongoing project is the extension of FSPVM-E to support EOS blockchain platform [32], and we will then verify our new framework in Coq. Besides, we will develop a general formal verification toolchain using HOL proof technology for blockchain smart contracts with the goal of automatic smart contract verification.

APPENDIX

A. Source Code of the Case Study

As shown in Algorithm 1, we give the partial source code of the case study contract.

```
Solidity ^0.4.8;
function example () public payable {
   uint index=indexes[msg.sender];
   uint open;
   uint close; …
   if (privileges[msg.sender]) {
      open=privilegeOpen;
      close=privilegeClose;
   …} else {
      open=ordinaryOpen;
      close=ordinaryClose;…}
   if (now<open || now>close) {
      throw(); }
   if (subscription+rate>TOKEN_TARGET_AMOUNT) {
      throw (); }
   …
   if (msg.value <= finalLimit) {
      safe.transfer(msg.value);
      deposits[index] +=msg.value;
      subscription +=msg.value / 1000000000000000000 * rate;
      Transfer(msg.sender, msg.value); } else {
      safe.transfer(finalLimit);
      deposits[index] +=finalLimit;
      subscription +=finalLimit / 1000000000000000000 * rate;
```

```
        Transfer(msg.sender, finalLimit);
        msg.sender.transfer(msg.value - finalLimit);
    }
}
```

Algorithm 1. Partial source code of case study contract.

B. Formal Version of the Case Study

As shown in Algorithm 2, we give the formal version of Algorithm 1 written in Lolisa.

```
Coq ^8.8;
Definition Example :=
  (Fun public payable (Efun (Some example) Tundef) pnil nil);;
    (Var (Some public) Evar (Some index) Tuint));;
    (Assignv (Evar (Some index) Tuint)
    (Econst (@Vmap Iaddress Tuint indexes (Mstr_id Iaddress msg (sender ~>>\\\)) None));;
      (Var (Some public) (Evar (Some open) Tuint));;
      (Var (Some public) (Evar (Some close) Tuint));;
      (Var (Some public) (Evar (Some quota) Tuint));;
      ...
      (If (Econst (@Vmap Iaddress Tbool priviledges
        (Mstr_id Iaddress msg (sender ~>>\\\)) None))
      ((Assignv (Evar (Some open) Tuint) (Evar (Some privilegeOpen) Tuint));;
        (Assignv (Evar (Some close) Tuint) (Evar (Some privilegeClose) Tuint));;
      ...;; nil)
      ((Assignv (Evar (Some open) Tuint) (Evar (Some ordinaryOpen) Tuint));;
        (Assignv (Evar (Some close) Tuint) (Evar (Some ordinaryClose) Tuint));;
      ...;; nil)
      (If ((Evar (Some now) Tuint) (<) (Evar (Some open) Tuint) (||)
      (Evar (Some now) Tuint) (>) (Evar (Some close) Tuint))
      (Throw;; nil) (Snil;; nil));;
```

	(If ((Evar (Some subscription) Tuint) (+) (Evar (Some rate) Tuint) (>)
	TOKEN_TARGET_AMOUNT)
	(Throw;; nil) (Snil;; nil));;
	…
	(If ((Econst (Vfield Tuint (Fstruct _0xmsg msg) (values ~> \\) None))
	(<=) (Evar (Some finalLimit) Tuint))
	((Fun_call (Econst (Vfield (Tfid (Some safe)) (Fstruct _0xaddress safe) (send ~> \\) None))
	(pccons (Econst (Vfield Tuint (Fstruct _0xmsg msg) (values ~> \\) None)) pcnil));;
	(Assignv (Econst (@Vmap Iuint Tuint deposits (Mvar_id Iuint index) None))
	((Econst (Vfield Tuint (Fstruct _0xmsg msg) (values ~> \\) None)) (+)
	((Econst (@Vmap Iuint Tuint deposits (Mar_id Iuint index) None))));;
	(Assignv (Evar (Some subscription) Tuint) ((Econst Vfield Tuint (Fstruct _0xmsg msg)
	(values ~> \\) None)) (+) (Evar (Some finalLimit) Tuint) (/)
	(Econst (Vint (INT I64 Unsigned 1000 000 000 000 000 000))) (x)
	(Evar (Some rate) Tuint))));; nil) …;; nil);; nil.

Algorithm 2. Formal version of Algorithm A written in Lolisa.

REFERENCES

1. S. B. Nakamoto, "A peer-to-peer electronic cash system," 2020, https://bitcoin.org/bitcoin.pdf.
2. A. Narayanan, J. Bonneau, E. Felten, A. Miller, and S. Goldfede, *Bitcoin and Cryptocurrency Technologies: A Comprehensive Introduction*, Princeton University Press, Princeton, NJ, USA, 1 edition, 2016.
3. V. E. Buterin, "A next-generation smart contract and decentralized application platform," 2020, https://github.com/ethereum/wiki/wiki/White-Paper.
4. S. Demirkan, I. Demirkan, and A. McKee, "Blockchain technology in the future of business cyber security and accounting," *Journal of Management Analytics*, vol. 7, no. 2, pp. 189–208, 2020.
5. 2020, Ethereum solidity documentation. https://Solidity.readthedocs.io/en/develop/.
6. J. McKee, J. Cheng, N. Xiong, L. Zhan, and Y. Zhang, "A distributed privacy preservation approach for big data in public health emergencies using smart contract and SGX," *Computers Materials & Continua*, vol. 65, no. 1, pp. 723–741, 2020.
7. 2020, The D. A. O. Attacked: Code issue leads to $60 million ether theft. https://www.coindesk. com/dao-attacked-code-issue-leads-60-million-ether-theft/.
8. L. Luu, D. H. Chu, H. Olickel, P. Saxena, and A. Hobor, "Making smart contracts smarter," in *Proceedings of ACM Conference on Computer and Communications Security*, pp. 24–28, Vienna, Austria, October 2016.
9. H. Xi, C. Chen, and G. Chen, "Guarded recursive datatype constructors," in *Proceedings of SIGPLAN-SIGACT Symposium on Principles of Programming Languages*, pp. 224–235, New Orleans, LA, USA, January 2003.
10. 2020, The Coq proof assistant reference manual. https://coq.inria.fr/distrib/current/refman/.
11. Z. Yang and H. Lei, "FEther: an extensible definitional interpreter for smart-contract verifications in Coq," *IEEE Access*, vol. 7, pp. 37770–37791, 2019.
12. T. M. Bits, 2020, https://github.com/trailofbits/manticore.
13. B. M. Mueller, 2020, https://github.com/b-mueller/mythril/.

14. E. Hildenbrandt, M. Saxena, X. Zhu, N. Rodrigues, and P. Daian, "KEVM: a complete semantics of the ethereum virtual machine," in *Proceedings of IEEE Computer Security Foundations Symposium*, pp. 204–217, Oxford, UK, July 2018.
15. Y. Hirai, "Defining the Ethereum Virtual Machine for interactive theorem provers," in *Proceedings of Financial Cryptography and Data Security*, pp. 35–47, Sliema, Malta, April 2017.
16. S. Amani, M. Bégel, and M. Bortin, "Towards verifying ethereum smart contract bytecode in Isabelle/HOL," in *Proceedings of Acm Sigplan International Conference on Certified Programs*, pp. 66–77, Los Angeles, CA, USA, January 2018.
17. C. Reitwiessner, "Dev update: formal methods," 2020, https://ethereum.org/2016/09/01/formal-methods-roadmap/.
18. P. Rizzo, "Ethereum seeks smart contract certainty," 2020, http://www.coindesk.com/ethereum-formal-verification-smart-contracts/.
19. C. Liu, H. Liu, Z. Cao, Z. Chen, B. Chen, and B. Roscoe, "Reguard: finding reentrancy bugs in smart contracts," in *Proceedings of IEEE/ACM International Conference on Software Engineering: Companion*, pp. 65–68, Gothenburg, Sweden, May 2018.
20. P. Tsankov, A. M. Dan, D. Drachsler-Cohen, A. Gervais, F. Bünzli, and M. T. Vechev, "Securify: practical security analysis of smart contracts," in *Proceedings of ACM SIGSAC Conference on Computer and Communications Security*, pp. 67–82, Toronto, ON, Canada, October 2018.
21. N. Grech, M. Kong, A. Jurisevic, L. Brent, B. Scholz, and Y. Smaragdakis, "Madmax: surviving out-of-gas conditions in ethereum smart contracts," in *Proceedings of the ACM on Programming Languages*, pp. 1–27, Philadelphia, PA, USA, November 2018.
22. E. Albert, P. Gordillo, B. Livshits, A. Rubio, and I. E. Sergey, "A framework for high-level analysis of ethereum bytecode," in *Proceedings of Automated Technology for Verification and Analysis*, pp. 513–520, Los Angeles, CA, USA, October 2018.
23. A. Mavridou, A. Laszka, E. Stachtiari, and A. Dubey, "Verisolid: correct-by-design smart contracts for ethereum," in *Proceedings of Financial Cryptography and Data Security*, pp. 446–465, Frigate Bay, St. Kitts and Nevis, February 2019.

24. A. Mavridou and A. Laszka, "Tool demonstration: fsolidm for designing secure ethereum smart contracts," *Lecture Notes in Computer Science*, Springer, Cham, Switzerland, pp. 270–277, 2018.
25. T. Abdellatif and K. Brousmiche, "Formal verification of smart contracts based on users and blockchain behaviors models," in *Proceedings of International Conference on New Technologies, Mobility and Security*, pp. 1–5, Paris, France, February 2018.
26. D. Park, A. Stefănescu, and G. Roşu, "KJS: a complete formal semantics of JavaScript," *Acm Sigplan Notices*, vol. 50, no. 6, pp. 346–356, 2015.
27. X. Leroy, S. Blazy, D. Kästner, B. Schommer, and M. Pister, "Compcert-a formally verified optimizing compiler," in *Proceedings of European Congress on Embedded Real Time Software and Systems*, pp. 35–62, Toulouse, France, January 2016.
28. A. W. Appel, *Verified Software Toolchain*, Springer Press, Berlin, Germany, 1 edition, 2011.
29. R. H. Gu, Z. Shao, H. Chen et al., "CertiKOS: an extensible architecture for building cerified concurrent OS kernels," in *Proceedings of the USENIX Symposium on Operating Systems Design and Implementation*, pp. 653–669, Savannah, GA, USA, November 2016.
30. S. Maffeis, J. C. Mitchell, and A. Taly, "An operational semantics for JavaScript," in *Proceedings of Asian Symposium on Programming Languages and Systems*, pp. 307–325, Bangalore, India, December 2008.
31. Z. Yang, H. Lei, and W. Qian, "A hybrid formal verification system in Coq for ensuring the reliability and security of ethereum-based service smart contracts," *IEEE Access*, vol. 8, pp. 21411–21436, 2020.
32. 2020, EOS blockchain documentation. https://eos.io/.

Chapter

7 ONTOLOGY OF DOMAINS. ONTOLOGICAL DESCRIPTION SOFTWARE ENGINEERING DOMAIN–THE STANDARD LIFE CYCLE

Ekaterina M. Lavrischeva

Moscow Physics-Technical Institute, Dolgoprudnuy, Russia

ABSTRACT

Basic concepts and notions of ontological description of domains are implemented in the conceptual model being understandable to ordinary users of this domain. Ontological approach is used for the presentation of software engineering domain—Life Cycle (LC) ISO/IEC 12207 with the aim to automate LC processes and to generate different variants of LC for development systems. And the second aim of Conceptual Model must teach

Citation: Lavrischeva, E. (2015), "Ontology of Domains. Ontological Description Software Engineering Domain - The Standard Life Cycle". Journal of Software Engineering and Applications, 8, 324-338. doi: 10.4236/jsea.2015.87033.

the student to standard process LC, which includes general, organizational and supported processes. These processes are presented in graphical terms of DSL, which are transformed to XML for processing systems in the modern environment (IBM, VS.Net, JAVA and so on). The testing process is given in ontology terms of Protégé systems, and semantics of tasks of this process is implemented in Ruby. Domain ontology LC is displayed by the several students of MIPT Russia and Kiev National University as laboratory practicum course "Software Engineering".

Keywords: Ontology, Life Cycle, Models, Processes, Actions, Tasks, Testing, DSL, XML, Protégé

INTRODUCTION

At the given work, new conception of automation in general processes of LC and generation of variants specialized are offered for their use in the modern programs, information systems and technologies and implementations in the distributed environments of Grid and Clouds processing, highly productive cluster systems and in web-semantic to the Internet.

This conception is formulated by the author for the students of MIPT and Kiev National University (KNU) at the basic course lections of "Software Engineering" (2010-2013). Standard LC ISO/IEC 12207-2007 is a general mechanism of construction of various program systems (PS) and program products (PP). The 17 processes enter into his composition, 74 under processes (actions) and 232 technological tasks. The automation of LC is a very thorny and heavy problem. Variants of the standard LC will be implemented by many companies in case of development of the different application systems. A submachine gun is absent. Offered by our conception of LC automation through the formal conceptual model LC is an attempt in development of the Case commons instruments for support LC to the future industry PP [1] - [5] .

In addition, for implementation of this conception we use new languages of description in conceptual models of knowledge: OWL (Web Ontology Language), ODSD (Ontology-Driven Software Development), XML (Extensible Markup Language), MBPN (Modeling Biasness Process Notation) and others like that. There are systems of design of domains—ODM (Organizational Domain Modeling), FODA (Feature-Oriented Domain Analysis), DSSA (Domain-Specific Software Architectures), DSL (Domain Specific Language) Tools VS.Net, Eclipse-DSL, Protégé and others like that.

That is terms are used for the formal specification of the LC processes and design from them of different PP. The ontological approach ODSD allows getting descriptions of classes from notions to the domain. Unlike previous, models to the domain can be used not only for the generation of code, but also can be "executable" artifacts.

An important aspect of design in different domains is the notion base and system of notions, by which all problems are formulated to the domain. The notion base is given by terminology, substantial relations between notions and their interpretations.

Among the relations the main are [1] :

- Concretization, as an union of notions in the new notion, the substantial signs of which can be a sum of signs of notions or substantially new;
- Association, that approves a presence of communication between notions without clarification of dependence of them from maintenance and volumes;
- Aggregation of terminology, notions, characters for their relations and paradigms of their interpretations in scopes to the domain is accepted to name the ontology of domain knowledge.

In the same general case, ontology is an agreement about the general use of notions, which includes facilities of subject knowledge and agreement about methods of reasoning representation.

The ontology appears by the semantic networks, knots of which are been by domain notion, and by arcs—copulas or relation, associations between them. On the given time the ontological approaches got the wide distribution in the decision of problems of knowledge, semantic integration of informative resources, informative searches and others like that representation. They allow getting descriptions of classes of objects domain, which are specific at notion and knowledge about them. Some ontology domains are given by knowledge, dictionaries of notions, concepts and relations between them. So, the XML become a standard language for marking of various data domains for their saving and exchange between different domains. It is a mean of automatic transformation of descriptions of model domains in the modern ontological languages to the charts, which are suitable for work in the different applied applications.

Offered conception ontology was considered in the different models of LC (spiral, interaction, incremental and so on) on student lections and on the scientific seminars of the Theory Programming and Information Systems departments of the Kiev National University (KNU), and also case of the discipline teaching the "Software engineering" [3] . Within the framework of this discipline the students learn modern methods and design facilities domains and PS constructions, and also learning standards of LC ISO/IEC 12207-2007 and General Data Types ISO/IEC 11404 GDT-2006. At the practical classes evaluation of facilities is conducted for general description and implementation of some experimental ontology on DSL Tools VS.Net and Protégé. For example of description of some fragments of science domains by the ontological facilities they built models PS with purpose of their use in case of PP construction [4] [5] . Some students of department of the IC and TP faculty of cybernetics defended diploma works on given topics with the use of ontological facilities and notions (classes, axioms, slots, facets and others like that) for description of calculable geometry, GDT and LC on the ITC of developing object and component and configuration them [6] . (http://sestudy.edu-ua.net) Approach to LC automation and its evaluation by students in curriculum of the model LC and standard ISO/IEC 12207-2007, namely to study LC structure, processes and actions, and also use of ontology facilities for their description and implementation in the open ontological instruments are offered—DSL Tools VS.Net and Protégé. Students received ontological knowledge may apply them in the implementation of other application areas.

ONTOLOGY AS A BASIC FORMAL DESCRIPTION OF SUBJECT AREAS

Ontology is a conceptual tool to describe base set of concepts and relations for some domains (or subject area- SA). The concept of the SA is classified and dictionary and thesaurus database schema knowledge is created. Domain Ontology—is a system of concepts or conceptual model which is supplied with a set of entities and relationships between them. Now, many anthologies for various scientific and applied areas are created. For example: ontology Census general knowledge of English natural concepts (70,000 more terms and their definitions); ontology concepts of e-commerce; global ontology products and services (UN); commercial ontology SCTG, Rosetta Net-traffic products from 400 companies. Medical ontology's: Galen-to determine the clinical condition; UMLS-US National Library of Medicine; ON9-famous for certification of health systems; chemical, biological ontology; all-Web portal mathematical resources; universal mathematical system Math Lab, Ret, etc.

A basic instrument of implementation of the subject description is the DSL Tools VS.Net and Eclipse-DSL [7] - [9] , the result of the tools is described in the XML (Extensible Markup Language) language, which actually became a standard of data marking for their saving and exchange of information between different applications. XML serves to transformation of the domain ontology model in XML-charts, suitable for work of applied applications.

Using the properties ontology for description of processes LC was given the subject-oriented DSL (Domain Specific Language), and also language the BPMN description of semantics of these processes, the author offers approach of implementation of suggested conception. In the example implementation of the given conception we select the process of testing by the ontological facilities and semantics description by language of programming. At developments to the domain LC used ontological instrumental facilities, DSL Tools VS.Net, DSL Eclipse and Protégé. Set of different methodologies, facilities of language for description of domains are shown in Figure 1.

Ontology of LC is absent. We consider two means of implementing domain LC-language OWL and tools DSL for the overall presentation of the model standard ISO/IEC 12207 and basic concepts of the testing process LC in the standard system Protégé.

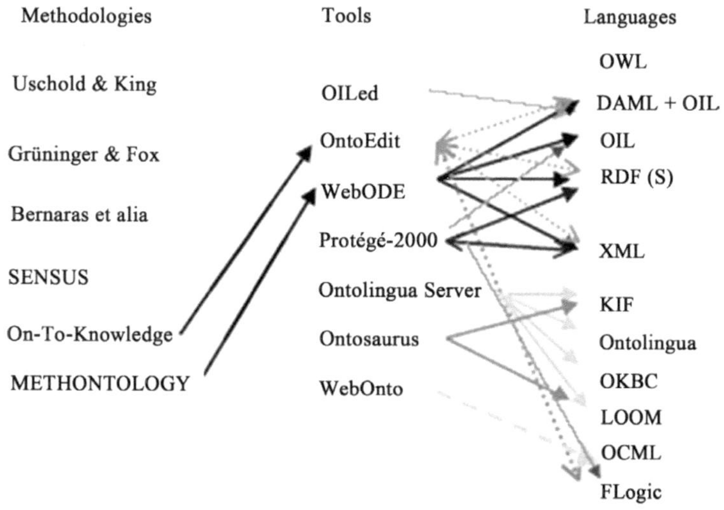

Figure 1. List of methodologies, facilities and the ontology language.

Ontology Means

The form of representation of ontology is a conceptual model (CM) on the reflecting system concepts with common properties (attributes), attitude and behavior rules. CM serves as a communication (between people, between computer systems), storage of information in a computer environment and the recycling of finished objects stored in libraries and repositories. To describe the use of ontology language OWL (Web Ontology Language) with a range of languages and markup languages RDF is to access and exchange ontological knowledge in the Internet. Description in ontology language OWL is a sequence of axioms and facts, information about classes, properties and resources for ID documents and Web imported URI in the form: :: = (see Figure 2).

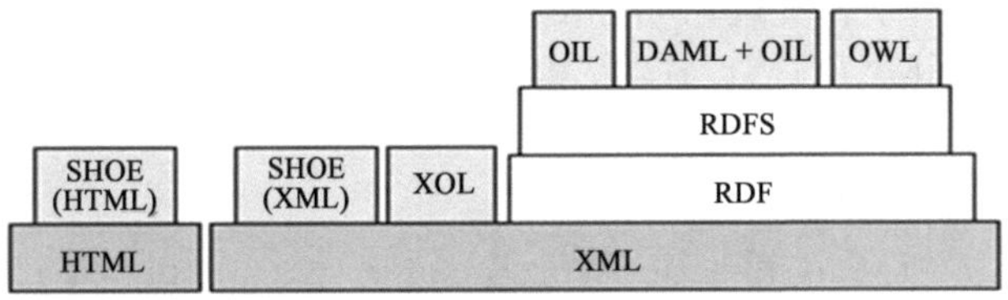

Figure 2. Languages definition of ontology.

Axiom class is a set of more general classes and restrictions on local properties of objects.

The class or a subset of the intersection of more general classes of constraints may be equivalent.

Axiom class in OWL is a set of specifications that can be in the form of generalized classes, restrictions, sets of resources, Boolean combinations of descriptions and more.

As ontology editor was used Protégé 3.4. It seems ontology classes, slots, facets and axioms. (http://protege.stanford.edu/)

Classes describe concepts of, and slots—properties (attributes) classes.

Facets describe the properties of slots (specific types and ranges of possible values).

The axioms define additional constraints (rules).

Classes can be abstract or concrete.

Abstract classes are classes and concrete containers may contain abstract attributes (which do not contain specific values).

Attributes of concepts in a domain called Protégé slots. Specific classes contain specific slots, which can be assigned a value (a copy attributes). To determine the types and limits on the value (like the rules of XML- Schema) used facets.

Protégé supports multiple inheritances but a class can have more than one superclass.

LIFE CYCLES ONTOLOGY OF SOFTWARE SYSTEMS

LC received evolution from the beginning of programming, from simplified life cycle for each application models to spiral, iterative and so on. They formed a separation in group development of various types PS. As the result the standard ISO/IEC 1996 (first edition) was introduced, and in 2007 the second edition of its life cycle appeared, which reflects the overall structure of processes that may be involved in the development of the different PS. These standards should be studied by students who will participate in the joint development of various applications and commercial systems [3] .

Presentation of Formal Specification of the LC Standard ISO/IEC-12207

The LC processes are given in standard by three categories (see Table 1).

In the each process it defines types of activity (actions-activity), tasks, aggregate of results (going out) of activity and decision of tasks design, testing, assembly and others, and also tracing some specific requirements. A list of works for the basic, organizational and support processes is led in standard, but method of their implementation and form of presentation not available. Next, we give a general description of the basic, organizational and support processes.

Table 1. The process is of standard life cycle.

№ п/п		**Process (subprocess)**
	1. Category the "Basic processes"	
1.1	Order (agreement)	
	1.1.1	Preparation of order, choice of supplier
	1.1.2	Monitoring supplier activity, acceptance by user
1.2	Delivery (acquisition)	
1.3	Development	

	1.3.1	Exposure of requirements
	1.3.2	Analysis of system requirements
	1.3.3	Planning system architecture
	1.3.4	Analysis of system requirements
	1.3.5	Planning the system
	1.3.6	Constructing (code) the system
	1.3.7	Integration of the system
	1.3.8	Testing the system
	1.3.9	System integration
	1.3.1	System testing
	1.3.1	Installation of the system
1.4	Exploitation	
	1.4.1	Functional application
	1.4.2	Support of user
1.5	Accompaniment	
	2. The support "Processes category"	
2.1	Documenting	
2.2	Management by configuration	
2.3	Providing a quality guarantee	
2.4	Verification	
2.5	Validation	
2.6	General review	
2.7	Audit	
2.8	Decision of problems	
2.9	Providing product applicability	
2.10	Evaluation of product	
	3. Category the "Organizational processes"	
3.1	Management	
	3.1.1	Management at level organization
	3.1.2	Management by project
	3.1.3	Management by quality
	3.1.4	Management by the risk
	3.1.5	Organizational providing
	3.1.6	Measuring
	3.1.7	Management by knowledge's
3.2	Improvement	
	3.2.1	Introduction of processes

	3.2.2	Evaluation of processes
	3.2.3	Improvement of processes

To the basic processes belong:

- Acquisition process determines actions of buyer at automated system or service. Actions are initiation and preparation of query, legalization of contract; monitoring and acceptance;
- Delivery process determines actions from the transmission of product or service to the buyer. It has preparation of suggestions, legalization of contract, planning, implementation and control of product, and also its estimation and delivery;
- Development process determines the processes and actions (development of requirements, planning, encoding, integration, testing, the system testing, and installation) for development of PP;
- Exploitation process (introduction, support of user, testing functions, exploitation of the system) determines actions of operator from maintenance of processes the system during its exploitation by users;
- Maintenance process (management by modifications, support of current status and functional fitness, PP in- stallation in the operating environment, accompaniment and modification, development of system modification plans, PP migrations on other and others like that), which determines actions of organization, that development PP.

The LC standard contains description of the ancillary proceeding, that regulate the additional actions from verification of product, management by project and his quality.

The support process contains: documenting, management by versions, verification and validation, revisions, audits, evaluation of product and etch. To the organizational processes belong: management by project (development management) and perfection of processes.

The management process includes the processes of management by configuration, project, quality, risks, changes and others like that. The perfection process includes introduction, project estimation and his perfection.

Quantity of processes, actions and LC standard tasks are shown in Table 2.

Table 2. Processes, under processes tasks and LC actions.

Classes	Process	Action	Task
Basic processes	5	35	135
Support processes	10	25	70
Organizational processes	2	14	27
All	17	74	232

Depending on the purpose of concrete project the main developer and project manager choose the processes, actions and tasks, line up the LC chart for application in the concrete project. Description of semantics of processes and methods of their implementation (objective, component, service and so on) written in kernel of SWEBOK knowledge and [3] . (www.swebok.com)

Theoretical, applied methods, quality standards, general and fundamental types of data (ISO/IEC 15404, ISO/IEC 9126, ISO/IEC 11404 and others), and also recommendations and methods of this standards are used at every technology of the PS programming with the use of the LC standard.

Task of automation of standard LC arose up in the students groups MIPT and KNU of course Software Engineering. Taking into account this task, the author discussed with the students the features of standards and machineries of their presentation in the modern operating environments. On the practical lessons the students learned LC processes and gave their description for DSL Tools VS.Net. The students executed LC ontology description in graphic (Figure 3) and the XML kinds within this framework. Then they used DSL Eclipse and Protégé.

XML description of general, support and organization processes are given on web-site ITC.

Formal Presentation of Conceptual Model Domain LC

Starting from Table 2, we give description to the conceptual model (CM) domain LC standard, described highly from terms: P—processes, A—actions and T—tasks.

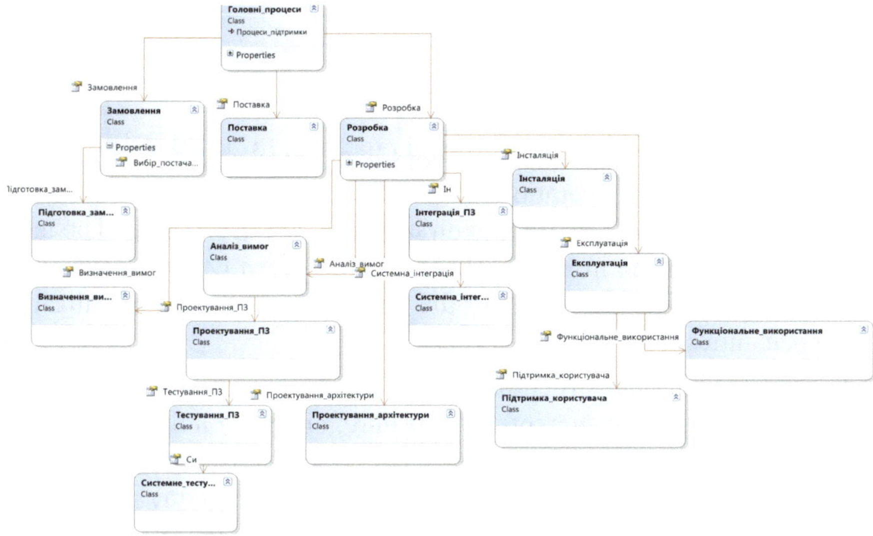

Figure 3. Graphical representation of the basic life cycle processes.

The LC model has such kind:

$$M_{\text{LC}} = \left(P_{\kappa}, A_{\scriptscriptstyle M}, T_n\right)$$

where $P\kappa = \left(P_{1\kappa}, P_{2\kappa 1}, P_{3\kappa 2}\right) - P_{1\kappa,\kappa\ =1-5}$—basic processes of the LC first category,

$P_{2\kappa 1,\ \kappa 1=1-8}$—processes of LC support of the second category;

$P_{3\kappa 2,\ \kappa 2=4}$—organizational processes of the LC second category;

$$A_{\scriptscriptstyle M} = \left(A_{\kappa r}, A_{\kappa 1l}, A_{\kappa 2j}\right)$$

where $A_{\mathbf{\kappa},\ r\ r=\ 1-35}$—action on the basic processes LC,

$A_{\kappa 1l,\ l=1-25}$—Action on the LC support processes,

$A_{\kappa 2\ j=\ 1-14}$—action on the processes LC;

$$T_n = \left(T_{n\kappa}, T_{nl}, T_{nj}\right)$$

where $T_{n\kappa},\ \kappa = 1\text{ - }135$—tasks of the basic processes LC;

$T_{nl, l=1-70}$ —tasks of LC support processes;

$T_{nj, j=1-27}$ —tasks of the organizational processes LC.

The goal processes, operations given is highly contained in essence, and description of maintenance of tasks on it is led them in standard. The tasks not formal and will be in the future at the first given description of their setting, then selection languages for the formal specification for realization to their semantics.

For presentation the structure of the CM LC is used graphic language DSL. This language has an expressive feature, directed on the reflection of the process specific LC, while languages of the general setting (Java, C++, C#, Ruby and others) oriented on description of actions of any programs of the data processing. The DSL contains general abstractions for the reflection of classes of objects domain type process, action, and also relations between them [2] [9] . On its maintenance this language near to the HTML, XML, WSDL and others like that.

Model LC it is described by one DSL, can be transformed in model by other DSL. It allows freely to integrate between itself different parts of system processes, written in the different DSL. That is domain LC can be described at one level of abstraction, and then regenerate with the additional going into detail on the more low level of abstraction, that allows complementing a model domain by the repeated components and objects. Main to the CM domain LC there is a model of general descriptions of processes as objects domain.

Processes of transformation of the LC models in the DSL at the different levels are given in Figure 4.

Transformation of description of the models LC in this language DSL is conducted by facilities of the model-guided development MDD (Model Driven Development). According to this model system architecture are designed at two levels—platform of level independent on the PIM (Platform Independent Model) model and platform of dependent level on the PSM model (Platform Specific Models).

The LC domain CM model can be automated with the use of specific languages, be tuned especially for processes and actions, which are in class language ontology. The models can contain information about the union of processes and actions, including artifacts, which participate in it, and also their dependence between itself. They can also contain information about the configuration structure of the programs of treatment of processes, vehicle and program resources, necessary in case of implementation of the programs of automation of processes and their development.

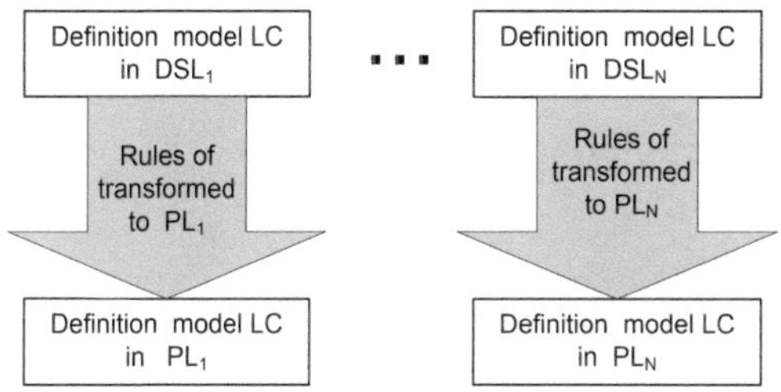

Figure 4. Transformation of description of models LC in DSL.

Ontology of Domain Characteristic Model

DSL development pre-condition is made by the detailed analysis and structured to the domain. Among the existent methodologies of domain analysis most knowing such: ODM (Organization Domain Modeling), FODA (Feature-Oriented Domain Analysis), characteristic analysis to the domain and DSSA (Domain-Specific Software Architectures) [9] - [11] .

In case of analysis to the domain is created a model of characteristics. This model secures generalization and disagreements of the PS domain processes by the indication of general characteristic for all processes and excellent characteristic each of the LC processes.

A model of characteristic is given by diagrams of characteristic with description of relations between them. Conception of diagrams is inherited from the FODA method, which gives possibility briefly to describe all possible configurations of processes within the limits of different categories of the LC processes, which are considered as instances, selecting general and alternative characteristics, which can be excellent for each configuration of the LC processes.

For the given time notation of characteristic diagram is executed by the DSL language under the FDL (Feature Definition Language) name, as languages of description of characteristic of notions to the domain and formal definite operations for treatment of FDL expressions.

The diagrams of characteristic is given system characteristic the different domains. In case of creation of automated instruments, intended for construction of diagrams of characteristic and their treatment, text presentation is necessary. It inflicts all information, which exists in the graphic diagram. The determination consists of great number of characteristic (feature definitions), names of characteristic and expression (feature expression), that includes:

- Atomic characteristics;
- Composition characteristics: names of which determination elsewhere;
- Optional characteristics (optional) of expression, is it completed by the "character"?
- Obligatory characteristics (mandatory) of expressions, what reserved in construction of all ();
- Alternative characteristic (exclusive-choice): y expression of one-of ();
- Exceptional set of descriptions ("or-features") from the list of characteristic expressions of more-of () and their combination;
- Value of characteristic by default (default)—atomic to description;
- Other (indefinite) characteristic in the form of ".".

The specification of FDL characteristic gives formals for determination of syntax, which it is possible to compare to the BNF (P.Naur) form for conducting a lexical and semantic analysis of described characteristic of model domain, which is used for creation of the different variants PS.

About Machineries of Dependence of Characteristic

Offered approach is contained on principle of inherences characteristic with such terms:

- Every characteristic answers class;
- Associations (copulas) between classes are noticed to so call , which marks a type of characteristic dependence:
- Obligatory (mandatory) dependence between aggregations in classes;
- Optional (optional) dependence between association and range of cardinal numbers (by power of great number or quantity of elements of great number) from 0 до1;
- Obligatory list of one-of and more-of in specified class each of alternatives.

The result of translation of description of characteristic in FDL can be given by the XMI language, as a format of exchange by information of Meta data (the XML Meta data Information Exchange format). The XML?documents can be imported in the UML design instruments, such as Rational and UML, and also for the generation of the Java classes. After creation of the DSL language to the domain it is necessary to use the FDL language. Approach to description of model of domain it is used for developed the LC processes of variants PS by configuring different processes for automation the PS. On the given model LC are solved the task of providing a generation of special variants from necessary processes for realization of the set PS. Every variant will be addition of semantics of some tasks for included processes. For the receipt of the working variant LC PS a use of Java facilities is planned [6] [7] .

Standard Life Cycle Ontology in DSL Eclipse

For description of ontology of domains there is other approach of Eclipse DSL [3] [6] . This development environment is used for presentation of the graphic models LC because it has effective instruments for description the object of this domain. On beginning it is necessary to develop a visual model of domain LC. Than it make description of classes of sections processes LC domain and relations between them (Figure 5 and Figure 6).

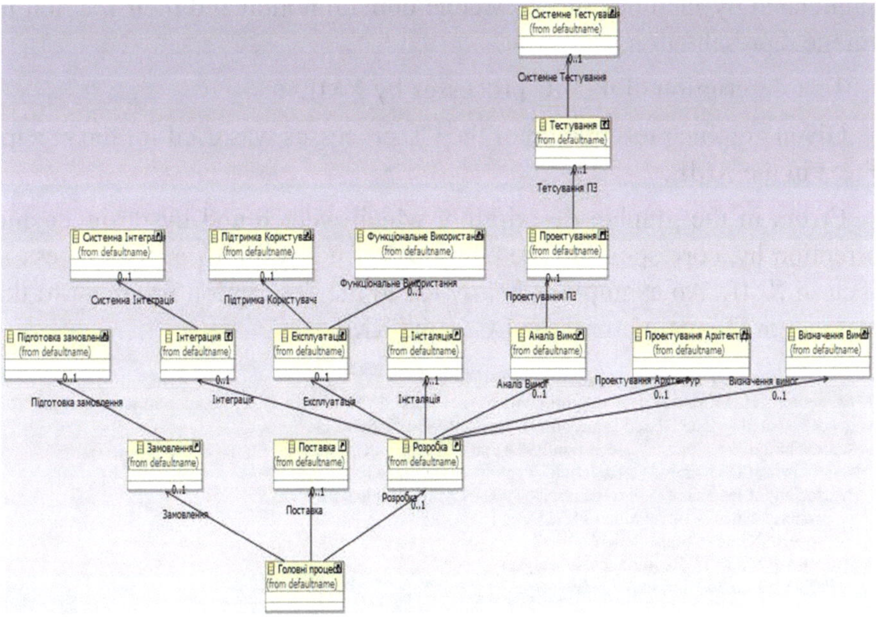

Figure 5. Structure organization process LC in DSL.

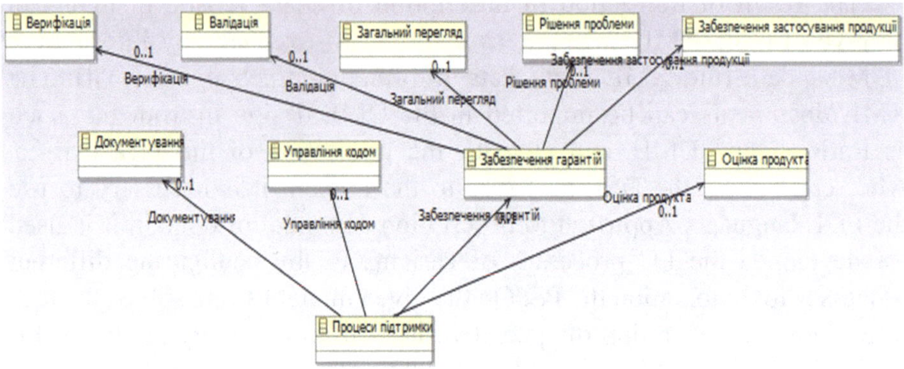

Figure 6. Ontology of the basic processes LC in DSL.

The types of relations allow realizing basic logic of project. Present methods and fields necessary are described in every class for functioning a project. The support processes contain all processes that are executed after the domain construction and support his capacity and actuality. Their ontological structure answers a structure of basic processes and is pointed it will not be.

A next step is been by the generation of text presentation of present graphic models, and then generation in XML. A process of the LC testing is annotated by facilities of knowledge domain notion and their relation in Protégé representation.

Text description of the LC processes by XML

Given graphic presentation of the CL processes was used for the receipt of text in the XML.

Errors in the graphic description, which were found by designer and correction by, correspond to the editor. After it a result of every process is given in XML. An example is below led to the description fragment to the fragment of the main processes LC in the XML.

```
<? Xml version="1.0" encoding="utf-8"?>
   <Association Line Name="Determination of requirements" Type="Main. Determination of requirements"
Manually Routed = "true" FixedFromPoint = "true"   </Association Line>
   <AssociationLine Name = "Integration PS" Type = "Main. Integration PS"
   <AssociationLine Name = "Installetion" Type = "Main. Installetion"
   <AssociationLine Name = "Exploatation" Type = "Main. Exploatation"
     <Property Name = " Integration PS" />
     <Property Name = " Installetion" />
     <Property Name = " Requirements Analyses" />
     <Property Name = "Exploitation" />...
```

For receipt descriptions of the LC processes in XML is given their semantic description. Annotating is executed on example of the LC testing processes by Protégé facilities [10] [11] .

Facilities of the Protégé for Description Ontology

To the basic facilities of the Protégé for description of ontology belong:

- classes (or notion);
- relation (or properties, attributes);
- functions;
- axioms,
- copies (or individuals).

Classes—it is abstract groups, collections or sets of objects. They can include the copies, other classes, or halving both that and second. The relations give a type to co-operation between notions domain. Functions—it is the special case of relations, in which an n-element of relation is simply determined by the n-1 previous elements. Axioms are used for determination of complex limitations on the value of attributes, arguments of relations, for verification of information correctness, or for inference of new information.

By these facilities Protégé forming an ontological model to the LC domain is conducted.

The classes answer the types of artifacts, which, in same queue, answer the roles of program components in system and in the functional properties product/Classes are reflected in Protégé as an inheritance (inheritance hierarchy) hierarchy, which is disposed in to the window navigator of classes (Class Browser). By root of tree of classes in Protégé, by default, appointed class THING (thing, something). All created classes are to be inherited immediately or mediocre.

The protégé will be use for presentation CL testing processes.

It are a new type of description LC and testing processes, which are very necessary for e-learning students for practice preparing some tests for testing the programs [3] [5] [11] .

DESCRIPTION OF ONTOLOGY OF PROCESS TESTING LC

The conceptual model of process testing of the PS has a kind [4] [5] [7] [8] :

$$SFT = \{TM, TD, TA, Env\}$$

where TM—subprocess of management by testing;

TD and TA—subprocesses of testing accordingly domains and applications;

Env—conceptual and informative environment of testing process of the PS.

To all three subprocesses will give the compatible formal presentation:

$$TM = \{Task(TM, TD, TA), CM(TA,\ TD), Env\}$$

where Task—tasks of correspond under process;

$$Env = En(TM) \cup En(TD) \cup En(TA)$$

En—conceptual and informative environment of correspond under process;

CM—under model of co-ordination of operations of correspond under process.

Environment composition is determined by expression:

$$Env = TG \cup SG \cup T \cup P \cup RG \cup RP$$

where TG and SG—test active voices and prepared programs;

T and P—tests and application for testing;

RG and RP—reports about implementation of the tests of programs.

Ontological description of testing process. For description of this process used ontological system Protégé. In her knowledge about the process model are given by classes, slots, facets and axioms. Similar possibility give also and other instruments. For example, diagrams of classes in the UML Rational Rose, which can translate in the program code of a few languages of programming.

For presentation of testing ontology use two groups of notions: simple and complex.

Testing—simple notions. It such: Tester (Tester), Context (Context), Action (Activity), Method (Method), Artifact and Environment.

Can have simple notion attributes. In quality attributes such are selectable under notions, which characterize base (paternal) notion and can accept the concrete values. Will give short maintenance of basic concepts.

Tester—the subject or object, which executes testing determines. The group of testing has a leader, which is a notion attribute, and him name—by the value of attribute. Attributes are been by name, type, duties. Tester attribute-duties-describes, that can do a tester in the process of testing. Notion duties?complex notion, which is determined on the basic of simple notions. For this notion it is possible to select next attributes: tester name, tester type, duties. Tester attribute-duties-describes actions, which can be done by tester in the process of testing.

Example to the tester XML-fragment:

```
<! -- TESTER -->
<xs: element name = "TESTER">
  MaxOccurs = "unbounded" ref = "TESTER"/>
     </xs: sequence>
     <xs:attributename = "TESTER_TYPE" use = "required">
     <xs:simpleType>
          </xs:attribute>
     <xs:attributename = "TESTER_NAME" use = "required"/>
```

Context determines the proper levels, methods of testing, entrances and going out tasks of testing. In ontology this notion determines one attribute: Context type (Level of testing) on form

Level of testing = {module, integration, system, regressive}.

Action consists of notions, that go into detail the steps of process of testing: planning testing, development (generation) of tests, implementation of tests, estimation of results, measuring test coverage, generation of reports and others like that. For this notion one attribute is inflicted-type of action (Activity type) with the possible values: type of action = {planning, development of tests, implementation of tests, and verification of results, coverage estimation, and preparation of report}.

Method—this notion, which is answered by a few methods of testing. For example, the module testing—methods of the structural and functional testing. Every method in relation to the initial code can be classified as a "white small box", "black small box" or based on specification of testing (specification-based). Fragment of method notion XML?chart:

```
<!-- METHOD -->
<xs:element name = "METHOD">
 <xs:complexType>
   <xs:attribute name = "METHOD_NAME" use = "required">
    <xs:simpleType>
      <xs:restriction base = "xs:token">
        <xs:enumeration value = "CONTROL_FLOW_TESTING"/>
        <xs:enumeration value = "DATA_FLOW_TESTING"/>
        <xs:enumeration ...
```

The methods based on coda subdivide on: structural; over seeding of errors; mutational. The structural methods subdivide on: testing a stream of management and testing a data flow. The methods of testing a management (control-flow methods) stream include coverage of operators, coverage of branches and different criteria of coverage of ways. This concrete methods testing is copies of different class of methods of testing.

By the similar rank methods of "black small box" are classified or based on specification: functional; on supposition about the errors; heuristic and so on.

From other side, in relation to process of search of errors and refusals, it is possible to divide all methods into systematic (search of errors) and stochastic (statistical)?exposure of refusals.

Thus, for description of method of testing will enter next attributes:

- Name (method name), "laying out on category";
- Method type (structural, based on the errors);
- Approach, based on code, on specifications, statistical.

Such lying out of methods allows simply classifying every method of testing and extending ontology.

Artifacts. Every action from testing can include a few artifacts, such as an object of testing, intermediate data, results of testing, plans, sets of tests, scripts and others like that. Name them "test active voices". The objects of testing can be different types: initial code, the HTML files, the XML files, built-in images, sound, video, documents and so on. All this artifacts mapping in ontology. Every artifacts is also associated with place of its saving, data, history of creation and revision (for example, creator, upgrade time, version number and others like that).

Environment. Program environments, where testing is executed, as a rule notion such given: name, type and product version. Given notion is broken up on two under notion: vehicle and program with attributes. Attributes of hardware environment are: device name, model, and producer.

Attributes of software environment are: product name, product type and version. The possible values of attribute can be seen by such:

Environment = {ОС, БД, Compiler, web-browser}.

Complex notions of process of testing. Such belong to the complex notions: tester (capability) duties and task (task). They are determined by simple notions.

In the distributed system co-operation between components is executed by interfaces (reports). After treatment of report, the component which got him returns the answer. That is why in ontology expediently to enter additional notions of report and answer. With every report it is possible to link the attributes Type and Value. With every answer it is possible to link its state, which is set as an attribute with two possible values:

The State answer = {Success, Refusal}.

Tricking into result to description of basic terms ontology of testing it will present an ontological model of process of testing with the use of led notions is given on the Figure 7.

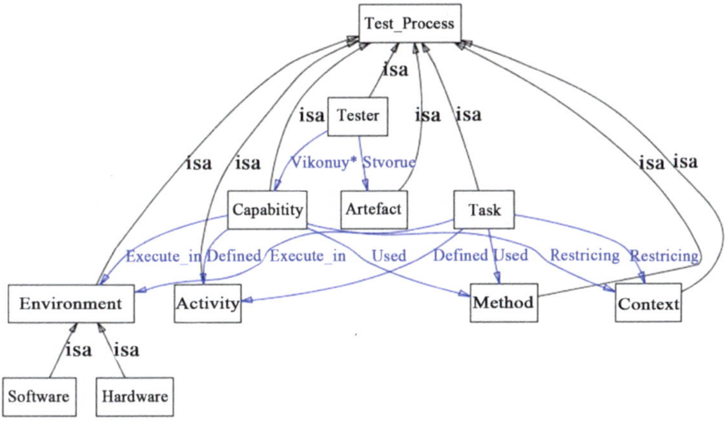

Figure 7. Count of ontology of testing.

For transformation this count to XML gets out format in windows Protege system (Figure 8).

Fragment of ontology of testing in the XML, automatically generates Protege 3.4 and is presented in the UNICODE terms:

```
<class>
        <name>: STANDARD-SLOT</name>
        <type>: STANDARD-CLASS</type>
        <own_slot_value>
                <slot_reference>: ROLE</slot_reference>
                <value value_type="string">Concrete</value>
        </own_slot_value>
        <superclass>:SLOT</superclass>
        <template_slot>: DOCUMENTATION</template_slot>
        <template_slot>: SLOT-CONSTRAINTS</template_slot>
        <template_slot>: SLOT-MAXIMUM-CARDINALITY</template_slot>
        <template_slot>: SLOT-MINIMUM-CARDINALITY</template_slot>
        <template_slot>: SLOT-NUMERIC-MAXIMUM</template_slot>
        <template_slot>: SLOT-NUMERIC-MINIMUM</template_slot>
        <template_slot>: SLOT-INVERSE</template_slot>
        <template_slot>: SLOT-DEFAULTS</template_slot>
        <template_slot>: SLOT-VALUES</template_slot>

        <template_slot>: ASSOCIATED-FACET</template_slot>
        <template_slot>: DIRECT-SUBSLOTS</template_slot>
        <template_slot>: DIRECT-SUPERSLOTS</template_slot>
        <template_slot>: DOCUMENTATION</template_slot>
        <template_slot>: SLOT-CONSTRAINTS</template_slot>
        <template_slot>: SLOT-MAXIMUM-CARDINALITY</template_slot>
        <template_slot>: SLOT-MINIMUM-CARDINALITY</template_slot>
        <template_slot>: SLOT-NUMERIC-MAXIMUM</template_slot>
        <template_slot>: SLOT-NUMERIC-MINIMUM</template_slot>
        <template_slot>: SLOT-INVERSE</template_slot>
        <template_slot>: SLOT-DEFAULTS</template_slot>
        <template_slot>: SLOT-VALUES</template_slot>
        <template_slot>: ASSOCIATED-FACET</template_slot>
        <template_slot>: DIRECT-SUBSLOTS</template_slot>
        <template_slot>: DIRECT-SUPERSLOTS</template_slot>
```


Using these notions, the KNU students on the practical getting by busy created a variant of ontology of process of testing and realization of the program of testing by the Ruby (Figure 9). Erroneous characters are marked on the checked program (Figure 9) on the right, and a correct record is on the left given.

Model of testing ontology and this program made two students in magister works and are placed on web-site http://sestudy.edu-ua.net. It is necessary to do the appeal to, which by pressing on the name "Ontology" word at the main panel of this site.

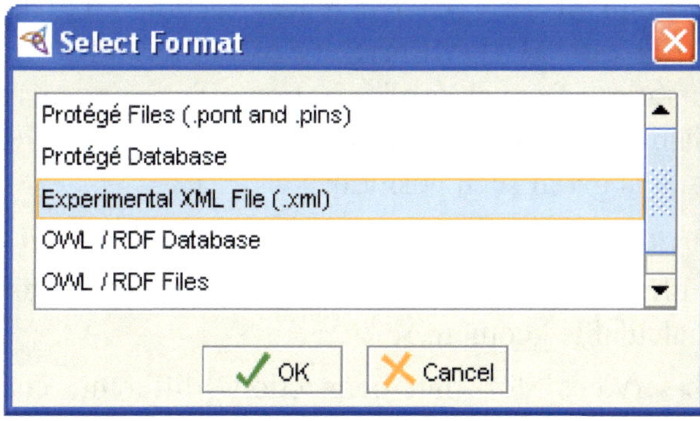

Figure 8. Saving ontology in the XML format.

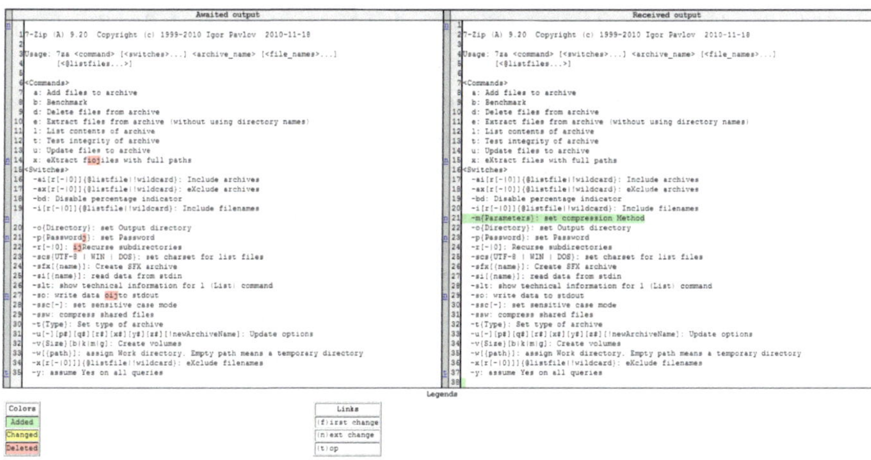

Figure 9. Testing program with the marked errors in it.

LIFE CYCLE ONTOLOGY ON SITE

Complex technology that includes a spectrum of technologies, facilities, instruments of planning and reuses specification is realized in ITK of website [6] . (http://sestudy.edu-ua.net)

This site is based on standard systems (Eclipse, Protégé, CORBA, MS.Net and others), systems of support of co-operation of the programs, systems and environments between itself VS.Net “Eclipse” Java [6] [11] -[15] .

The main menu of web-site has a few sections: TECHNOLOGIES, INTERPRABILITY, INSTRUMENTS, TEACHING and others. Realization of specified operations from class of operations of components development, assembling, change and their configuring is led in the "Technology" section. On this section it is given such position:

- Generation of DSL description to the LC domain;
- Ontology of presentation of the standard LC domain and domain of calculable geometry;
- Beb-services for interconnection different components in environment MS.Net, IBM, Eclipse;
- Transformation of general types of GDT data to fundamental FDT and others.

Web-site is oriented on realization by LC ontological facilities with the use of the Protege system. After its help by the student T Litho of departments the "Informative systems" KNU are developed ontology of calculable geometry, which behaves toward the normative course. Web-site is developed by three languages (Ukrainian, Russian, and Eng.). As Google statistics show, to web-site apply from the different countries (more 35,000 users?teachers and students). This site contains a textbook the "Software Engineering" and is used by author for the E-teaching to all aspects of this discipline. By me lecture at the ICTERI-2012 conference were done, in which mapping new approaches to teaching students of SE.

CONCLUSIONS

The essence of this work focuses on the automation LC by ontological description conceptual model. It is new approaches to the description of domain SE Standard ISO/IEC 12207-2007. Perform three basic tasks: to develop a conceptual model LC and describe this model in terms of language (DSL, OWL); to generate variants LC for development different systems; to consider the training of students scheme using LC by described ontology.

The formal terms for describing the conceptual model LC of the domain ontology are given. The table description of general, organizational processes and support processes of the standard LC is used for presentation processes in the language of DSL. The characteristic domain model and process model LC in DSL is done. A scheme describing LC in DSL is transformed to a lower level XML for processing systems in the environment (IBM, VS.Net, JAVA and so on).

Submission LC processes graphically DSL Tool VS.Net, and in the language XML is described. Conception of automation of the LC and realization testing process is discussed. As a practical implementation process is selected by process testing. A formal description of the conceptual model testing in terms of Protégé systems and algorithm testing in the language Ruby is realized.

It is noted that the ontological model of LC and computational geometry is implemented by the MIPT, KNU students. Technology of work with that ontology is presented on the website, which gives an access to realize ontology. (http://sestudy.edu-ua.net)

REFERENCES

1. Gomes-Perez, A., Fernandez-Lopez, M. and Corcho, O. (2004) Ontological Engineering. Springer-Verlag, London, 403 p.
2. Mernik, M., Heering, J. and Sloane, A.M. (2006) When and How to Develop Domain—Specific Languages. ACM Computing Surveys, 37, 316-344.
3. Lavrischeva, E.M. (2014) Software Engineering Computer Systems. Paradigms, Technologies, CASE-Tools Programming. Nauk, Dumka, 284 p. (In Russian)
4. Lavrischeva, E.M. (2013) Ontological Representation for the Total Life Cycle of AC Line Production of Software Product. Proceedings Conf.TAAPSD'2012, Theoretical and Applied Aspects of Building Software Systems, Yalta, 25 May-2 June 2013, 81-90.
5. Lavrischeva, E.M. (2013) The Approach to the Formal Submission of Ontology Life Cycle of Software Systems, Vesnik KNU, a Series of Physical and Math. Sciences, 4, 140-149.
6. Lavrischeva, E.M., Zinkovich, V., Kolesnik, A., et al. (2012) Instrumental and Technological Complex for Development and Learning Design Patterns of Software Systems. State Intellectual Property Service of Ukraine, Copyright Registration Certificate No. 45292, 103 p. (In Ukrainian)
7. Korotun, T.M. and Lavrischeva, E.M. (2002) Construction of the Testing Process of Software Systems. Problems of Programming, 2, 272-281. (In Ukrainian)
8. Korotun, T.M. (2005) Models and Methods Testing Engineering Programs Systems in Resource-Limited Settings. Autoref Dissertation, Kiev, 23 p. (In Ukrainian)
9. (2005) Walkthrough. Domain–Specific Language (DSL) Tools.
10. Protégé—Frames User's Guide. http://protege.stanford.edu/doc/index.php/PrF_UG
11. Mens, C., Van Gorp, P. and Czarnecki, K.A. Taxonomy of Model Transformation. http://drops.dagstuhl.de/2–5/11
12. Lavrischeva, E.M. (2013) Generative and Composition Programming: Aspects of Developing Software System Families. Cybernetics and Systems Analysis, 49, 110-123.

13. Lavrischeva, E.M. and Ostrovski, A. (2013) New Theoretical Aspects of Software Engineering for Development Application and E-Learning. Journal of Software Engineering and Application, 6, 34-40. http://www.crirp.org/journal/jsea
14. Lavrischeva, E.M., Stenyashin, A. and Kolesnyk, A. (2014) Object-Component Development of Application and Systems. Theory and Practice. Journal of Software Engineering and Applications, 7, 14. http://www.scirp.org/journal/jsea
15. Lavrischeva, E.M. (2013) Conception of Programs Factory for Presentating and E-Learning Disciplines Software Engineering. 10th International Conference on ICT in Education, Research and Industrial Applications, Ukraine, 16 June 2013, 15. http://senldogo0039.springer-sbm.com/ocs/
16. Lavrischeva, E., Ostrovski, A. and Radetskyi, I. (2012) Approach to E-Learning Fundamental Aspects of Software Engineering. 8th international Conf. ICTERI—ICT in Education, Research and Industrial Applications, Kherson, 6-10 June 2012. http://ceur-ws.org/Vol-848/ICTERI-2012-CEUR-WS-p-176-187

Chapter

GUIDELINES BASED SOFTWARE ENGINEERING FOR DEVELOPING SOFTWARE COMPONENTS

8

Muthu Ramachandran

Faculty of Arts, Environment and Technology, School of Computing and Creative Technologies, Leeds Metropolitan University, Leeds, UK.

ABSTRACT

Software guidelines have been with us in many forms within Software Engineering community such as knowledge, experiences, domain expertise, laws, software design principles, rules, design heuristics, hypothesis, experimental results, programming rules, best practices, observations, skills, algorithms have played major role in software development. This paper presents a new discipline known as Guidelines Based Software Engineering

Citation: M. Ramachandran, "Guidelines Based Software Engineering for Developing Software Components," Journal of Software Engineering and Applications, Vol. 5 No. 1, 2012, pp. 1-6. doi: 10.4236/jsea.2012.51001.

where the aim is to learn from well-known best practices and documenting newly developed and successful best practices as a knowledge based (could be part of the overall KM strategies) when developing software systems across the life cycle. Thereby it allows reuse of knowledge and experiences.

Keywords: Software Reuse, Software Guidelines, Software Design Knowledge, CBSE, GSE

INTRODUCTION

The term Software Engineering was coined by F. L. Bauer the chairman of 1968 NATO Software Engineering conference held in Garmisch, Germany to promote a disciplined approach to developing software. The term Software is meant a list of machine instructions where as the Engineering is meant the use of disciplined approaches and laws when building software systems. This paper would argue that the term Software should include best practices which are the laws due to the nature and the age of software as a science compared with Science and Engineering where the laws have been proved and established. In the world of software our principles are out current practices and are continue to emerge as we speak. Later, the term algorithm has emerged to provide a structured step by step programmable instructions/solution to a software problem. Best practices provide a step by step instructions/solution to software problem across the life cycle and are based on the successful use in real world.

Guidelines provide a precise set of steps based on underlying software design principles which help us to follow any course of disciplined set of activities. The term guidelines are defined in the dictionary as follow:

- A recommended approach, parameter, etc. for conducting an activity or task, utilizing a product, etc.;
- A statement of desired, good or best practice;
- Advice about how to design an interface;
- A document used to communicate the recommended procedures, processes, or usage of a particular business practice;
- A recommendation that leads or directs a course of action to achieve a certain goal;
- A written statement or outline of a policy, practice or conduct. Guidelines may propose options to enable a user to satisfy provisions of a code, standard, regulation or recommendation.

Software Engineering is a set of disciplined activities that are based on well defines standards and procedures. In Software Design we use guidelines that help us to identify a suitable design criterion when faced with design decisions. Therefore software guidelines summarises expert knowledge as a collection of design judgements, rationales, and principles. This can be used by students/ engineers when learning about new principles with examples and experts alike.

GUIDELINES BASED SOFTWARE ENGINEERING

The very definition of Software Engineering deals with best practices, disciplined & systematic approaches to software development and management. These best practices have been found throughout software development life cycle. Starts from good program design by Parnas [1], Algorithms design by Dijkstra [2], concurrent programs by Hoare and they all have provided good design guidelines which are applicable until now. The term best practices should support knowledge and wisdom that has emerged from many years of successful use across several projects, products, programs, and portfolios. Software as a profession, we must also include a list of recommended conduct and ethical activities when developing software product or research. Once we accept the term Software Guidelines as a new discipline that provide well established principles and rules that are successful in practice and thus also provide knowledge and wisdom. This way we can also tell the world proudly, we are Engineers since we follow principles strictly and ethically. Where do we start?

In practice we are not sure of the process by which to apply those principles. Therefore, our work on software guidelines have started on specifically on software components [3-5], extended to concurrency, software process improvement, agile methods, and software product line based development (aimed on good practice requirements guidelines). Therefore, we prefer to call Guidelines Based Software Engineering (GSE) which aimed to collect best practices and experiences as guidelines from many years of wealth of knowledge and wisdom in Software Engineering and apply them wherever possible across all artefacts of software development. Guidelines provide rationale for making a solution that has worked well and successfully in previous applications, environment, and in people. **Figure 1** shows the process view of guidelines based software engineering.

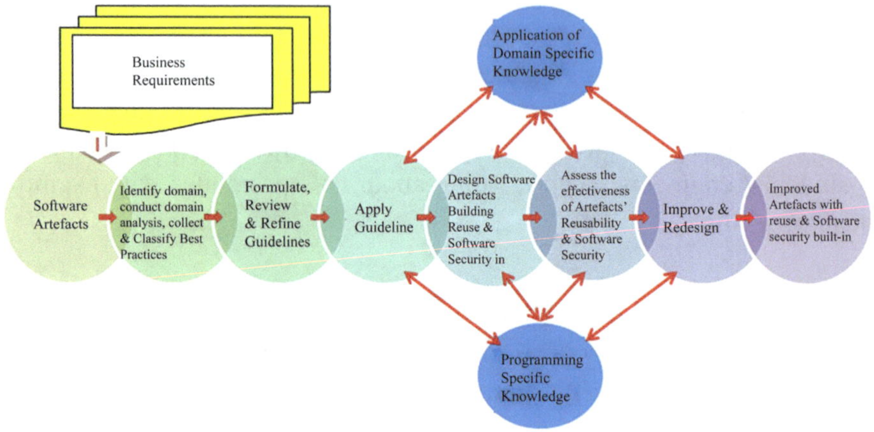

Figure 1. The process of guidelines based software engineering.

The process states start with gathering domain knowledge, classify domain, classify best practice design, identify artefacts (components, patterns, frameworks), identify and classify best practice design guidelines on various aspect of their design (for example how well requirements have been represented as use cases and how well use case have been used effectively and their features, how well OCL specifications have been used to document and describe the model). Building the domain knowledge is crucial for success of using software guidelines or GSE. We can define domain analysis is an activity for identifying a key set of software artefacts that can be ready-made for reuse. There are numerous approaches to this end which can conclude by summarising a common set of domain analysis process as follow:

- Setting Domain principles: Select a domain, definitions, business analysis, scope and boundaries and planning.
- Data collection—learn more about the domain, discover success and failures, and collect guidelines, discover abstractions, review literature extensively, interview and discuss with domain experts, and develop scenarios.
- Data analysis—the aim is to identify entities, objects, models, sub-domains, related classes and models, events, operations, relationships amongst all of them, tacit knowledge, analyse similarities and variabilities, analyse combinations and trade-offs, cost-benefit analysis, modular decompositions and design decisions.

- Classification—the aim during this phase is to describe domain classes, models, and components, conduct cluster analysis and HIPO chart, describe artefacts, classify models and components, generalize artefacts descriptions, conduct domain vocabulary.
- Evaluation of domain models—the aim in the last phase is to evaluate the findings systematically—use expert meeting, reviews, discussions, and review interviews.

In case if the artefacts are represented in any programming language, then identify and classify best design constructs that can be used for expressing various design factors such as reuse, flexibility, security, and so on. Guidelines fall into several categories such as good practice guidelines on requirements engineering (Sommerville and Sawyer [6]), RE methods-specific guidelines such on UML, Use Case driven modelling, design (OO, generic design principles), quality and SQA procedures and best practices, software development (good program design and language-specific guidelines), and good test process guidelines, and guidelines on software process improvement. The first step in building guidelines based SE is to devise a classification system/mechanism for collating guidelines which the useful for finding an appropriate guideline. A number of guidelines, best practices, projects, and knowledge engineering support for software development life cycles are presented by Ramachandran [7].

Best practice guidelines on components based software engineering (CBSE) fall into a number categories such as definitions, process, methods, techniques, models, design, implementation, domain engineering, and development for component reuse, component security, component testing, validation, certification, and QSA. Identifying software components from your application models is a human intensive activity. This comes from domain expertise. However, Pressman [8] has identified a few self-assessment questions to identify components from your design abstracts as given below:

- Is component functionality required on future implementations?
- How common is the component's function within the domain?
- Is there duplication of the component's function within the domain?
- Is the component hardware-dependent?
- Does the hardware remain unchanged between implementations?
- Can the hardware specifics be removed to another component?

- Is the design optimized enough for the next implementation?
- Can we parameterize a non-reusable component so that it becomes reusable?
- Is the component reusable in many implementations with only minor changes?
- Is reuse through modification feasible?
- Can a non-reusable component be decomposed to yield reusable components?
- How valid is component decomposition for reuse?

Example of a Process Guideline for Component Identification: One rule of thumb can be use here is to identify a group of related object classes to make up a selfindependent component. UML view of component identification process is depicted in the following diagram (**Figure 2**). UML process starts with identifying use cases, class modeling, dynamic modeling (state transition and message sequence models), collaboration models (grouping related classes), packaging, components, and deployment/implementation models (processors and network architectures) where components and packages will be placed in the expected processors.

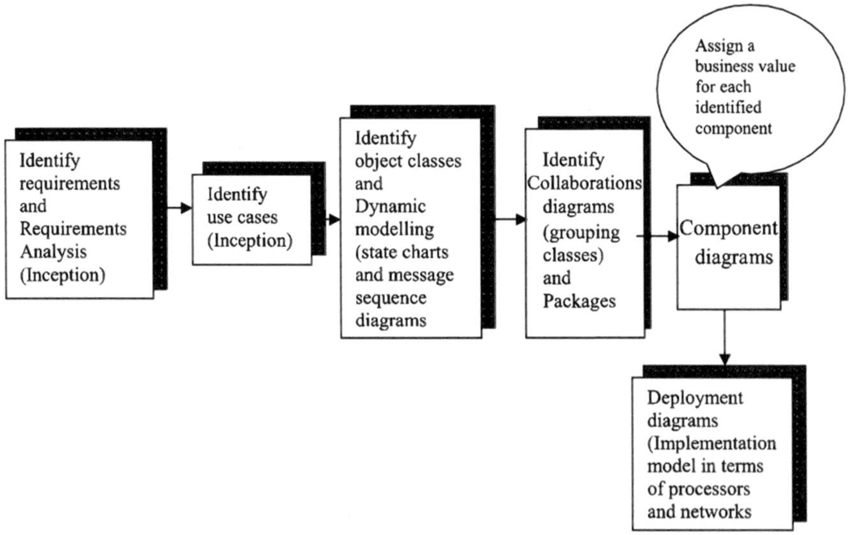

Figure 2. UML view of component identification.

Implementation effort and Return on Investment (RoI): This is an initial step in CBSE and it is therefore vital to identify a component which will have a longer life in your application domain and hence high returns on investment. Therefore it is absolutely essential to have a business view to each identified components with domain experts.

Process guidelines have also helped us to identify common processes and patterns across CBSE and reuse. Knowledge about commonly occurring patterns in a process helps to save cost. Therefore, for each guideline, it is important to present a description, illustration, return on investment (RoI), and possible implementation effort required along with cost-benefit analysis.

GUIDELINES, OBSERVATIONS, EMPIRICAL STUDIES TO LAWS AND THEORIES

Guidelines form principles from observations, laws, and theories. Observations, in software terms, mean to visually able to see changes or results of an experiment/software tools used by people, etc. However, these observations may not be a repeatable event. A law can be defined as repeatable observations according to Endres and Rombach [9]. For example, a rainy season, symptoms of a widespread disease, etc. Theories can help to explain and order our guidelines, observations, and laws. Theories can also help it predict new facts from existing guidelines, observations and laws. The diagram shown in **Figure 3** illustrates the relationships amongst guidelines, observations, law, and theory. Guidelines also add human perspective to observations, laws, and theories as it adds knowledge and experiences.

We have used similar approach to domain-specific modelling to generate reusable software components automatically for several application domains. An example of a CBSE guidelines classification system has been shown in **Figure 4** and their relevant guidelines have been adopted when designing software components [5]. Best practice guidelines on components based software engineering (CBSE) fall into a number categories such as definitions, process, methods, techniques, models, design, implementation, domain engineering, and development for component reuse, component security, component testing, validation, certification, and QSA.

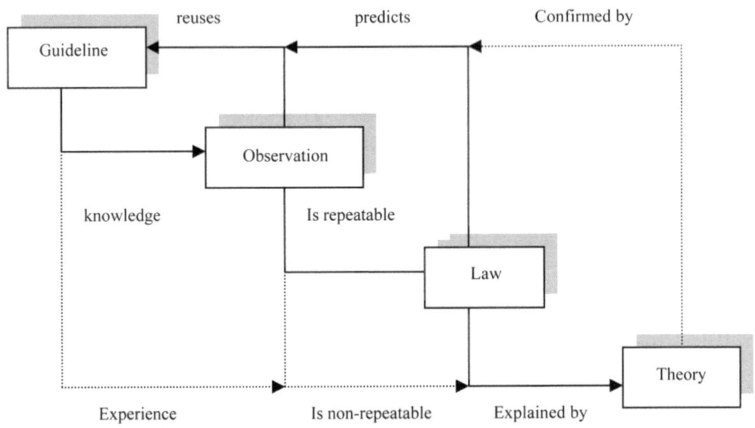

Figure 3. Guidelines, observations, laws, and theories.

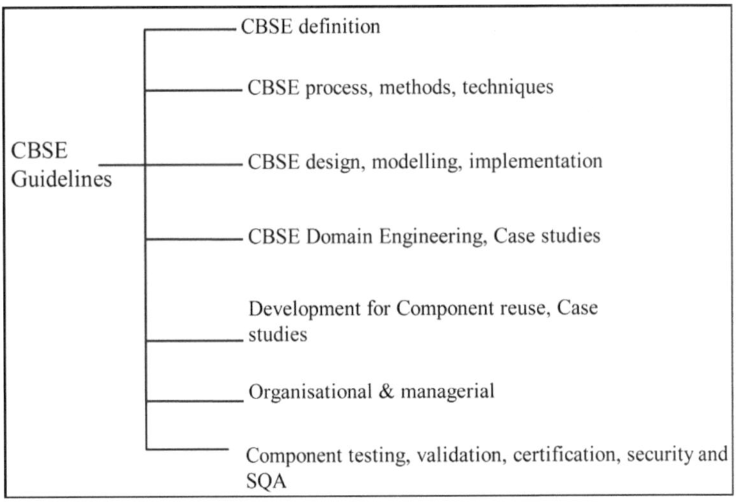

Figure 4. Classification of best practice CBSE guidelines.

Each of these guidelines has been followed against various models for Helpdesk management systems. There were 15 software component identified and their relevant interfaces. Each of these guidelines can also be used to conduct a systematic inspection against use case models, class diagrams, and component diagrams. Therefore, it allows us to achieve fine tuned models that can be further checked against guidelines during implementation as there are plenty of guidelines developed for JavaBeans and C# components. Similar best practices have been presented by many authors [10-21], all of them can be encoded as a knowledge base.

Our earlier results have shown components designed with guidelines seem to have improved reuse and easy to re-design (more than 70% reusability gain has been achieved) for a simple help desk management system. The **Table 1** shows an example of a list of components and their reusability gain in percent.

Table 1. Component reusability gain & security guidelines met.

Helpdesk System Components	Reuse Gain	Security Guidelines
GUI component 1	50%	92%
GUI component 2	40%	99%
User records management	70%	95%
Job status	55%	90%
Test components	65%	99.99%

Reuse gain represents the percent of reusability which is measured against percent of guidelines met. The GUI component 1 consists of a large component for Helpdesk system for the front-end consisting of more than 100 interfaces that can be served to other components. This component has met 50% of the best practice guidelines therefore reusability gain is 50%. Guidelines become highly useful for building software security. This is a new are for research and hence formulating best practice guidelines can help to achieve software security early in the life cycle. According the above data we can see the percent of security-specific design guidelines that have been met. The security design guidelines are further classified into a set of language-specific features (when not to use some features found in most programming practices) and design principles that help to design components for software security built in rather than as add ons.

Our future work includes designing automated tool to predict developing high quality software components that are designed for reuse and quality. This can be achieved by encoding guidelines as knowledge to assess, review and improve components development right from analysis. This will improve component based development with less effort and cost and can be manufactured as a mass production that has been seen in other industry. Due to current improvement in knowledge based technologies, this is will be possible to encode domain knowledge thereby best practice guidelines can be implemented efficiently.

CONCLUSION

Muthu Ramachandran Guidelines based SE can create best practices as guidelines to be followed when developing software artefacts. Guidelines provide knowledge and wisdom that has emerged from several years of best practice and experiences in previous projects successfully. This can save time, cost, and effort with quality that we all seek. Our work has shown increase in reuse gains to the maximum of 70%. The security factor can be achieved up to 99%. Thus, we believe, attributes such as reuse and security factors can be improved significantly which results in achieving high quality of the software systems and reducing software development costs.

REFERENCES

1. L. Parnas, "Good Program Design," Prentice-Hall, Upper Saddle River, 1979.
2. E. W. Dijkstra, "Selected Writings on Computing: A Personal Perspective," ACM Classic Books Series, 1982.
3. T. Hoare, "Concurrent Programs," Prentice-Hall, Upper Saddle River, 1979.
4. M. Ramachandran and Sommerville, "Software Reuse Assessment," First International Workshop on Software Reuse, Germany, 1992.
5. M. Ramachandran, "Software Components: Guidelines and Applications," Nova Publishers, New York, 2008. https://www.novapublishers.com/catalog/product_info.php?products_id=7577
6. I. Sommerville and P. Sawyer, "Requirements Engineering: Good Practice Guide," Addison Wesley, Boston, 1999.
7. M. Ramachandran, "Knowledge Engineering for Software Development Life Cycle," IGI Global, Hershey, 2011. doi:10.4018/978-1-60960-509-4
8. Pressman, "Software Engineering," 6th Edition, McGraw Hill, New York, 2005.
9. A. Endres and D. Rombach, "A Handbook of Software and Systems Engineering," Addison Wesley, Boston, 2003.
10. W. A. Brown and K. C. Wallnau, "The Current State of CBSE," The Current State of CBSE, IEEE Software, Vol. 15, No. 5, 1998.
11. M. Broy, et al., "What Characterizes a Software Component?" Software—Concepts and Tools, Vol. 19, No. 1, 1998, pp. 49-56. doi:10.1007/s003780050007
12. J. Cheesman and J. Daniels, "UML Components," Addison Wesley, Boston, 2000.
13. D'Souza and Wills, "Objects, Components and Frameworks with UML," Addison Wesley, Boston, 1999.
14. G. Eddon and H. Eddon, "Inside Distributed COM," Microsoft Press, Washington, 1998.
15. G. T. Heineman and W. T. Councill, "Component-Based Software Engineering," Addison Wesley, Boston, 2001.
16. IEEE SW, "Special Issue on Software Components," IEEE Software, Vol. 15, No. 5, 1998.

17. I. Jacobson, et al., "Software Reuse: Architecture, Process and Organisation for Business Success," Addison Wesley, Boston, 1997.
18. K.-K. Lau and Z. Wang, "A Taxonomy of Software Component Models," Proceedings of the 31st EUROMICRO Conference on Software Engineering and Advanced Applications, 2005.
19. O. Rob Van, et al., "The Koala Component Model for Consumer Electronics Software," IEEE Computer, 2000.
20. R. Sessions, "COM and DCOM," Wiley, New York, 1998.
21. C. Szyperski, "Component Software," Addison Wesley, Boston, 1998.

Chapter 9

INTELLIGENT AGENT BASED MAPPING OF SOFTWARE REQUIREMENT SPECIFICATION TO DESIGN MODEL

Emdad Khan and Mohammed Alawairdhi

College of Computer and Information Sciences, Al-Imam Muhammad Ibn Saud Islamic University, Riyadh, KSA.

ABSTRACT

Automatically mapping a requirement specification to design model in Software Engineering is an open complex problem. Existing methods use a complex manual process that use the knowledge from the requirement specification/modeling and the design, and try to find a good match between them. The key task done by designers is to convert a natural language based requirement specification (or corresponding UML based representation) into

Citation: E. Khan and M. Alawairdhi, "Intelligent Agent Based Mapping of Software Requirement Specification to Design Model," Journal of Software Engineering and Applications, Vol. 6 No. 12, 2013, pp. 630-637. doi: 10.4236/jsea.2013.612075.

a predominantly computer language based design model—thus the process is very complex as there is a very large gap between our natural language and computer language. Moreover, this is not just a simple language conversion, but rather a complex knowledge conversion that can lead to meaningful design implementation. In this paper, we describe an automated method to map Requirement Model to Design Model and thus automate/ partially automate the Structured Design (SD) process. We believe, this is the first logical step in mapping a more complex requirement specification to design model. We call it IRTDM (Intelligent Agent based requirement model to design model mapping). The main theme of IRTDM is to use some AI (Artificial Intelligence) based algorithms, semantic representation using Ontology or Predicate Logic, design structures using some well known design framework and Machine Learning algorithms for learning over time. Semantics help convert natural language based requirement specification (and associated UML representation) into high level design model followed by mapping to design structures. AI method can also be used to convert high level design structures into lower level design which then can be refined further by some manual and/or semi automated process. *We emphasize that automation is one of the key ways to minimize the software cost, and is very important for all, especially, for the "Design for the Bottom* 90% *People" or BOP (Base of the Pyramid People).*

Keywords: Software Engineering, Artificial Intelligence, Ontology, Intelligent Agent, Requirements Specification, Requirements Modeling, Design Modeling, Semantics, Natural Language Understanding, Machine Learning, Universal Modeling Language (UML), ICT (Information and Communication Technology and BOP (Base of the Pyramid People)

INTRODUCTION

Converting requirement specification or model to design model followed by an implementation is an important part of software engineering, especially for a large scale software. It is both information conversion and knowledge conversion, and it involves both art and science. Hence the process is complex. In fact, the various levels of abstractions involved in such mapping (e.g. from requirement model to design model, to architecture, to implementation) make the process even more complex. Designers use their expertise and various available tools to successfully complete the process. Since software cost is an important factor for many organizations (in fact, it is a key factor for almost all countries as it is a significant part of GDP, Gross Domestic Products), it is important that we keep the software cost minimal. This is even more true for underdeveloped and developing countries dominated by BOP (Base of the Pyramid People)—many of them are poor i.e. income is less than $2 per day. Minimizing software cost will help such countries afford ICT (Information and Communication Technologies) and associated software; and thus will provide the benefits of the Information Age to such population. This fits well, with "Design for the bottom 90% people". Automation is one of the key ways to minimize the software cost [1].

Many researchers have been working on automating various parts of the software engineering including software development process. e.g. to help architectural design, and various models have been proposed like Structural Models, Framework Models, Dynamic Models, Process Models and Functional Models ([2-5]). A number of different Architectural Description Languages (ADLs) have been developed to represent these models ([6,7]). Similarly, to help requirement modeling, various languages have been developed e.g. Requirement Modeling Language, RML ([8,9]). However, we could not find any citation regarding automatically mapping a Requirement Model to a Design Model. A few somewhat related researches are covered in ([10,11]).

In this paper, we present an Intelligent Agent (IA) based automated method to map Requirement Model to a Design Model. It is called IRTDM (Intelligent Agent based requirement model to design model mapping). The IA uses Artificial Intelligence (AI), semantic representation using Ontology or Predicate Logic, Design Structures (DS) using some well known design framework and Machine Learning algorithms for learning over time. We specifically focus on mapping Requirement Model to Architecture. Mapping to other key software areas/steps (e.g. converting the architecture into operational software) is also possible using similar approach but not covered in this paper.

Section 2 provides a brief high level overview of IRTDM (Intelligent Agent based requirement model to design model mapping). Section 3 describes the basics of the Flow-Oriented Requirement modeling to Data-Flow architecture mapping method as done by experienced designers. Section 4 describes an automated version of Section 3 using Natural Language Processing/Understanding, Artificial Intelligence and an Intelligent Agent. Section 5 describes the Architecture and Algorithms for more general and versatile Intelligent Agent. It also briefly discusses how to apply the concept for other types of mapping, Section 6 describes future work and Section 7 provides conclusions.

HIGH LEVEL OVERVIEW OF IRTDM

There is a good correspondence between requirement model and design model (Figure 1). Various parts of the Requirement Model have corresponding mapped parts in the design model. E.g. class-based elements map to data/ class, architecture and component design parts in the design model. In fact, designers use such basic mapping as a basis to come up with an architecture. Designers also use various levels of architectural abstractions (e.g. Architectural Genre, Architectural Styles, Archetypes) to come up with the structure showing key blocks or components. Our main theme is to use designers approach to come up with an automated approach. It is important to note that for some cases there is no practical mapping from requirement model to some architectural styles. But for many cases such mapping exists. A good example is mapping Flow-Oriented Requirement modeling to DataFlow architecture style. Since enough abstractions already exist and the manual method is understood reasonably well, we can convert the same into appropriate steps that can be done by an Intelligent Agent (IA) i.e. IA in IRTDM. First we discuss a simple IA to automatically handle Flow-Oriented Requirement modeling to Data-Flow architecture. Then we discuss more general IA.

The key issues a general IA needs to address are:

- Use of proper rules in doing the mapping.
- Use of semantics to ensure correct mapping.
- Use of appropriate rules and semantics to help map/transform one architectural style to another (e.g. Dataflow architecture to Layered architecture).
- Use of Learning to improve the outcome.
- Use of Verification to ensure correctness.
- Help Ensure that Implementation (coding) can also be automated in a similar way.
- Other key issues as appropriate (e.g. refactoring, generating test vectors and performing basic tests).

FLOW-ORIENTED REQUIREMENT MODELING TO DATA-FLOW ARCHITECTURE MAPPING

A mapping technique called Structured Design (SD) is often characterized as a data flow-oriented design method [10] as it provides a convenient transition from a data flow diagram (DFD) to software architecture. Such transformation involves the following 6 steps:

- The type of data (information) flow is established.
- Flow boundaries are determined.
- The DFD is mapped into the program structure.
- Control hierarchy is defined.
- Resultant structure is refined using design measures and heuristics, and
- The architectural description is refined and elaborated.

In order to design optimal module structure and interfaces two principles are crucial [12]:

- Cohesion which is "concerned with the grouping of functionally related processes into a particular module" and
- Coupling relates to "the flow of information, or parameters, passed between modules. Optimal coupling reduces the interfaces of modules, and the resulting complexity of the software".

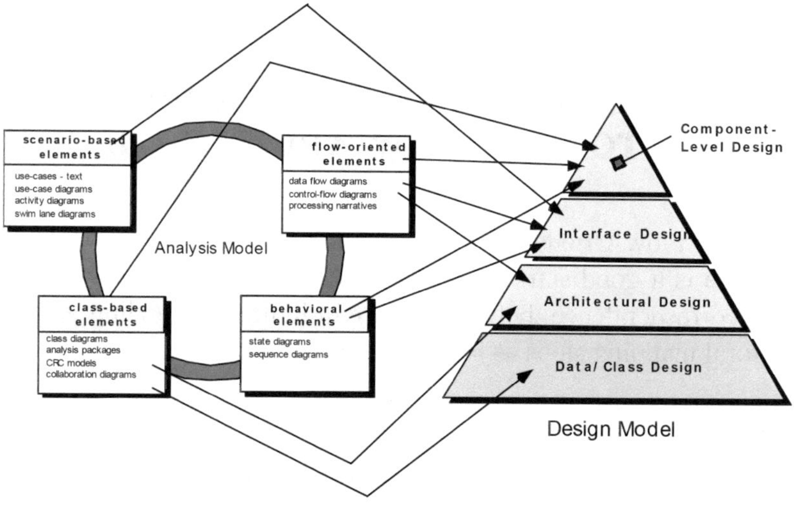

Figure 1. Flow-Oriented Requirement Modeling to Data-Flow Architecture Mapping (Courtesy [12]).

[Note: In general, Structured Design (SD) and Structured Analysis (SA) are methods for analyzing and converting business requirements into specifications and ultimately, computer programs, hardware configurations and related manual procedures. SA includes Context Diagram, Data Dictionary, DFD, Structure Chart, Structured Design and Structured Query Language (SQL)].

One form of information mapping is called Transform mapping where incoming data is transformed into an internal form by a transform center. The transformed data then flows to external world using outgoing flow. Another form of information mapping is called Transaction mapping in which a single data item triggers one or a number of information flows that effect a function implied by the triggering data item. The data item is called a transaction.

The above mentioned steps are done by designers (all types of designers including database and data warehouse designers and system architects) using the Requirement Model (in this case the Flow-oriented model) and the design structures including Design Genre, Design Styles (in this case data flow architecture), set of archetypes (e.g. Controller, Detector, Indicators, Node), basic classes (some of which are described in the Requirement Model) and some basic design guidelines. Refer to "Software Engineering: A Practitioner's Approach" by Roger Pressman [12] for a detailed example. We basically automate these steps using NLU, AI and an Intelligent Agent as described below in Sections 4 and 5.

AUTOMATING FLOW-ORIENTED REQUIREMENT MODELING TO DATA-FLOW ARCHITECTURE MAPPING

Converting Flow-Oriented Requirement Modeling to Data-Flow Architecture is a good start because of its simplicity. In this case there is a direct correspondence between the requirement modeling steps and architectural mapping steps as both use the same DFD.

Basic Ideas

Use the requirement modeling flow information and match it using AI rules to the corresponding Data-Flow Architecture. Since there is 1-1 correspondence (refer to Figure 1), Flow-Oriented elements have 1-1 correspondence with the Design Model blocks like Architectural Design), developing such rules are straight forward (refer to Sections 4.2, 4.3 and the example in Section 5). The rules are needed mainly to map DFD to the program structure, determine control hierarchy, complete refinement and elaboration.

Referring to Figure 1, there is a 1-1 correspondence from the DFD Requirement Model to Architectural Design, Interface Design and Component Level Design. Thus, we need appropriate rules to map to all such design levels. Cohesion and coupling are appropriately used to ensure optimal design module structures and interfaces. Any standard automatic/ semi-automatic technique can be used to determine the optimal design module structures and interfaces. All these key steps can be iterated during the refinement process (steps #e and #f in Section 3).

Requirement Modeling and Natural Language Processing (NLP)

Requirement Modeling methods usually use natural language words or equivalent methods. For example, in a Use Case diagram, the concept is expressed using natural language type concept. Class based, Behavioral based and DFD approaches also use natural language type concept. Thus, it is important to use Natural Language semantics and Natural Language Processing (NLP) in automating the mapping of Requirement Modeling to Design process. In case of DFD based modeling (as already mentioned), we would need semantics and NLP to map DFD to the program structure, determine control hierarchy, complete refinement and elaboration.

Besides, in a typical design,

- The software must be placed into context i.e. the design should define the external entities (other systems, devices, people) that the software interacts with and the nature of the interaction.
- A set of architectural archetypes should be identified—an archetype is an abstraction (similar to a class) that represents one element of system behavior.
- The designer specifies the structure of the system by defining and refining software components that implement each archetype.

NLP becomes handy in automating all these activities. Let's use an example to demonstrate the use of semantics and NLP:

Refer to Figure 2—it shows a simple DFD with reasonable details (i.e. say level 3 DFD). An analog signal is input to an Analog to Digital conversion unit (the Transform center circle or bubble #2) after doing some filtering operation by circle #1. The transform center outputs the digital signal in two format—binary (bubble #3) and hexadecimal (bubble #4). All bubbles are labeled with words that are easily understandable to human being as these are natural language words. Our goal is to use the semantic meaning of these words to come up with a design structure as designers usually do.

Consider the words "Analog to Digital Conversion" in bubble #2. The semantic meaning of this is "Conversion from an analog signal to digital signal takes place here" (see Section 4.3 below how such semantics is derived/ programmed). Once the program knows this semantics, it can determine the corresponding design archetypes and top level design box using AI rules which are based on the domain knowledge, semantics, and the DFD itself. Figure 3 shows the corresponding design structure. Such as structure is achieved using the following concept (the corresponding rules are given in Section 4.3):

- The boundaries shown in Figure 2 are used to focus on the design of bubble #2. This is as per standard DFD based design process as outlined in Section 3.
- Such boundaries can easily be done by representing the DFD using a Graph which can be implemented using netlist.
- Since bubble #2 is taking one input and producing 2 outputs of different data formats, bubble #2 is doing a "Transform flow".
- The outputs of the transform flow are detailed out in the DFD itself. So, corresponding design blocks can easily be constructed (Figure 3 shows this using DFD based mapping to a Call and Return architecture).
- As bubble #2 is doing a transform operation, it needs to do a "control function" in addition to do the main "transform function". This again is part of the standard design process that de signers use in a Structured Design.
- Netlist of the DFD is used to move and identify the new boundaries (by the automation software i.e. IA), find the new transform center and complete the design for new transform center, e.g. Binary Format-3 bubble and Hex Format bubble (Figure 3).

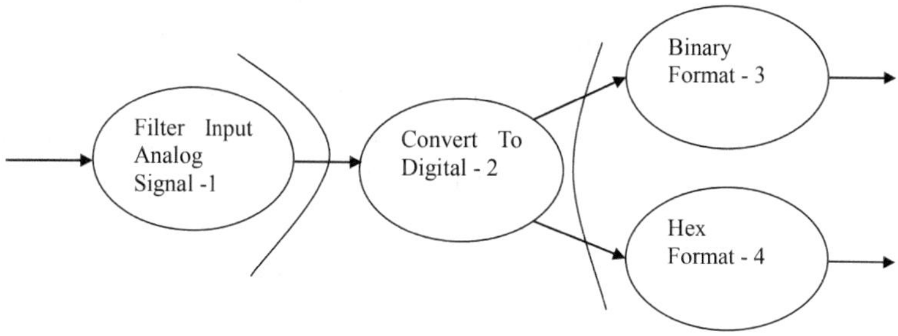

Figure 2. A Simple Transform Flow DFD. "Convert to Digital" circle (bubble) is the Transform Center. Input is an Analog signal which is converted by the Transform Center into Digital signal with two formats—Binary and Hexadecimal. The semantics of the "label" words of each bubble are used to automate the Design Process—see texts in Section 4.2 for details.

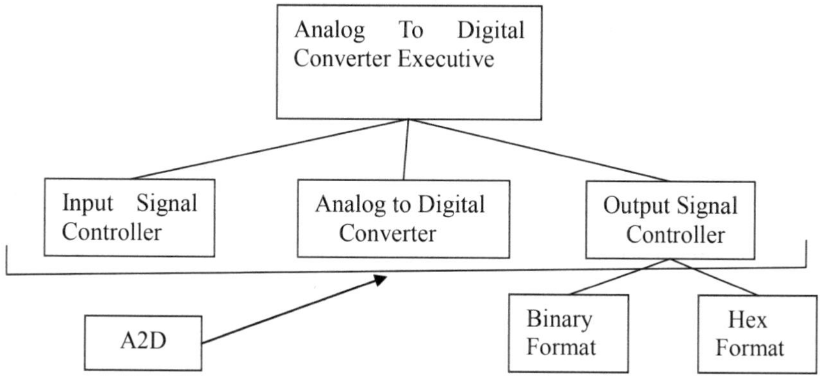

Figure 3. Design structure constructed by using the DFD in Figure 2. Semantics of the bubbles 2, 3 and 4 in Figure 2 and corresponding rules are used to make the construction. Semantics and all associated rules are implemented using First Order Logic (FOL). See Section 4.2 and Section 4.3 for details.

The following Section implements these concepts using semantics, NLP and AI. And all these are part of the Intelligent Agent, IA.

Predicate Calculus and Mapping Rules

The rules mentioned above can be represented by Predicate Calculus rules. Predicate Calculus can also be used to define semantics. We can also use

Ontology to define semantics. In this paper, we are using Predicate Calculus to describe the rules and semantics.

Consider the words "Analog to Digital Conversion" in bubble #2 in Figure 2 (as described in Section 4.2). The semantic meaning of this is "Conversion from an analog signal to digital signal takes place here" or simply "Conversion from an analog signal to digital". In predicate calculus (or First order logic, FOL), we can use the following to represent this semantics:

Converts (Convert to Digital-2, AnalogToDigital) …(1)

AnalogToDigitalConverter (Convert to Digital-2) …(2)

Converts (AnalogToDigitalConverter, AnalogToDigital) …………………………………………..…………….(2a)

When "Convert to Digital-2" label is seen in DFD bubble #2, the semantics determines that this is an analog to digital converter. Hence, all the design structures have the key blocks needed to implement the function of an analog to digital converter (Figure 3).

To make it more general, we use universal quantifier "for all" i.e. ∀ to say,

"All analog to digital converters convert analog signal to digital signal" …………………………..…………..(3a)

Which can be written in FOL

∀x AnalogtoDigitalConverter (x) ⇒ Converts (x, AnalogToDigital) …………………………………….(3b)

Using the universal quantifier, we allow to use any analog to digital converter in our knowledgebase or library.

[Note: mathematically, x can be any variable, including an instance of a non-AnalogToDigitalConverter [13]. This, however, can be avoided in various ways. We take care of this by only allowing analog to digital converters in the corresponding library].

In addition, an Executive control block (Analog To Digital Converter Executive) and a few other associated control blocks (e.g. input signal controller and output signal controller) are generated (Figure 3) as per standard design technique used in DFD model. Similarly, using the semantics of other bubbles, blocks to handle the binary and hex format are constructed. The FOL rules are used to describe all these as shown below:

If x is AnalogToDigitalConverter then Blocks are
"Analog To Digital Converter Executive"
AND "Analog To Digital Converter"
AND "Input Signal Controller"
AND "Output Signal Controller" ....(4)

If x is Binary Format then Blocks are "Binary Format" ....(5)

If x is Hex Format then Blocks are "Hex Format" ....(6)

The actual blocks for the analog to digital converter can have more than one block and also multi-level blocks as appropriate. But the whole thing can be labeled in the knowledge base as one block (e.g. A2D as shown in Figure 3) so that it is placed properly when such a rule (i.e. Equation (4)) is fired (see Section 4.3 for more details). The same is true for all other blocks and associated rules (e.g. Binary and Hex format blocks in Figure 3). Note, in a rule (e.g. Equation (4)), the semantics that it is an Analog To Digital Converter is derived using Equations (1) and (2) [see Section 4.4 for more details].

It may seem trivial that we could just use the label directly to construct the design structure using appropriate blocks. Yes, it is true for simple cases. But label may be more complex (can have more words and mean multiple operations), the format and words may vary considerably and the like. Use of NLP & FOL can define the meaning in a more flexible and reliable way, especially for complex cases. NLP & FOL become more important for refining the resultant structure (step #e in Section 3), and when the architectural description is refined and elaborated (step #f in Section 3). See Section 5 and Section 6 for more details.

Design Structures

In order to properly execute steps (#c to #f) in Section 3, namely,

c) The DFD is mapped into the program structure.
d) Control hierarchy is defined.
e) Resultant structure is refined using design measures and heuristics, and
f) The architectural description is refined and elaborated.

Designers follow various policies and processes. An architectural genre (e.g. Operating System or Artificial Intelligence), architectural style (e.g. Data-centric or Call and Return) and a set of Archetypes (e.g. Nodes, Detector, Indicator, Controller) need to be selected/defined. These are heavily influenced by designer's experience and knowledge. Such knowledge and experience need to be put in the knowledgebase using appropriate rules and predefined structures and blocks. Here, the designers have the option to make the automated system very efficient. Such structures and blocks need to be refined on a regular basis for continuous improvement.

To make the design modeling & construction of the design structure flexible and efficient, and to better support refinement and elaborations, design structures/ blocks needs to be configurable via some parameters. This scheme will better support the flexibility in the A2D implementation as mentioned in Section 4.3.

The Automation Process

The automation process involves the following key steps:

1) Create a good knowledgebase (KB) that has key information that designers follow in converting a requirement model to design model or structure. Designers use various policies and processes. Such a knowledgebase need to include all architectural genre, architectural styles, and set of archetypes.
2) The KB also would need to include all rules to convert a DFD (other representations used for Requirement Modeling) to design structures and blocks.
3) Design library needs to have all the key structures, blocks, components with appropriate parameterization.
4) Establish mechanism to continuously improve the library and the design process based on learning from previous design structures. This part can be automated using separate rules and semantics.

Once the above keys steps are completed, the IA (see Section 5), can take a DFD directly and produce a design structure as shown in Figure 3. IA accomplishes this by taking the DFD netlist and implementing (i.e. converting) each bubbles using the semantics of the bubbles and the rules.

The facts and the rules are combined using an inference mechanism, like Modus-Ponens.

Multiple rules can be fired and Forward Chaining, or Backward Chaining can be used to derive the final design structure. A short example is shown below using the AnalogToDigitalConverter example discussed in Sections 4.2 and 4.3:

AnalogToDigitalConverter (Convert to Digital-2) ....(2)

[a Fact—Convert to Digital-2 is an AnalogToDigitalConverter]

∀x AnalogtoDigitalConverter (x) ⇒ Converts (x, AnalogToDigital) ....(3b)

[Rule—for all x, if x is an AnalogtoDigitalConverter, then it converts AnalogToDigital]

[Using Modus-Ponen] Converts (Convert to Digital -2, AnalogToDigital) [New derived fact]

Note that the new derived fact by using Modus-Ponens is already shown in Equation (1). But it is shown there to express the semantics of the bubble #2 in Figure 2. But it is not used to represent a fact there. When it is derived as a fact, then Equation (4) will fire and will create the design structure (Rule represented by Equation (4) is not an implication as used in Equation 3(b). However, it can be converted to an implication form). Also, while Forward and Backward Chaining are sound, neither is complete. This means that there are valid inferences that cannot be found using these methods alone. An alternative inference technique called Resolution is sound and complete but computationally expensive [13].

INTELLIGENT AGENT

An intelligent Agent, IA implements the automation described in Section 4.5. It also performs other functions including some advanced functions needed to handle requirement models other than DFD i.e. Class based, Use Case based and State based models or their combinations that may include DFD. The key functions of IA are mentioned in Section 2. The implementation of key functions are described in Sections 3 & 4 for DFD based mapping to a Call and Return architecture. Such implementations are, in general, applicable for all other mappings with some refinements. Figure 4 shows the architecture of a general IA. A few key functions not yet described are:

- Use of appropriate rules and semantics to help map/ transform one architectural style to another (e.g. Dataflow architecture to Layered architecture).

- Use of Learning to improve the outcome.
- Use of Verification to ensure correctness.

Architectures for which direct mapping does not exists, the mapping process becomes complex. The designers approach the translation of requirements to design for such cases using their knowledge, more analyses and considering more architectural tradeoffs. Although there is no simple steps like steps #a to steps #f as mentioned in Section 2 for DFD based mapping, the designer's approach can be captured into similar flow and steps but with more natural language descriptions. Thus, for such cases, the issue of using NLP becomes more important and semantics & rules become more complex.

The learning over time can be implemented using any standard good learning algorithms. The verification process can be implemented by allowing performing some basic tests on the constructed system. Each component will have netlist or behavioral model representation which can take input vectors and verify the outputs with some predefined expected outputs (in compliance with the specification). In some cases, formal verification can be done using formal mathematical specification of the software.

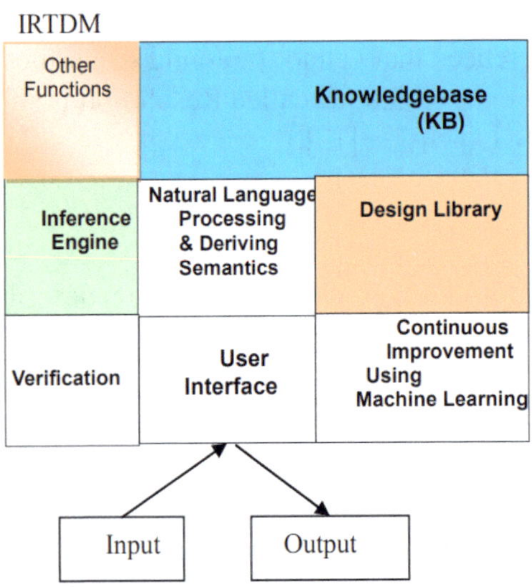

Figure 4. IRTDM—Intelligent Agent for requirement model to design model mapping. Shows all the key blocks. The KB and Design Library can reside outside. Input is mainly the requirement model and output is mainly the design structure and blocks.

FUTURE WORKS

The semantics represented by FOL and other similar techniques are good but they work satisfactorily mainly for small domain. As shown in Section 4.3, we need to define semantics for almost everything i.e. existing schemes do not allow to automatically derive new semantics from semantics of existing words. In ([14,15]) we have mentioned that while traditional approaches to Natural Language Understanding (NLU) have been applied over the past 50 years and have had some good successes mainly in a small domain, results show insignificant advancement, in general, and NLU remains a complex open problem. NLU complexity is mainly related to semantics: abstraction, representation, real meaning, and computational complexity. We argued that while existing approaches are great in solving some specific problems, they do not seem to address key Natural Language problems in a practical and natural way. In [16], we proposed a Semantic Engine using Brain-Like approach (SEBLA) that uses Brain-Like algorithms to solve the key NLU problem (i.e. the semantic problem) as well as its sub-problems.

SEBLA can calculate semantics of sentences using the semantics of words and the semantics of a paragraph using the semantics of the sentences. Enhanced semantics capability is needed to handle complex mapping cases mentioned in Section 5. We plan to use SEBLA for such cases.

We also plan to use SEBLA to automate/partially automate the implementation of the architecture into final software form (i.e. converting the architecture into operational software). Note that the automation presented in this paper is not the implementation in final software form; it is rather automating the mapping to design structure or architecture or blueprint of the desired system.

CONCLUSIONS

IRTDM (Intelligent Agent based requirement model to design model mapping) will significantly help today's large software development process. It takes long time to manually map the requirement model to a design model. As the software size gets bigger and bigger (a common trend in the industry), this process will become much more complex, and need for an automation of this process will become mandatory. In fact, automation is already mandatory to handle existing software design/development if we focus on the design for the bottom 90% people (the so-called Base of the pyramid people, BOP).

IRTDM will also increase the reliability and correctness of the said mapping and associated software. Moreover, with Natural Language Processing/Understanding and Artificial Intelligence (AI), the IA (Intelligent Agent) can map the design model to high level design components, thus further providing significant help in already very complex software engineering process.

Thus, our IRTDM will save significant cost for software which is a key component of the total yearly expense of most countries. Lower software cost implies lower price for buying new software; thus allowing many more people in the world to enjoy the benefits of the Information Age.

We have emphasized the need for enhanced Natural Language Processing/Understanding to better handle semantics, especially, for the complex software development cases. Use of natural semantics (e.g. SEBLA [16]) is the key to achieve this which we plan to do next.

REFERENCES

1. E. Khan, "Internet for Everyone: Reshaping the Glob-al Economy by Bridging the Digital Divide," 2011.
2. G. Abowd et al., "Structural Modeling: An Application Framework and Development Process for Flight Simulators," CMU Technical Report CMU/SEI-93-TR-014, 1993.
3. Structured Analysis. http://en.wikipedia.org/wiki/Structured_analysis
4. D. Garlan and M. Shaw, "An Introduction to Software Architecture," Advances in Software Engineering and Knowledge Engineering, Vol. I, World Scientific Publishing Company, 1995.
5. F. Buschmann, et al., "Pattern-Oriented Software Architecture, A System of Patterns," Wiley, 2007.
6. "Architecture Analysis and Design Language, Software (AADL)," Engineering Institute, Carnegie-Mellon University, 2004.
7. P. Clements, "A Survey of Architectural Description Languages," Paul C. Clements, Software Architecture, Software Engineering Institute, 1996.
8. S. Greenspan, et al., "A Requirements Modeling Language and Its Logic," Information Systems, Vol. 11, No. 1, 1986, pp. 9-23. http://dx.doi.org/10.1016/0306-4379(86)90020-7
9. J. Rumbaugh, et al., "The Unified Modeling Language Reference Manual," 2nd Edition, Addison-Wesley, 2004.
10. "Process Model Requirements Gap Analyzer," 2012. http://www.accenture.com/SiteCollection Documents/PDF/Accenture-Process-Model-Requirements-Gap-Analyzer.pdf
11. H. E. Okud, et al., "Experimental Development Based on Mapping Rule between Requirements Analysis Model and Web Framework Specific Design Model," SpringerPlus Journal, Vol. 2, 2013, p. 123. http://dx.doi.org/10.1186/2193-1801-2-123
12. R. Pressman, "Software Engineering: A Practitioner's Approach," McGrawHill, 2010.
13. D. Jurafsky, et al., "Speech and Language Processing: An Introduction to Natural Language Processing, Computational Linguistics and Speech Recognition," Pearson/ Prentice Hall, 2009.

14. E. Khan, "Natural Language Based Human Computer Interaction: A Necessity for Mobile Devices," International Journal of Computers and Communications, 2012.
15. E. Khan, "Addressing Big Data Problems using Semantics and Natural Language Understanding," 12th Wseas International Conference on Telecommunications and Informatics (Tele-Info '13), Baltimore, September 17-19, 2013.
16. E. Khan, "Natural Language Understanding Using Brain-Like Approach: Word Objects and Word Semantics Based Approaches help Sentence Level Understanding," Applied to US Patent Office, 2012.

SECTION 3 - FINITE AUTOMATA

Chapter

THE EQUIVALENT CONVERSION BETWEEN REGULAR GRAMMAR AND FINITE AUTOMATA

10

Jielan Zhang[1] and Zhongsheng Qian[2]

[1]Department of Information Technology, Yingtan Vocational and Technical College, Yingtan, China

[2]School of Information Technology, Jiangxi University of Finance and Economics, Nanchang, China.

ABSTRACT

The equivalence exists between regular grammar and finite automata in accepting languages. Some complicated conversion algorithms have also been in existence. The simplified forms of the algorithms and their proofs are given. And the construction algorithm 5 of the equivalent conversion from

Citation: J. Zhang and Z. Qian, "The Equivalent Conversion between Regular Grammar and Finite Automata," Journal of Software Engineering and Applications, Vol. 6 No. 1, 2013, pp. 33-37. doi: 10.4236/jsea.2013.61005.

finite automata to left linear grammar is presented as well as its correctness proof. Additionally, a relevant example is expounded.

Keywords: Regular Grammar, Finite Automata, NFA, DFA

INTRODUCTION

A rapid development in formal languages has made a profound influence on computer science, especially played a greater role in the design of programming languages, compiling theory and computational complexity since formal language system was established by Chomsky in 1956. Chomsky's Conversion Generative Grammar was classified into phase grammar, context-sensitive grammar, context-free grammar and linear grammar (or regular grammar) that includes left linear grammar and right linear grammar. All these are just a simple introduction to grammar, and automata theory, which plays an important role in compiling theory and technology, has another far-reaching impact on computer science.

A regular grammar G, applied to formal representation and theoretical research on regular language, is the formal description of regular language, mainly describes symbolic letters and often identifies words in compiler. A finite automata M including NFA (Non-deterministic Finite Automata) and DFA (Deterministic Finite Automata), applied to the formal model representation and research on digital computer, image recognition, information coding and neural process etc., is the formal model of discrete and dynamic system that have finite memory, and is applied to word identification and the model representation and realization of generation process during the course of word analysis in compiler. As far as language representation is concerned, the equivalence exists between the language regular grammar G describes and that finite automata M identifies.

SOME EQUIVALENT CONVERSION ALGORITHMS BETWEEN REGULAR GRAMMAR AND FINITE AUTOMATA

The definition of DFA where some notations in the remainder of this paper are shown is given first. The definition of NFA and regular grammar as well as the subset-based construction algorithm from NFA to DFA can be easily found in [1-4].

Definition 1. A DFA M is an automatic recognition device that is a quintuple denoted by (S,Σ,δ,s_0,F), where each element in S indicates one state in present system; $\sum$ denotes the set of conditions under which the system may happen; δ is a single valued function from $S\times\Sigma$ to S with $\delta(s_1,a)=s_2$ indicating if the state of the current system is s_1 with an input a, there will be a transition from the current state to the successive one named s_2; s_0 is the very unique start state and F the set of final states.

With δ, one can easily identify whether the condition in Σ can be accepted by DFA or not. Now, we extend the definition domain of δ to $S\times\Sigma^*$ meaning that for any $s\in S$, $a\in\Sigma$ and $w\in\Sigma^*$, $\delta(s,\varepsilon)=s$ and $\delta(s,aw)=\delta(\delta(s,a),w)$ hold. That is to say, if the condition is ε, the current state is unchanged; if the state is s and the condition aw, the system will first map δ(s, a) to s_1, then continue to map from s_1 until the last one. For some set ω where $\omega\in\Sigma^*$, if $\delta(s_0,\omega)=f$ where $f\in F$ holds, then we say that DFA can accept the condition set ω.

Definition 2. If a regular grammar G describes the same language as that a finite automata M identifies, viz., $L(G)=L(M)$, then G is equivalent to M.

The following theorems are concerned about the equivalence between regular grammar and finite automata.

Theorem 1. For each right linear grammar G_R (or left linear grammar G_L), there is one finite automata M where $L(M)=L(G_R)$ (or $L(M)=L(G_L)$).

Here is the construction algorithm from regular grammar to finite automata, and the proof of correctness. It contains two cases, viz., one from right linear grammar and another from left linear grammar to finite automata.

Construction Algorithm 1.

For a given right linear grammar $G_R=\langle V_T,V_N,S,\Psi\rangle$, there is a corresponding NFA

$$M=(V_N\cup\{f\},V_T,\delta,\{S\},\{f\})$$

where f is a newly added final state with $f\notin V_N$ holding, the transition function δ is defined by the following rules.

1) For any $A \in V_N$ and $a \in V_T^*$, if $A \to a \in \Psi$ holds, then let $\delta(A,a) = f$ hold; or 2) For any $A, B \in V_N$ and $a \in V_T^*$, if $A \to aB \in \Psi$ holds, then let $\delta(A,a) = B$ hold.

Proof. For a right linear grammar G_R, in the leftmost derivation of S =>*ω (ω ∈ ∑*), using A→aB once is equal to the case that the current state A meeting with a will be transited to the successive state B in M. In the last derivation, using A→a once is equal to the case that the current state A meeting with a will be transited to f, the final state in M. Here we let $\omega = \omega_1 \cdots \omega_n$ where $\omega_i \in \Sigma \ (i = 1, \cdots, n)$, then S =>*ω if and only if

$$\delta(S,\ \omega) = \delta(\delta(S,\omega_1),\omega_2,\cdots,\omega_n)$$
$$= \delta(\delta(\delta(S,\omega_1),\omega_2),\omega_3,\cdots,\omega_n)$$
$$= \cdots = \delta\Big(\delta\Big(\delta\big(\cdots\delta(\delta(S,\omega_1),\omega_2),\cdots\big),\omega_i\Big),\omega_{i+1},\cdots\omega_n\Big)$$
$$= \cdots = \delta\Big(\delta\Big(\delta\big(\cdots\delta(\delta(S,\omega_1),\omega_2),\cdots\big),\omega_{n-1}\Big),\omega_n\Big) = f$$

holds.

For G_R, therefore, the enough and necessary conditions of S =>*ω are that there is one path from S, the start state to f, the final state in M. During the course of the transition, all the conditions met following one by one are just equal to ω, viz., $\omega \in L(G_R)$ if and only if $\omega \in L(M)$. Therefore, it is evident that $L(M) = L(G_R)$ holds.

Construction Algorithm 2.

For a given left linear grammar $G_L = \langle V_T, V_N, S, \Psi \rangle$, there is a corresponding NFA

$$M = (V_N \cup \{q\}, V_T, \delta, \{q\}, \{S\})$$

where q is a newly added start state with $q \notin V_N$ holding, the transition function δ is defined by the following rules.

1) For any $A \in V_N$ and $a \in V_T^*$, if $A \to a \in \Psi$ holds, then let $\delta(q,a) = A$ hold; or 2) for any $A, B \in V_N$ and $a \in V_T^*$, if $A \to Ba \in \Psi$ holds, then let $\delta(B,a) = A$ hold.

The proof of construction Algorithm 2 is similar to that of construction algorithm 1 and we obtain

$L(M)=L(G_L)$

Theorem 2. For each finite automata M, there is one right linear grammar G_R or left linear grammar G_L where $L(G_R)=L(G_L)=L(M)$.

Construction Algorithm 3.

For a given finite automata $M=(S,\Sigma,\delta,s_0,F)$, a corresponding right linear grammar $G_R=\langle\Sigma,S,s_0,\Psi\rangle$ can be constructed. We discuss this in two cases.

1) If $s_0 \notin F$ holds, then Ψ is defined by the following rules.

For any $a\in\Sigma$ and $A,B\in S$, if $\delta(A,a)=B$ holds, then a) if $B\notin F$ holds, let A→aB hold; or b) if $B\in F$ holds, let A→a|aB hold. Or 2) if $s_0\in F$ holds, then $\varepsilon\in L(M)$ holds because of $\delta(s_0,\varepsilon)=s_0$. From step 1) we know that

$L(G_R)=L(M)-\{\varepsilon\}$ holds. So, a new generation rule $s_1 \to s_0|\varepsilon$ is added to G_R created from step 1) where s_1 is a newly added start symbol with the original symbol s_0 being no longer the start symbol any more and $s_1\notin S$ holding. Such a right linear grammar obtained is still named G_R, viz. $G_R=\langle\Sigma,S\cup\{s_1\},s_1,\Psi\cup\{s_1\to s_0|\varepsilon\}\rangle$.

THE IMPROVED VERSION FOR CONSTRUCTION ALGORITHM 3

Construction Algorithm 3 discussed above is complex in some sort. The following one named as Construction Algorithm 4, more easily understood, is its simplified version.

Construction Algorithm 4.

For a given finite automata $M=(S,\Sigma,\delta,s_0,F)$, a corresponding right linear grammar $G_R=\langle\Sigma,S,s_0,\Psi\rangle$ can be constructed. For any $a\in\Sigma$ and $A,B\in S$ 1) If $\delta(A,a)=B$ holds, then let A→aB hold;

2) If $B \in F$ holds, then we add a generation rule B→ε. Here B may be equal to s_0, and as long as B is a member of the set of final states, B→ε must be added.

Proof. For any $\omega \in \Sigma^*$ in G_R, if s_0 =>*ω holds, let $\omega = \omega_1 \cdots \omega_n$ hold where $\omega_i \in \Sigma \ (i = 1, \cdots, n)$, we have s_0=> $\omega_1 s_1$ = > $\omega_1 \omega_2 s_2$ => ··· => ω_1 ··· $\omega_i s_i$ => ··· => ω_1 ··· ω_n.

That's to say, s_0 = > *ω holds if and only if there is a path from s_0 meeting $\omega_1, \cdots, \omega_n$ one by one to final states in M. Therefore, $\omega \in L(G_R)$ if and only if $\omega \in L(M)$ holds, viz., $L(G_R) = L(M)$.

It is obvious that Construction Algorithm 4 is much simpler than Construction Algorithm 3.

THE PROPOSED CONSTRUCTION ALGORITHM

The following Construction Algorithm 5 presented in this work as much as I know so far is an effective algorithm about the equivalent conversion from finite automata M to left linear grammar G_L according to construction algorithm 4; its proof of correctness is also given.

Construction Algorithm 5.

Let a given finite automata be $M = (S, \Sigma, \delta, s_0, F)$, adding q, a new symbol, as the start symbol with $q \notin S$ holding. Let $G_L = \langle \Sigma, S \cup \{q\}, q, \Psi \rangle$ hold where Ψ is defined by the following rules.

For any $a \in \Sigma$ and $A, B, f \in S$ 1) If $\delta(A, a) = B$ holds, then let B→Aa hold;

2) Add a generation rule s_0→ε; and 3) For any $f \in F$, add a generation rule q→f.

The rule 3) means that we add a new state q as the final state, and then link all the original final states which are no longer final ones to q through ε respectively in the state transition diagram of M.

In particular, we can let $G_L = \langle \Sigma, S, f, \Psi \rangle$ hold when F, the set of final states, contains only one final state f where Ψ is defined by the following rules.

For any $a \in \Sigma$ and $A, B \in S$ 1) If $\delta(A,a) = B$, let B→Aa hold;

2) Add a generation rule $s_0 \to \varepsilon$.

Proof. For left linear grammar G_L, using q→f once is equivalent to the case one of the original states meeting ε will be transited to q in M in the very beginning of the rightmost derivation of q =>*ω where $\omega \in \Sigma^*$; during the course of the derivation, using B→Aa once is equivalent to the case the state A meeting a will be transited to the successive state B in M; in the final step of the derivation, using $s_0 \to \varepsilon$ once is equivalent to the case that the state s_0 meeting ε stops in s_0 in M. Therefore, the rightmost derivation of q = > *ω is just the inverse chain of the path M transits from the very start state s_0 to the very final state f with all the conditions linked together in the path are just identical with ω.

Let $\omega = \omega_1 \cdots \omega_n$ hold without thought where $\omega_i \in \Sigma\ (i = 1, \cdots, n)$. If q => *ω holds, we have q => f => $s_{n-1}\omega_n$ => $s_{n-2}\omega_{n-1}\omega_n$ => ⋯ => $s_{i-1}\omega_i \cdots \omega_n$ => ⋯ => $s_0\omega_1 \cdots \omega_n$ => $\omega_1 \cdots \omega_n$, and there is a transition

$$\delta(s_0,\omega) = \delta(\delta(s_0,\omega_1),\omega_2,\cdots,\omega_n)$$
$$= \delta(\delta(\delta(s_0,\omega_1),\omega_2),\omega_3,\cdots,\omega_n)$$
$$= \cdots = \delta\Big(\delta\big(\delta(\cdots\delta(\delta(s_0,\omega_1),\omega_2),\cdots),\omega_i\big),\omega_{i+1},\cdots,\omega_n\Big)$$
$$= \cdots = \delta\Big(\delta\big(\delta(\cdots\delta(\delta(s_0,\omega_1),\omega_2),\cdots),\omega_{n-1}\big),\omega_n\Big) = f$$

of which each inverse step is corresponding to the one of the rightmost derivation above.

There, $\omega \in L(G_L)$ holds if and only if $\omega \in L(M)$ holds, viz. $L(G_L) = L(M)$ holds.

According to all of the above discussed and the equivalence between NFA and DFA, Theorem 2 is proved.

An example expatriated for Construction Algorithm 5 is taken as follows.

Example 1. Let DFA be

$M = \langle \{A,B,f\},\{0,1,2\},\delta,A,\{f\} \rangle$ which is equivalent to regular expression 02(102)* where δ satisfies $\delta(A,0) = B$, $\delta(B,2) = f$ and $\delta(f,1) = A$. The state transition diagram of M is shown in **Figure 1**. Now we can construct a left linear grammar

$G_L = \langle \{0,1,2\}, \{A,B,f,q\}, q, \Psi \rangle$ equivalent to M where

$\Psi = \{q \to f, f \to B2,\ B \to A0,\ A \to f1,\ A \to \varepsilon\}$

holds.

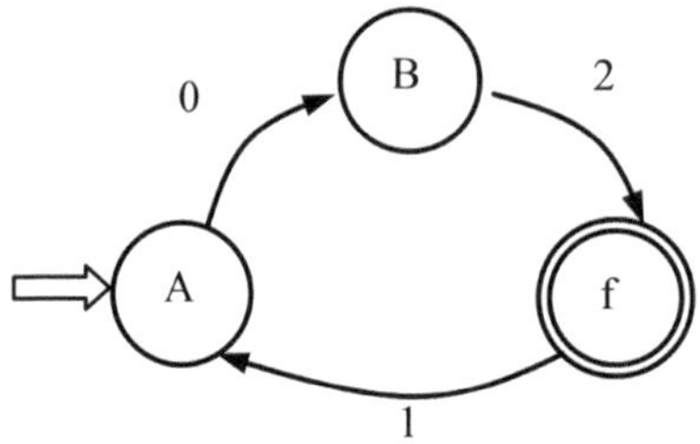

Figure 1. The state transition diagram of M.

In **Figure 1**, we can reduce G_L to $\langle \{0,1,2\}, \{A,B,f\}, f, \Psi \rangle$ because of only one final state f here where $\Psi = \{f \to B2,\ B \to A0,\ A \to f1,\ A \to \varepsilon\}$ holds. Furthermore, we can also get rid of ε from A→ε for A is not a start symbol in G_L, and then

$\Psi = \{f \to B2, B \to 0|A0, A \to f1\}$ is obtained.

RELATED WORK

The known proofs that the equivalence and containment problems for regular expressions, regular grammars and nondeterministic finite automata are PSPACE-complete that depends upon consideration of highly unambiguous expressions, grammars and automata. R. E. Stearns and H. B. Hunt III [5] proved that such dependence is inherent. Deterministic polynomial-time algorithms are presented for the equivalence and containment problems for unambiguous regular expressions, unambiguous regular grammars and unambiguous finite automata. The algorithms are then extended to ambiguity bounded by a fixed k. Their algorithms depend upon several elementary observations on the solutions of systems of homogeneous linear difference equations with constant coefficients and their relationship with the number of derivations of strings of a given length n by a regular grammar.

V. Laurikari [6] proposed a conservative extension to traditional nondeterministic finite automata (NFAs) to keep track of the positions in the input string for the last uses of selected transitions, by adding "tags" to transitions. The resulting automata are reminiscent of nondeterministic Mealy machines. A formal semantics of auto- mata with tagged transitions is given. An algorithm is given to convert these augmented automata to the corresponding deterministic automata, which can be used to process strings efficiently. The application to regular expressions is discussed, explaining how the algorithms can be used to implement, for example, substring addressing and a look ahead operator, and an informal comparison to other widely-used algorithms is made.

Cyril Allauzen, et al. [7] presented a general weighted grammar software library, the GRM Library, that can be used in a variety of applications in text, speech, and biosequence processing. The underlying algorithms were designed to support a wide variety of semirings and the representation and use of very large grammars and automata of several hundred million rules or transitions. They described several algorithms and utilities of this library and pointed out in each case their application to several text and speech processing tasks.

Several observations were presented on the computational complexity of regular expression problems [8]. The equivalence and containment problems were shown to require more than linear time on any multiple tape deterministic Turing machine. The complexity of the equivalence and containment problems was shown to be "essentially" independent of the structure of the languages represented. Subclasses of the regular grammars, that generated all regular sets but for which equivalence and containment were provably decidable deterministically in polynomial time, were also presented. As corollaries several program scheme problems studied in the literature were shown to be decidable deterministically in polynomial time.

Anne Brüggemann-Klein [9] showed that the Glushkov automaton can be constructed in a time quadratic in the size of the expression, and that this is worst-case optimal. For deterministic expressions, their algorithm has even linear run time. This improves on the cubic time methods.

Motivated by Li and Pedrycz's work on fuzzy finite automata and fuzzy regular expressions with membership values in lattice-ordered monoids and inspired by the close relationship between the automata theory and the theory of formal grammars, Xiuhong Guo [10] established a fundamental framework of L-valued grammar. It was shown that the set of L-valued regular languages coincides with the set of L-languages recognized by nondeterministic L-fuzzy finite automata and every L-language recognized by a deterministic L-fuzzy finite automaton is an L-valued regular language.

Formal construction of deterministic finite automata (DFA) based on regular expression was presented [11] as a part of lexical analyzer. At first, syntax tree is described based on the augmented regular expression. Then formal description of important operators, checking nullability and computing first and last positions of internal nodes of the tree is described. Next, the transition diagram is described from the follow positions and converted into deterministic finite automata by defining a relationship among syntax tree, transition diagram and DFA. Formal specification of the procedure is described using Z notation and model analysis is provided using Z/Eves toolset.

Sanjay Bhargava, et al. [12] described a method for constructing a minimal deterministic finite automaton (DFA) from a regular expression. It is based on a set of graph grammar rules for combining many graphs (DFA) to obtain another desired graph (DFA). The graph grammar rules are presented in the form of a parsing algorithm that converts a regular expression R into a minimal deterministic finite automaton M such that the language accepted by DFA M is same as the language described by regular expression R.

CONCLUDING REMARKS

The conversion algorithm can be realized from regular grammar to finite automata for the equivalence exists between the language regular grammar G describes and that finite automata M identifies and vice versa. In fact, the conversion between them is the very conversion between generation rules of grammar and mapping function of finite automata. The simplified forms of the conversion algorithms which are a little complicated and their proofs are given. And an algorithm about the equivalent conversion from finite automata to left linear grammar is presented as well as its correctness proof. Additionally, a relevant example is expounded.

ACKNOWLEDGEMENTS

Jielan Zhang, Zhongsheng Qian (NSFC) under grant No. 61262010 and the Jiangxi Provincial Natural Science Foundation of China under Grant No. 2010GQS 0048.

REFERENCES

1. H. W. Chen, C. L. Liu, Q. P. Tang, K. J. Zhao and Y. Liu, "Programming Language: Compiling Principle," 3rd Edition, National Defense Industry Press, Beijing, 2009, pp. 51-53.
2. A. V. Aho, M. S. Lam, R. Sethi and J. D. Ullman, "Compilers: Principles, Techniques, and Tools," 2nd Edition, Addison-Wesley, New York, 2007.
3. J. E. Hopcroft, R. Motwani and J. D. Ullman, "Introduction to Automata Theory, Languages, and Computation," Addison-Wesley, New York, 2007.
4. P. Linz, "An Introduction to Formal Languages and Automata," 5th Edition, Jones and Bartlett Publishers, Inc., Burlington, 2011.
5. R. E. Stearns and H. B. Hunt III, "On the Equivalence and Containment Problems for Unambiguous Regular Expressions, Regular Grammars and Finite Automata," SIAM Journal on Computing, Vol. 14, No. 3, 1985, pp. 598-611. doi:10.1137/0214044
6. V. Laurikari, "NFAs with Tagged Transitions, Their Conversion to Deterministic Automata and Application to Regular Expressions," Proceedings of the 7th International Symposium on String Processing Information Retrieval, IEEE CS Press, New York, 2000, pp. 181-187.
7. C. Allauzen, M. Mohri and B. Roark, "A General Weighted Grammar Library," Implementation and Application of Automata, LNCS 3317, 2005, pp. 23-34. doi:10.1007/978-3-540-30500-2_3
8. H.B. Hunt III, "Observations on the Complexity of Regular Expression Problems," Journal of Computer and System Sciences, Vol. 19, No. 3, 1979, pp. 222-236. doi:10.1016/0022-0000(79)90002-3
9. A. Brüggemann-Klein, "Regular Expressions into Finite Automata," Theoretical Computer Science, Vol. 120, No. 2, 1993, pp. 197-213. doi:10.1016/0304-3975(93)90287-4
10. X. H. Guo, "Grammar Theory Based on Lattice-ordered Monoid," Fuzzy Sets and Systems, Vol. 160, No. 8, 2009, pp. 1152-1161. doi:10.1016/j.fss.2008.07.009
11. N. A. Zafar and F. Alsaade, "Syntax-Tree Regular Expression Based DFA Formal Construction," Intelligent Information Management, Vol. 4, No. 4, 2012, pp. 138- 146. doi:10.4236/iim.2012.44021

12. S. Bhargava and G. N. Purohit, "Construction of a Minimal Deterministic Finite Automaton from a Regular Expression," International Journal of Computer Applications, Vol. 15, No. 4, 2011, pp. 16-27.

Chapter 11

CONTROLLABILITY, REACHABILITY, AND STABILIZABILITY OF FINITE AUTOMATA: A CONTROLLABILITY MATRIX METHOD

Yalu Li,[1] Wenhui Dou,[1] Haitao Li,[1,2] and Xin Liu[1]

[1]School of Mathematics and Statistics, Shandong Normal University, Jinan 250014, China

[2]Institute of Data Science and Technology, Shandong Normal University, Jinan 250014, China

ABSTRACT

This paper investigates the controllability, reachability, and stabilizability of finite automata by using the semitensor product of matrices. Firstly, by expressing the states, inputs, and outputs as vector forms, an algebraic form is obtained for finite automata. Secondly, based on the algebraic form, a

Citation: Yalu Li, Wenhui Dou, Haitao Li, Xin Liu, "Controllability, Reachability, and Stabilizability of Finite Automata: A Controllability Matrix Method", Mathematical Problems in Engineering, vol. 2018, Article ID 6719319, 6 pages, 2018. https://doi.org/10.1155/2018/6719319.

controllability matrix is constructed for finite automata. Thirdly, some necessary and sufficient conditions are presented for the controllability, reachability, and stabilizability of finite automata by using the controllability matrix. Finally, an illustrative example is given to support the obtained new results.

INTRODUCTION

In the research field of theoretical computer science, finite automaton is one of the simplest models of computation. Finite automaton is a device whose states take values from a finite set. It receives a discrete sequence of inputs from the outside world and changes its state according to the inputs. The study of finite automata has received many scholars' research interest in the last century [1–5] due to its wide applications in engineering, computer science, and so on.

As we all know, controllability and stabilizability analysis of finite automata are fundamental topics, which are important and necessary to the solvability of many related problems [1, 4, 6]. The concepts of controllability, reachability, and stabilizability of finite automata were defined in [2] by resorting to the classic control theory. The controllability of a deterministic Rabin automaton was studied in [7] by defining the "controllability subset." Kobayashi et al. [8] investigated the state feedback stabilization of a deterministic finite automaton and presented some new results.

Recently, a new matrix product, namely, the semitensor product (STP) of matrices, has been proposed by Cheng et al. [9]. Up to now, STP has been successfully applied to many research fields related to finite-valued systems like Boolean networks [10–20], multivalued logical networks [21–23], game theory [24, 25], finite automata [5, 26], and so on [27–35]. The main feature of STP is to convert a finite-valued system into an equivalent algebraic form [22]. Thus, STP provides a convenient way for the construction and analysis of finite automata [5, 26]. Xu and Hong [5] provided a matrix-based algebraic approach for the reachability analysis of finite automata with the help of STP. Yan et al. [26] studied the controllability and stabilizability analysis of finite automata based on STP and presented some novel results. It should be pointed out that although the concepts of controllability, reachability, and stabilizability of finite automata come from classic control theory, there exist fewer results on the construction of controllability matrix for finite automata.

In this paper, we investigate the controllability, reachability, and stabilizability of deterministic finite automata by using STP. The main contribution of this paper is to construct a controllability matrix for finite automata based on the algebraic form. Using the controllability matrix, we present some necessary and sufficient conditions for the controllability, reachability, and stabilizability of finite automata. Compared with the existing results [5, 26], our results are more easily verified via MATLAB.

The rest of this paper is organized as follows. Section 2 contains some necessary preliminaries on the semitensor product of matrices and finite automata. Section 3 studies the controllability, reachability, and stabilizability of finite automata and presents the main results of this paper. In Section 4, an illustrative example is given to support our new results, which is followed by a brief conclusion in Section 5.

Notations. $\mathbb{R}$, $\mathbb{N}$, and $\mathbb{Z}_+$ denote the set of real numbers, the set of natural numbers, and the set of positive integers, respectively. $\Delta_n := \{\delta_n^k : k = 1, \ldots, n\}$, where δ_n^k denotes the *k*th column of the $n \times n$ identity matrix I_n. An $n \times t$ matrix M is called a logical matrix, if $M = \begin{bmatrix}\delta_n^{i_1} & \delta_n^{i_2} & \cdots & \delta_n^{i_t}\end{bmatrix}$, which is briefly denoted by $M = \delta_n \begin{bmatrix}i_1 & i_2 & \cdots & i_t\end{bmatrix}$. The set of $n \times t$ logical matrices is denoted by $\mathscr{L}_{n \times t}$. Given a real matrix A, $\mathrm{Col}_i(A)$, $\mathrm{Row}_j(A)$, , and $(A)_{i,j}$ denote the *i*th column, the *j*th row, and the (i, j)th element of , respectively. $A > 0$ if and only if $(A)_{i,j} > 0$ holds for any i, j. $\mathrm{Blk}_i(A)$ denote the *i*th block of an $n \times mn$ matrix A.

PRELIMINARIES

Semitensor Product of Matrices

In this part, we recall some necessary preliminaries on STP. For details, please refer to [9].

Definition 1. Given two matrices $A \in \mathbb{R}^{m \times n}$ and $B \in \mathbb{R}^{p \times q}$, the semitensor product of A and B is defined as

$$A \ltimes B = \left(A \otimes I_{\alpha/n}\right)\left(B \otimes I_{\alpha/p}\right), \quad (1)$$

where $\alpha = \operatorname{lcm}(n, p)$ is the least common multiple of n and p and $\otimes$ is the Kronecker product of matrices.

Lemma 2. *STP has the following properties:*

- *Let* $X \in \mathbb{R}^{t\times 1}$ *be a column vector and* $A \in \mathbb{R}^{m\times n}$. *Then*

$$X \ltimes A = (I_t \otimes A) \ltimes X. \tag{2}$$

- *Let* $X \in \mathbb{R}^{m\times 1}$ *and* $Y \in \mathbb{R}^{n\times 1}$ *be two column vectors. Then*

$$Y \ltimes X = W_{[m,n]} \ltimes X \ltimes Y, \tag{3}$$

where $W_{[m,n]} \in \mathcal{L}_{mn\times mn}$ *is called the swap matrix.*

Finite Automata

In this subsection, we recall some definitions of finite automata.

A finite automaton is a seven-tuple $A = (X, U, Y, f, g, x_0, X_m)$, in which X, U, and Y are finite sets of states, input symbols, and outputs, respectively; x_0 and $X_m \subset X$ are the initial state and the set of accepted states; f and g are transition and output functions, which are defined as $f: X \times U \to 2^X$ and $g: X \times U \to 2^Y$, where 2^X and 2^Y denote the power set of X and Y, respectively; that is, $f(x,u) \subset X$, $g(x,u) \subset Y$. U^* represents the finite string set on U, which does not include the empty transition. Given an initial state $x_0 \in X$ and an input symbol $u \in U$, the function *f* uniquely determines the next subset of states, that is, $f(x_0, u) \subset X$, while the function g uniquely determines the next subset of outputs; that is, $g(x_0, u) \subset Y$.

Throughout this paper, we only consider the deterministic finite automata; that is, $|f(x,u)| \leq 1$ holds for any $x \in X$ and $u \in U$. In addition, we only investigate the controllability, reachability, and stabilizability of deterministic finite automata, and thus we do not use Y and g in the seven-tuple $A = (X, U, Y, f, g, x_0, X_m)$.

In the following, we recall the definitions of controllability, reachability, and stabilizability for deterministic finite automata.

Definition 3. (i) A state $x_p \in X$ is said to be controllable to $x_q \in X$, if there exists a control sequence $u^t \in U^*$ such that $f(x_p, u^t) = x_q$.

(ii) A state $x_p \in X$ is said to be controllable, if $x_p \in X$ is controllable to any state $x_q \in X$.

Definition 4. (i) A state $x_q \in X$ is said to be reachable from $x_p \in X$, if there exists a control sequence $u^t \in U^*$ such that $f(x_p, u^t) = x_q$.

(ii) A state $x_q \in X$ is said to be reachable, if $x_q \subset X$ is reachable from any state $x_p \in X$.

Given two nonempty sets $X_1 \subseteq X$ and $X_2 \subseteq X$ satisfying $X_1 \cup X_2 = X$ and $X_1 \cap X_2 = \emptyset$, we have the following definitions.

Definition 5. A nonempty set of state $X_1 \subseteq X$ is said to be controllable, if, for any state $x_q \in X_2$, there exist an $x_p \in X_1$ and a control sequence $u^t \in U^*$ such that $f(x_p, u^t) = x_q$.

Definition 6. A nonempty set of state $X_2 \subseteq X$ is said to be reachable, if, for any state $x_p \in X_1$, there exist an $x_q \in X_2$ and a control sequence $u^t \in U^*$ such that $f(x_p, u^t) = x_q$.

Definition 7. A nonempty set of state $X_1 \subseteq X$ is said to be 1-step returnable, if, for any state $x_0 = x_p \in X_1$, there exists an input $u_i \in U$ such that $f(x_p, u_i) = x_s \in X_1$.

Definition 8. A nonempty set of state $X_1 \subseteq X$ is said to be stabilizable, if X_1 is reachable and 1-step returnable.

MAIN RESULTS

In this section, we investigate the controllability, reachability, and stabilizability of deterministic finite automata by constructing a controllability matrix.

Controllability Matrix

For a deterministic finite automaton $A = (X, U, Y, f, g, x_0, X_m)$, where $X = \{x_1, \ldots, x_n\}$ and $U = \{u_1, \ldots, u_m\}$, we identify x_i as δ_n^i $(i = 1, \ldots, n)$ and call δ_n^i the vector form of x_i. Then, X can be denoted as Δ_n; that is, $X = \{\delta_n^1, \ldots, \delta_n^n\}$. Similarly, for U, we identify u_j with δ_m^j $(j = 1, \ldots, m)$ and call δ_m^j the vector form of u_j. Then, $U = \{\delta_m^1, \ldots, \delta_m^m\}$.

Using the vector form of elements in X and U, Yan et al. [26] construct the transition structure matrix (TSM) of $A = (X, U, Y, f, g, x_0, X_m)$ as

$F = [F_1 \ \cdots \ F_m]$. One can see that if there exists a control $u_j \in U$ which moves state x_p to state x_q, then

$$\delta_n^q = F \ltimes \delta_m^j \ltimes \delta_n^p. \tag{4}$$

In this case, $(F_j)_{q,p} = 1$. Otherwise, $(F_j)_{q,p} = 0$. Thus, setting

$$M = \sum_{j=1}^{m} F_j \in \mathbb{R}^{n \times n}, \tag{5}$$

then one can use M to judge whether or not state x_p is controllable to state x_q in one step. Precisely, state x_p is controllable to state x_q in one step, if and only if $(M)_{q,p} > 0$.

Now, we show that, for any $t \in \mathbb{Z}_+$, state x_p is controllable to state x_q at the tth step, if and only if $(M^t)_{q,p} > 0$. We prove it by induction. Obviously, when $t = 1$, the conclusion holds. Assume that the conclusion holds for some $t \in \mathbb{Z}_+$. Then, for the case of $t + 1$, state x_p is controllable to state at the (t + 1)th step, if and only if there exists some state $x_r \in X$ such that state x_p is controllable to state x_r at the tth step and state x_r is controllable to state x_q in one step. Hence,

$$\left(M^{t+1}\right)_{q,p} = \sum_{i=1}^{n} (M)_{q,i} \left(M^t\right)_{i,p} \geq (M)_{q,r} \left(M^t\right)_{r,p} > 0. \tag{6}$$

By induction, for any $t \in \mathbb{Z}_+$, state x_n is controllable to state x_q at the tth step, if and only if $(M^t)_{q,p} > 0$. Thus, $\sum_{t=1}^{\infty} M^t$ contains all the controllability information of the finite automata. Noticing that M is an $n \times n$ square matrix, by Cayley-Hamilton theorem, we only need to consider $t \leq n$. Then, we define the controllability matrix for finite automata as follows.

Definition 9. Set $M = \sum_{j=1}^{m} F_j \in \mathbb{R}^{n \times n}$. The controllability matrix of finite automata is $C = \sum_{t=1}^{n} M^t$.

Based on the controllability matrix, we have the following result.

Algorithm 10. Consider the finite automata $A = (X, U, Y, f, g, x_0, X_m)$. Then, the controls which force x_p to x_q in the shortest time can be designed by the following steps:

- Find the smallest integer l such that, for

$$\mathscr{F}_0^l := M^{l-1}F = \left[\mathrm{Blk}_1\left(\mathscr{F}_0^l\right)\ \mathrm{Blk}_2\left(\mathscr{F}_0^l\right)\ \cdots\ \mathrm{Blk}_m\left(\mathscr{F}_0^l\right)\right], \tag{7}$$

there exists a block, say, $\mathrm{Blk}_\theta(\mathscr{F}_0^l)$, satisfying $[\mathrm{Blk}_\theta(\mathscr{F}_0^l)]_{q,p} > 0$.

- Set $u(0) = \delta_m^\theta$ and $x_q = \delta_n^q$. If $l = 1$, stop. Otherwise, go to Step (3).
- Find r and η such that $[\mathrm{Blk}_\eta(\mathscr{F}_0^1)]_{q,r} > 0$ and $[\mathrm{Blk}_\theta(\mathscr{F}_0^{l-1})]_{r,p} > 0$, where $\mathscr{F}_0^1 = F$ and $\mathscr{F}_0^{l-1} = M^{l-2}F$. Set $u(l-1) = \delta_m^\eta$ and $x(l-1) = \delta_n^r$.
- If $l - 1 = 1$, stop. Otherwise, replace l and q by $l = 1$ and r, respectively, and go to Step (3).

Example 11. Consider a finite automaton $A = (X, U, Y, f, g, x_0, X_m)$ given in Figure 1, where $X = \{x_1, x_2, x_3\}$, $U = \{1, 2\}$. Suppose that $x_0 = x_1$ and $X_m = \{x_3\}$. Then, X can be denoted as $\Delta_3 = \{\delta_3^1, \delta_3^2, \delta_3^3\}$. Similarly, $U = \Delta_2 = \{\delta_2^1, \delta_2^2\} x_0 = \delta_3^3$, $X_m = \{\delta_3^2\}$.

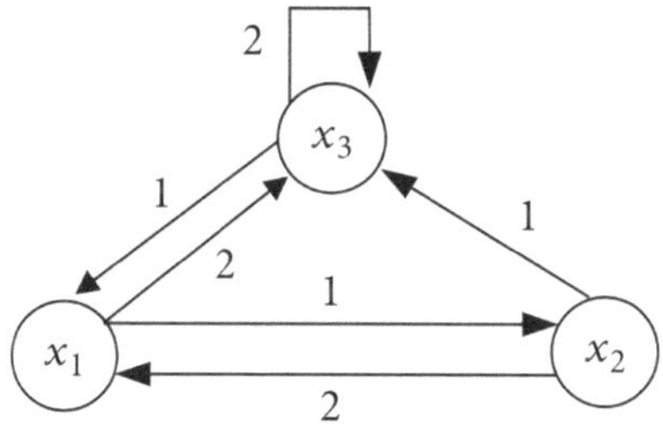

Figure 1. A finite automata.

The transition structure matrix of the finite automata A is

$$F = \delta_3\,[2\ \ 3\ \ 1\ \ 3\ \ 1\ \ 3]. \tag{8}$$

Split $F = [F_1\ \ F_2]$, where $F_1 = \delta_3\,[2\ \ 3\ \ 1]$ and $F_2 = \delta_3\,[3\ \ 1\ \ 3]$. Then,

$$M = \begin{bmatrix} 0 & 1 & 1 \\ 1 & 0 & 0 \\ 1 & 1 & 1 \end{bmatrix}. \tag{9}$$

Thus, the controllability matrix is

$$C = \sum_{t=1}^{3} M^t = \begin{bmatrix} 4 & 5 & 5 \\ 3 & 2 & 2 \\ 7 & 7 & 7 \end{bmatrix}. \tag{10}$$

By Algorithm 10, one can obtain that $[\mathrm{Blk}_1(\mathscr{F}_0^2)]_{2,3} > 0$. Setting $x_p = \delta_3^3$, $u(0) = \delta_2^1$, and $x_q = \delta_3^2$, one can find $r = 1$ and $\eta = 1$ such that $[\mathrm{Blk}_1(\mathscr{F}_0^1)]_{2,1} > 0$ and $[\mathrm{Blk}_1(\mathscr{F}_0^1)]_{1,3} > 0$. Let $u(1) = \delta_2^1$ and $x_1 = \delta_3^1$. Hence, state x_3 is controllable to state x_2 at the 2nd step.

Controllability, Reachability, and Stabilizability

In this part, we study the controllability, reachability, and stabilizability of deterministic finite automata based on the controllability matrix.

According to the meaning of controllability matrix, we have the following results.

Theorem 12. *The state* $x_p \in X$ *is controllable, if and only if*

$$Col_p(C) > 0. \tag{11}$$

Proof.

Necessity. Suppose that the state $x_p \in X$ is controllable. By Definition 3, $x_p \in X$ is controllable to any state $x_q \in X$. Based on (4), one can see that there exists a control sequence $\{u_{i_\alpha} : \alpha = 0, \ldots, s-1\} \subseteq U \ (1 \le s \le n)$ satisfying $\delta_n^q = F_{i_{s-1}} \cdots F_{i_0} \ltimes \delta_n^p$. Thus, $(F_{i_{s-1}} \cdots F_{i_0})_{q,p} > 0$, which implies that

$$(C)_{q,p} \ge \left(F_{i_{s-1}} \cdots F_{i_0}\right)_{q,p} > 0. \tag{12}$$

From the arbitrariness of q, we have $\mathrm{Col}_p(C) > 0$.

Sufficiency. Suppose that $\mathrm{Col}_p(C) > 0$ holds. Then, for any state $x_q \in X$, one can find some $(F_{i_{s-1}} \cdots F_{i_0})_{q,p} > 0$. Therefore, under the control sequence $\{u_{i_\alpha} : \alpha = 0, \ldots, s-1\} \subseteq U$, the state $x_p \subset X$ is controllable to $x_q \in X$. From the arbitrariness of q, the state $x_p \in X$ is controllable.

Theorem 13. *The state $x_q \in X$ is reachable, if and only if*

$$Row_q(C) > 0. \tag{13}$$

Proof.

Necessity. Suppose that the state $x_q \in X$ is reachable. By Definition 4, $x_q \in X$ is reachable to any state $x_p \in X$. One can obtain from (4) that there exists a control sequence $\{u_{i_\alpha} : \alpha = 0, \ldots, s-1\} \subseteq U \ (1 \leq s \leq n)$ satisfying $\delta_n^q = F_{i_{s-1}} \cdots F_{i_0} \ltimes \delta_n^p$. Thus, $(F_{i_{s-1}} \cdots F_{i_0})_{q,p} > 0$, which shows that

$$(C)_{q,p} \geq \left(F_{i_{s-1}} \cdots F_{i_0}\right)_{q,p} > 0. \tag{14}$$

From the arbitrariness of p, one can conclude that $\mathrm{Row}_q(C) > 0$.

Sufficiency. Suppose that $\mathrm{Row}_q(C) > 0$. Then, for any state $x_p \in X$, there exists some $(F_{i_{s-1}} \cdots F_{i_0})_{q,p} > 0$. Hence, under the control sequence $\{u_{i_\alpha} : \alpha = 0, \ldots, s-1\} \subseteq U$, the state $x_q \in X$ is reachable to $x_p \in X$. By Definition 4, the state $x_q \in X$ is reachable.

Given two nonempty sets $X_1 = \{\delta_n^{p_1}, \ldots, \delta_n^{p_s}\}$ and $X_2 = \{\delta_n^{p_{s+1}}, \ldots, \delta_n^{p_n}\}$, where $X_1 \cap X_2 = \emptyset$ and $X_1 \cup X_2 = X$, define

$$C_{X_1} = \left[\sum_{\beta=1}^{s} (C)_{p_{s+1},p_\beta} \cdots \sum_{\beta=1}^{s} (C)_{p_n,p_\beta}\right],$$

$$R_{X_2} = \left[\sum_{\gamma=s+1}^{n} (C)_{p_\gamma,p_1} \cdots \sum_{\gamma=s+1}^{n} (C)_{p_\gamma,p_s}\right]. \tag{15}$$

Based on Theorems 12 and 13, we have the following result.

Theorem 14. *(i) The nonempty set* $X_1 \subseteq X$ *is controllable, if and only if* $C_{X_1} > 0$.

(ii) The nonempty set $X_2 \subseteq X$ *is reachable, if and only if* $R_{X_2} > 0$.

Proof.

(i) *Necessity*. Suppose that the nonempty set $X_1 \subseteq X$ is controllable. By Definition 5, for any state $\delta_n^{p_\gamma} \in X_2$, there exist a $\delta_n^{p_\beta} \in X_1$ and a control sequence $u^t \in U^*$ such that $f(\delta_n^{p_\beta}, u^t) = \delta_n^{p_\gamma}$. Based on Theorems 12 and 13, for a fixed $\delta_n^{p_\gamma} \in X_2, \gamma = s+1, s+2, \ldots, n$, at least one of the following cases is true: $(C)_{p_\gamma, p_1} > 0, (C)_{p_\gamma, p_2} > 0, \ldots, (C)_{p_\gamma, p_s} > 0$. Therefore, for a fixed $\delta_n^{p_\gamma} \in X_2$, one can conclude that $\sum_{\beta=1}^{s} (C)_{p_\gamma, p_\beta} > 0$. From the arbitrariness of p_γ, one can see that

$$C_{X_1} = \left[\sum_{\beta=1}^{s} (C)_{p_{s+1}, p_\beta} \quad \cdots \quad \sum_{\beta=1}^{s} (C)_{p_n, p_\beta} \right] > 0. \tag{16}$$

Sufficiency. Suppose that $C_{X_1} = [\sum_{\beta=1}^{s} (C)_{p_{s+1}, p_\beta} \cdots \sum_{\beta=1}^{s} (C)_{p_n, p_\beta}] > 0$. Then, for any $\gamma = s+1, \ldots, n$, we have $\sum_{\beta=1}^{s} (C)_{p_\gamma, p_\beta} > 0$. It means that, for any state $\delta_n^{p_\gamma} \in X_2$, there exist a $\delta_n^{p_\beta} \in X_1$ and a control sequence $u^t \in U^*$ such that $f(\delta_n^{p_\beta}, u^t) = \delta_n^{p_\gamma}$. By Definition 5, the nonempty set $X_1 \subseteq X$ is controllable.

(ii) *Necessity*. Suppose that the nonempty set $X_2 \subseteq X$ is reachable. By Definition 6, for any state $\delta_n^{p_\beta} \in X_1$, there exist a $\delta_n^{p_\gamma} \in X_2$ and a control sequence $u^t \in U^*$ such that $f(\delta_n^{p_\beta}, u^t) = \delta_n^{p_\gamma}$. Based on Theorems 12 and 13, for a fixed $\delta_n^{p_\beta} \in X_1$, $\beta = 1, 2, \ldots, s$, at least one of the following cases is true: $(C)_{p_{s+1}, p_\beta} > 0, (C)_{p_{s+2}, p_\beta} > 0, \ldots, (C)_{p_n, p_\beta} > 0$. Therefore, for a fixed $\delta_n^{p_\beta} \in X_1$, one can see that $\sum_{\gamma=s+1}^{n} (C)_{p_\gamma, p_\beta} > 0$. From the arbitrariness of p_β, we have

$$\left[\sum_{\gamma=s+1}^{n} (C)_{p_\gamma, p_1} \quad \cdots \quad \sum_{\gamma=s+1}^{n} (C)_{p_\gamma, p_s} \right] > 0. \tag{17}$$

Sufficiency. Suppose that $\left[\sum_{\gamma=s+1}^{n} (C)_{p_\gamma, p_1} \quad \cdots \quad \sum_{\gamma=s+1}^{n} (C)_{p_\gamma, p_s} \right] > 0$. Then, for any $\beta = 1, 2, \ldots, s$, we have $\sum_{\gamma=s+1}^{n} (C)_{p_\gamma, p_\beta} > 0$. It means that, for any state $\delta_n^{p_\beta} \in X_1$, there exist a $\delta_n^{p_\gamma} \in X_2$ and a control sequence $u^t \in U^*$ such that $f(\delta_n^{p_\beta}, u^t) = \delta_n^{p_\gamma}$. By Definition 6, the nonempty set $X_2 \subseteq X$ is reachable.

Finally, we study the stabilizability of deterministic finite automata.

For $X_1 = \{\delta_n^{p_1}, \ldots, \delta_n^{p_s}\}$ and $X_2 = \{\delta_n^{p_{s+1}}, \ldots, \delta_n^{p_n}\}$, define

$$M_{X_1} = \left[\sum_{i=1}^{s} (M)_{p_i, p_1} \quad \cdots \quad \sum_{i=1}^{s} (M)_{p_i, p_s} \right]. \tag{18}$$

Theorem 15. *The nonempty set* $X_1 \subseteq X$ *is 1-step returnable, if and only if* $M_{X_1} > 0$.

Proof. By Definition 7, one can see that the nonempty set $X_1 \subseteq X$ is 1-step returnable, if and only if, for any state $\delta_n^{p_\beta} \in X_1$, there exist an input $u_i \in U$ and some $\delta_n^{p_i} \in X_1$ such that $f(\delta_n^{p_\beta}, u_i) = \delta_n^{p_i}$, that is, for a fixed $\delta_n^{p_\beta} \in X_1, \beta = 1, 2, \ldots, s,$ at least one of the following cases is true: $(M)_{p_1,p_\beta} > 0, (M)_{p_2,p_\beta} > 0, \ldots, (M)_{p_s,p_\beta} > 0$. Hence, $\sum_{i=1}^{s}(M)_{p_i,p_\beta} > 0, \forall \beta = 1, \ldots, s$. From the arbitrariness of p_β, one can obtain that $M_{X_1} > 0$.

Based on Theorems 14 and 15, we have the following result.

Corollary 16. *The nonempty set* $X_1 \subseteq X$ *is stabilizable, if and only if* $M_{X_1} > 0$ *and* $R_{X_1} > 0$.

Proof. By Definition 8, $X_1 \subseteq X$ is stabilizable, if and only if X_1 is reachable and 1-step returnable. Based on Theorems 14 and 15, the conclusion follows.

Remark 17. Compared with the existing results on the controllability and stabilizability of deterministic finite automata [5, 26], the main advantage of our results is to propose a unified tool, that is, controllability matrix, for the study of deterministic finite automata. The new conditions are more easily verified via MATLAB.

AN ILLUSTRATIVE EXAMPLE

Consider the finite automata $A = (X, U, Y, f, g, x^0, X^m)$ given in Figure 2, where $X = \{x_1, x_2, x_3, x_4\}$ and $U = \{1, 2\}$.

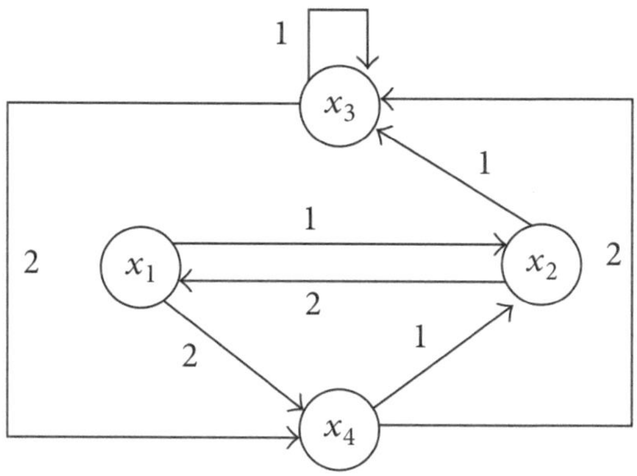

Figure 2. A finite automata.

From Figure 2, we can see that $x_1 \xrightarrow{1} x_2 \xrightarrow{1} x_3 \xrightarrow{2} x_4$ and $x_1 \xrightarrow{1} x_2 \xrightarrow{2} x_1$. Therefore, by Definition 3, one can obtain that x_1 is controllable. Similarly, by Definition 3, we conclude that x_2, x_3, and x_4 are also controllable. By Definition 4, we can also find that all the states are reachable.

Assume that $X_1 = \{x_1, x_2\}$ and $X_2 = \{x_3, x_4\}$. Since $x_1 \in X_1 \xrightarrow{1} x_2 \xrightarrow{1} x_3 \in X_2$ and $x_1 \in X_1 \xrightarrow{2} x_4 \in X_2$, by Definition 5, one can see that X_1 is controllable. Similarly, we also obtain that X_2 is controllable. Since $x_3 \xrightarrow{2} x_4 \xrightarrow{1} x_2 \xrightarrow{2} x_1$ and $x_1 \xrightarrow{1} x_2 \xrightarrow{1} x_3 \xrightarrow{2} x_4$, by Definition 6, we can obtain that X_1 and X_2 are reachable. From Figure 2, we can see that $x_1 \xrightarrow{1} x_2$, $x_2 \xrightarrow{2} x_1$, $x_3 \xrightarrow{2} x_4$ and $x_4 \xrightarrow{2} x_3$. Hence, by Definition 7, X_1 and X_2 are 1-step returnable. By Definition 8, the sets X_1 and X_2 are stabilizable.

Now, we check the above properties based on the controllability matrix.

The transition structure matrix of the finite automata A is

$$F = \delta_4 [2 \; 3 \; 3 \; 2 \; 4 \; 1 \; 4 \; 3]. \tag{19}$$

Split $F = [F_1 \;\; F_2]$, where $F_1 = \delta_4[2 \;\; 3 \;\; 3 \;\; 2]$ and $F_2 = \delta_4[4 \;\; 1 \;\; 4 \;\; 3]$. Then,

$$M = \begin{bmatrix} 0 & 1 & 0 & 0 \\ 1 & 0 & 0 & 1 \\ 0 & 1 & 1 & 1 \\ 1 & 0 & 1 & 0 \end{bmatrix}. \tag{20}$$

Thus, the controllability matrix is

$$C = \sum_{t=1}^{4} M^t = \begin{bmatrix} 3 & 4 & 2 & 3 \\ 7 & 5 & 5 & 6 \\ 12 & 13 & 14 & 14 \\ 8 & 8 & 9 & 7 \end{bmatrix}. \tag{21}$$

Since all rows and columns of C are positive, by Theorems 12 and 13, any state $x_i \in X$ is controllable and reachable, i = 1, 2, 3, 4.

A simple calculation gives $C_{X_1} = [25 \;\; 16]$, $R_{X_1} = [7 \;\; 9]$, $C_{X_2} = [5 \;\; 11]$, $R_{X_2} = [20 \;\; 21]$, $M_{X_1} = [10 \;\; 9]$ and $M_{X_2} = [23 \;\; 21]$. By Theorems 14 and 15 and Corollary 16, X_1 and X_2 are controllable, reachable, 1-step returnable, and stabilizable, respectively.

CONCLUSION

In this paper, we have investigated the controllability, reachability, and stabilizability of deterministic finite automata by using the semitensor product of matrices. We have obtained the algebraic form of finite automata by expressing the states, inputs, and outputs as vector forms. Based on the algebraic form, we have defined the controllability matrix for deterministic finite automata. In addition, using the controllability matrix, we have presented several necessary and sufficient conditions for the controllability, reachability, and stabilizability of finite automata. The study of an illustrative example has shown that the obtained new results are effective.

ACKNOWLEDGMENTS

The research was supported by the National Natural Science Foundation of China under Grants 61374065 and 61503225, the Natural Science Foundation of Shandong Province under Grant ZR2015FQ003, and the Natural Science Fund for Distinguished Young Scholars of Shandong Province under Grant JQ201613.

REFERENCES

1. S. Abdelwahed and W. M. Wonham, "Blocking Detection in Discrete Event Systems," in *Proceedings of the American Control Conference*, pp. 1673–1678, USA, 2003.
2. M. Dogruel and U. Ozguner, "Controllability, reachability, stabilizability and state reduction in automata," in *Proceedings of the IEEE International Symposium on Intelligent Control*, pp. 192–197, Glasgow, UK, 1992.
3. Y. Gang, "Decomposing a kind of weakly invertible finite automata with delay 2," *Journal of Computer Science and Technology*, vol. 18, no. 3, pp. 354–360, 2003.
4. J. Lygeros, C. Tomlin, and S. Sastry, "Controllers for reachability specifications for hybrid systems," *Automatica*, vol. 35, no. 3, pp. 349–370, 1999.
5. X. Xu and Y. Hong, "Matrix expression and reachability analysis of finite automata," *Control Theory and Technology*, vol. 10, no. 2, pp. 210–215, 2012.
6. A. Casagrande, A. Balluchi, L. Benvenuti, A. Policriti, T. Villa, and A. Sangiovanni-Vincentelli, "Improving reachability analysis of hybrid automata for engine control," in *Proceedings of the 43rd IEEE Conference on Decision and Control (CDC)*, pp. 2322–2327, 2004.
7. J. Thistle and W. Wonham, "Control of infinite behavior of finite automata," *SIAM Journal on Control and Optimization*, vol. 32, no. 4, pp. 1075–1097, 1994.
8. K. Kobayashi, J. Imura, and K. Hiraishi, "Stabilization of finite automata with application to hybrid systems control," *Discrete Event Dynamic Systems*, vol. 21, no. 4, pp. 519–545, 2011.
9. D. Cheng, H. Qi, and Z. Li, *Analysis and Control of Boolean Network: A Semi-Tensor Product Approach*, Communications and Control Engineering Series, Springer, London, UK, 2011.
10. E. Fornasini and M. E. Valcher, "On the periodic trajectories of Boolean control networks," *Automatica*, vol. 49, no. 5, pp. 1506–1509, 2013.
11. Y. Guo, P. Wang, W. Gui, and C. Yang, "Set stability and set stabilization of Boolean control networks based on invariant subsets," *Automatica*, vol. 61, pp. 106–112, 2015.

12. D. Laschov and M. Margaliot, "Minimum-time control of Boolean networks," *SIAM Journal on Control and Optimization*, vol. 51, no. 4, pp. 2869–2892, 2013.
13. F. Li and J. Sun, "Controllability and optimal control of a temporal Boolean network," *Neural Networks*, vol. 34, pp. 10–17, 2012.
14. F. Li, "Pinning control design for the synchronization of two coupled boolean networks," *IEEE Transactions on Circuits and Systems II: Express Briefs*, vol. 63, no. 3, pp. 309–313, 2016.
15. H. Li, L. Xie, and Y. Wang, "On robust control invariance of Boolean control networks," *Automatica*, vol. 68, pp. 392–396, 2016.
16. H. Li, Y. Wang, and L. Xie, "Output tracking control of Boolean control networks via state feedback: Constant reference signal case," *Automatica*, vol. 59, article 6422, pp. 54–59, 2015.
17. H. Li and Y. Wang, "Controllability analysis and control design for switched Boolean networks with state and input constraints," *SIAM Journal on Control and Optimization*, vol. 53, no. 5, pp. 2955–2979, 2015.
18. H. Li, L. Xie, and Y. Wang, "Output regulation of Boolean control networks," *Institute of Electrical and Electronics Engineers Transactions on Automatic Control*, vol. 62, no. 6, pp. 2993–2998, 2017.
19. J. Lu, J. Zhong, C. Huang, and J. Cao, "On pinning controllability of Boolean control networks," *Institute of Electrical and Electronics Engineers Transactions on Automatic Control*, vol. 61, no. 6, pp. 1658–1663, 2016.
20. M. Meng, L. Liu, and G. Feng, "Stability and gain analysis of Boolean networks with Markovian jump parameters," *Institute of Electrical and Electronics Engineers Transactions on Automatic Control*, vol. 62, no. 8, pp. 4222–4228, 2017.
21. Z. Liu, Y. Wang, and H. Li, "New approach to derivative calculation of multi-valued logical functions with application to fault detection of digital circuits," *IET Control Theory & Applications*, vol. 8, no. 8, pp. 554–560, 2014.
22. J. Lu, H. Li, Y. Liu, and F. Li, "Survey on semi-tensor product method with its applications in logical networks and other finite-valued systems," *IET Control Theory & Applications*, vol. 11, no. 13, pp. 2040–2047, 2017.

23. Y. Wu and T. Shen, "An algebraic expression of finite horizon optimal control algorithm for stochastic logical dynamical systems," *Systems & Control Letters*, vol. 82, article 3915, pp. 108–114, 2015.
24. D. Cheng, F. He, H. Qi, and T. Xu, "Modeling, analysis and control of networked evolutionary games," *Institute of Electrical and Electronics Engineers Transactions on Automatic Control*, vol. 60, no. 9, pp. 2402–2415, 2015.
25. P. Guo, H. Zhang, F. E. Alsaadi, and T. Hayat, "Semi-tensor product method to a class of event-triggered control for finite evolutionary networked games," *IET Control Theory & Applications*, vol. 11, no. 13, pp. 2140–2145, 2017.
26. Y. Yan, Z. Chen, and Z. Liu, "Semi-tensor product approach to controllability and stabilizability of finite automata," *Journal of Systems Engineering and Electronics*, vol. 26, no. 1, pp. 134–141, 2015.
27. D. Cheng and H. Qi, "Non-regular feedback linearization of nonlinear systems via a normal form algoithm," *Automatica*, vol. 40, pp. 439–447, 2004.
28. H. Li, G. Zhao, M. Meng, and J. Feng, "A survey on applications of semi-tensor product method in engineering," *Science China Information Sciences*, vol. 61, no. 1, Article ID 010202, 2018.
29. Z. Li, Y. Qiao, H. Qi, and D. Cheng, "Stability of switched polynomial systems," *Journal of Systems Science and Complexity*, vol. 21, no. 3, pp. 362–377, 2008.
30. Y. Liu, H. Chen, J. Lu, and B. Wu, "Controllability of probabilistic Boolean control networks based on transition probability matrices," *Automatica*, vol. 52, pp. 340–345, 2015.
31. Y. Wang, C. Zhang, and Z. Liu, "A matrix approach to graph maximum stable set and coloring problems with application to multi-agent systems," *Automatica*, vol. 48, no. 7, pp. 1227–1236, 2012.
32. Y. Yan, Z. Chen, and Z. Liu, "Solving type-2 fuzzy relation equations via semi-tensor product of matrices," *Control Theory and Technology*, vol. 12, no. 2, pp. 173–186, 2014.
33. K. Zhang, L. Zhang, and L. Xie, "Invertibility and nonsingularity of Boolean control networks," *Automatica*, vol. 60, article 6475, pp. 155–164, 2015.

34. J. Zhong, J. Lu, Y. Liu, and J. Cao, "Synchronization in an array of output-coupled boolean networks with time delay," *IEEE Transactions on Neural Networks and Learning Systems*, vol. 25, no. 12, pp. 2288–2294, 2014.
35. Y. Zou and J. Zhu, "System decomposition with respect to inputs for Boolean control networks," *Automatica*, vol. 50, no. 4, pp. 1304–1309, 2014.

Chapter

BOUNDED MODEL CHECKING OF ETL COOPERATING WITH FINITE AND LOOPING AUTOMATA CONNECTIVES

12

Rui Wang, Wanwei Liu, Tun Li, Xiaoguang Mao, and Ji Wang

College of Computer Science, National University of Defense Technology, Changsha, Hunan 410073, China

ABSTRACT

As a complementary technique of the BDD-based approach, bounded model checking (BMC) has been successfully applied to LTL symbolic model checking. However, the expressiveness of LTL is rather limited, and some important properties cannot be captured by such logic. In this paper, we present a semantic BMC encoding approach to deal with the mixture of ETL_f and ETL . Since such kind of temporal logic involves both finite and

Citation: Rui Wang, Wanwei Liu, Tun Li, Xiaoguang Mao, Ji Wang, "Bounded Model Checking of ETL Cooperating with Finite and Looping Automata Connectives", Journal of Applied Mathematics, vol. 2013, Article ID 462532, 12 pages, 2013. https://doi.org/10.1155/2013/462532.

looping automata as connectives, all regular properties can be succinctly specified with it. The presented algorithm is integrated into the model checker ENuSMV, and the approach is evaluated via conducting a series of imperial experiments.

INTRODUCTION

A crucial bottleneck of model checking is the state-explosion problem, and the symbolic model checking technique has proven to be an applicable approach to alleviate it. In the early 1990s, McMillan presented the BDD [1] based model checking technique [2]. It is first applied to CTL model checking and is later adapted to deal with LTL. With the rapid evolvement of SAT solvers, an entirely new approach, namely, bounded model checking (BMC), is presented in [3]. It rerpresents the problem "there is a path (with bounded length) violating the specification in the model" with a Boolean formula and then tests its satisfiability via a SAT solver. Usually, BMC is considered to be a complementary approach of the BDDbased approach: BMC is normally used for hunting bugs not for proving their absence. It performs better when handling a model having a large reachable state set but involving (relatively) shallow error runnings.

BMC has been successfully employed in LTL model checking. However, LTL has the drawback of limited expressiveness. Wolper was the first to complain about this by addressing the fact that some counting properties such as "p holds at every even moment" cannot be expressed by any LTL formula [4]. Indeed, LTL formulae are just as expressive as star-free ω-expressions, that is, ω-regular expressions disallowing arbitrary (in a star-free expression, Kleene-closure operators ($\bullet^*$ and $\bullet^\omega$) can only be applied upon Σ, which is the whole set of alphabet) use of Kleene-closure operators.

As pointed in [5, 6], it is of great importance for a specification language to have the power to express all ω-regular properties—as an example, it is a necessary requirement to support modular model checking. Actually, such specification language like PSL [7] has been accepted as industrial standard. For temporal logics within linear framework, there are several ways to pursue such an expressiveness.

(1) The first way is to add fixed-point operators or propositional quantifiers to the logic, such as linear μ- calculus [8] and QLTL [9].

(2) An alternative choice is to add regular expressions to LTL-like logics, as done in RLTL [10], FTL [11, 12], and PSL [7].

(3) The third approach is to cooperate infinitely many temporal connectives with the logic, just like various of ETLs [4, 9, 13].

The first extension requires finitely many operators in defining formulae. Meanwhile, the use of fixed-point operators and higher-order quantifiers tends to rise difficulties in understanding. In contrast, using regular expressions or automata as syntactical ingredients is much more intuitive in comprehension. To some extent, since nesting of automata connectives is allowed, the third approach generalizes the second one.

In [4], Wolper suggested using right linear grammars as connectives. Later, Wolper, Vardi, and Sistla consider taking various ω-automata [9, 13]. Depending on the type of automata used as temporal connectives, we may obtain various ETLs. As a result, ETLs employing ω-automata with looping, finite, and repeating (alternatively, Buchi [¨ 14]) acceptance are, respectively, named ETL_l , ETL_f, and ETL_r, and all of them are known to be as expressive as ω-regular expressions [13].

We have presented a BDD-based model checking algorithm for ETL_f in [15] and an algorithm for BDD-based model checking of an invariant of PSL in [16]. Jehle et al. present a bounded model checking algorithm for linear μ- calculus in [17]. And in [18], a tester based symbolic model checking approach is proposed by Pnueli and Zacks to deal with PSL properties. Meanwhile, a modular symbolic Buchi ¨ automata construction is presented in [19] by Cimatti et al.

In this paper, we present a semantic BMC encoding for ETL employing both finite acceptance and looping acceptance automata connectives (we in the following refer to it as ETLl+f). The reason that we study BMC algorithm for such kind of logic is for the following considerations.

(1) The BDD-based symbolic model checking technique for ETLf has been established in [15] by extending LTL construction [20]. Nevertheless, in a pure theoretical perspective, looping and finite acceptance, respectively, correspond to safety and liveness properties, and looping acceptance automata can be viewed as the counterparts of finite acceptance automata. Actually, both similarities and differences could be found in compiling the semantic models and translating Boolean representations when dealing with these two types of connectives. Since ETL_{l+f} has

a rich set of fragments, such as LTL, it is hopeful to develop a unified semantic BMC framework of such logics.

(2) Practically, things would usually be much more succinct when employing both types of automata connectives, in comparison to merely using finite or looping ones. As an example, there is no direct encoding for the temporal operator G just with finite acceptance automata—to do this with ETL_f, we need to use a two-state and two-letter connective to represent the operator F and then to dualize it. In contrast, with looping automata, we just need to define a one-state and one-letter connective. It would save much space overhead in building tableaux.

(3) Lastly, unlike syntactic BMC encodings (such kind of encodings give inductive Boolean translations with the formulae's structure, cf. [21, 22] for a survey), the semantic fashion [22] yields a natural completeness threshold computation approach, and it describes the fair path finding problem over the product model with Boolean formulae. In this paper, we give a linear semantic encoding approach (opposing to the original quadratic semantic encoding) for ETL_{l+f}. Moreover, the technique can also be tailored to semantic LTL BMC.

We have implemented the presented algorithm with our model checker ENuSMV (Ver. 1.2), and this tool allows end users to customize temporal connectives by defining automata. We have also justified the algorithm by conducting a series of comparative experiments.

The paper is structured as follows: Section 2 briefly revisits basic notions. Section 3 introduces semantic BMC encoding technique for ETL_{l+f}. In Section 4, experimental results of ETL_{l+f} BMC are given. Finally, we conclude the whole paper with Section 5.

PRELIMINARIES

An infinite word w over the alphabet Σ is a mapping from $\mathbb{N}$ to Σ; hence we may use (i) to denote the ith letter of w. For the sake of simplicity, we usually write w as the sequence $(0)w(1)\cdots$. A finite prefix of w with length n is a restriction of w to the domain $\{0, \ldots, n-1\}$, denoted by $w[n]$.

A (nondeterministic) automaton is a tuple $\mathscr{A} = \langle\Sigma, Q, \delta, q, F\rangle$, where:

- Σ is a finite alphabet,

- Q is a finite set of states,
- $\delta: Q \times \Sigma \rightarrow 2^Q$ is a transition function,
- $q \in Q$ is an initial state, and
- $F \subseteq Q$ is a set of accepting states.

An infinite run of $\mathscr{A} = \langle \Sigma, Q, \delta, q, F \rangle$ over an infinite word w is an infinite sequence $\sigma = q_0 q_1 \cdots \in Q^\omega$, where $q_0 = q$ and $q_{i+1} \in (q_i, w(i))$ for each i. In addition, we say that each prefix $q_0 \cdots q_{n+1}$ is a finite run over $[n]$.

In this paper, we are concerned with two acceptance types for ω-automata.

Looping. An infinite word w is accepted if it has an infinite run over w.

Finite. An infinite word w is accepted if it has a finite prefix $w[n]$, over which there is a finite run $q_0 \cdots q_{n+1}$ and q_{n+1} is an accepting state (call such a prefix accepting prefix).

In both cases, we denote by $\mathscr{L}(\mathscr{A})$ the set of infinite words accepted by $\mathscr{A}$.

Given an automaton $\mathscr{A} = \langle \Sigma, Q, \delta, q, F \rangle$ and a state $r \in Q$, we denote by $\mathscr{A}^r$ the automaton $\langle \Sigma, Q, \delta, r, F \rangle$. That is, $\mathscr{A}^r$ is almost identical to $\mathscr{A}$, except for that its initial state is replaced by r. Hence, $\mathscr{A}$ and $\mathscr{A}^q$ are the same.

Given a set of atomic propositions AP, the class of ETL_{l+f} formulae can be inductively defined as follows.

- Both $\top$ and $\bot$ are ETL_{l+f} formulae.
- Each proposition $p \in AP$ is an ETL_{l+f} formula.
- If φ is an ETL_{l+f} formula, then $\mathrm{o}\varphi$ and $\mathrm{I}\varphi$ are ETL_{l+f} formulae.
- If φ_1, φ_2 are ETL_{l+f} formulae, then both $\varphi_1 \wedge \varphi_2$ and $\varphi_1 \vee \varphi_2$ are ETL_{l+f} formulae.
- If A is an automaton with the alphabet $\Sigma = \{a_1, \ldots,_n\}$ and $\varphi_1,\ldots,\varphi_n$ are ETL_{l+f} formulae, then $\mathrm{A}(\varphi_1,\ldots,\varphi_n)$ is also an ETL_{l+f} formula.

Remark 1. In the original definition of various ETLs (say ETL, ETL_f, and ETL_r), the "next operator" (o) is not explicitly declared. However, this operator is extremely important in building the semantic BMC encodings for ETL_{l+f}. Hence, we explicitly use this operator in our definition, and it would not change the expressiveness of the logic.

Remark 2. Since we employ both finite and looping acceptance automata connectives, our logic is a mixture of ETL_l and ETL_f. On the one hand, ETL_{l+f} generalizes both of these two logics; on the other hand, it can be embedded into ETL_r; hence this logic is also as expressive as omega-regular expressions.

The satisfaction relation of an ETL_{l+f} formula φ with respect to an infinite word $\pi \in (2^{AP})^{\omega}$ and a position $i \in \mathbb{N}$ is inductively given as follows.

- $\pi, i \models \top$ and $\pi, i \not\models \bot$.
- $\pi, i \models p$ if and only if $p \in (i)$.
- $\pi, i \models \neg\varphi$ if and only if $\pi, i \not\models \varphi$.
- $\pi, i \models \mathrm{I}\varphi$ if and only if $\pi, i + 1 \models \varphi$.
- $\pi, i \models \varphi_1 \wedge \varphi_2$ if and only if $\pi, i \models \varphi_1$ and $\pi, i \models \varphi_2$.
- $\pi, i \models \varphi_1 \vee \varphi_2$ if and only if $\pi, i \models \varphi_1$ or $\pi, i \models \varphi_2$.
- If $\mathscr{A}$ is a looping acceptance automaton with the alphabet $\{a_1,...,a_n\}$, then $\pi, i \models \mathscr{A}(\varphi_1,...,\varphi_n)$ if and only if: there is an infinite word $w \in \mathscr{L}(\mathscr{A})$, and, for each $j \in \mathbb{N}$, $w(j) = a_k$ implies $\pi, i + j \models \varphi_k$.
- If $\mathscr{A}$ is a finite acceptance automaton with the alphabet $\{a_1,...,a_n\}$, then $\pi, i \models \mathscr{A}(\varphi_1,...,\varphi_n)$ if and only if: there is an infinite word $w \in \mathscr{L}(\mathscr{A})$ with an accepting prefix $w[n]$, such that, for each j

As usual, we directly use $\pi \models \varphi$ in place of $\pi, 0 \models \varphi$.

To make a better understanding of ETL_{l+f} formulas, we here give some examples of the use of automata connectives.

(1) Considering the LTL formula $\varphi_1 \mathrm{U} \varphi_2$, it can be described with an ETL_{l+f} formula $\mathscr{A}_{\mathrm{U}}(\varphi_1, \varphi_2)$, where $\mathscr{A}_{\mathrm{U}}$ is the finite acceptance automaton $\langle \{a_1, a_2\}, \{q_1, q_2\}, \delta_{\mathrm{U}}, q_1, \{q_2\}\rangle$, and we let $\delta_{\mathrm{U}}(q_1, a_1) = \{q_1\}$, $\delta_{\mathrm{U}}(q_1, a_2) = \{q_2\}$, and $\delta\mathrm{U}(q_2, a_1) = \delta\mathrm{U}(q_2, a_2) = 0$.

(2) The LTL formula $\mathrm{G}\varphi$ is equivalent to the ETL_{l+f} formula $\mathscr{A}_{\mathrm{G}}(\varphi)$, where $\mathscr{A}_{\mathrm{G}} = \langle \{a\}, \{q\}, \delta_{\mathrm{G}}, q, 0\rangle$ is a looping acceptance automaton and $\delta_{\mathrm{G}}(q, a) = \{q\}$.

Remark 3. The order of letters is important in defining automata connectives. Hence, the alphabet should be considered as a vector, rather than a set.

We use sub(φ)to denote the set of subformulae of φ. A formula φ is in negation normal form (NNF) if all negations in φ are adjacent to atomic propositions or automata connectives. One can achieve this by repeatedly using De Morgan's law and the schemas of $\neg o\varphi \equiv o\neg\varphi$ and $\neg\neg\varphi \equiv \varphi$. In addition, we call a formula φ being of the form $\mathscr{A}$ $(\varphi_1,\ldots,_n)$ an automaton formula.

Given a formula φ (in NNF), we use a two-letter-acronym to designate the type of an automaton subformula of φ: the first letter is either "P" or "N," which means "positive" or "negative"; and the second letter can be "F" or "L," which describes the acceptance type. For example, NL-subformulae stand for "negative automata formulae with looping automata connectives," such as $\neg\mathscr{A}^q$ (φ_1, φ_2), where $\mathscr{A}$ is a two-letter looping automaton.

A model or interchangeably a labeled transition system (LTS) is a tuple $\mathscr{M} = \langle S, \rho, I, L, \mathscr{F} \rangle$, where:

- S is a finite set of states,
- $\rho \subseteq S \times S$ is a transition relation (usually, we require ρ to be total; that is, for each $s \in S$, there is some $s' \in S$ having $(s, s') \in \rho$),
- $I \subseteq S$ is the set of initial states,
- $L{:}S \rightarrow 2AP$ is the labeling function, and
- $\mathrm{F} \subseteq 2S$ is a set of fairness constraints.

A path of $\mathscr{M}$ is an infinite sequence $\sigma = s_0 s_1 \cdots \in S^\omega$, where $s_0 \in I$ and $(s_i, s_{i+1}) \in \rho$ for each $i \in \mathbb{N}$. In addition, σ is a fair path if σ visits each $F \in \mathscr{F}$ infinitely often. Formally, σ is a fair path if $\inf(\sigma) \cap F \neq \emptyset$ for each $F \in \mathscr{F}$, where $\inf(\sigma)$ denotes the set of states occurring infinitely many times in σ.

An infinite word $\pi = a_0 a_1 \cdots$ is derived from a path σ of $\mathscr{M}$ (denoted by $\pi = (\sigma)$) if $a_i = L(s_i)$ for each $i \in \mathrm{N}$. We use $\mathscr{L}(\mathscr{M})$ to denote the set of infinite words derived from fair paths of $\mathscr{M}$.

Given an ETL_{l+f} formula φ and an LTS $\mathscr{M}$, we denote by $\mathscr{M} \vDash \varphi$ if $\pi \vDash \varphi$ for each $\pi \in \mathscr{L}(\mathscr{M})$. The model checking problem of ETL_{l+f} is just to verify if $\mathscr{M} \vDash \varphi$ holds for the given LTS $\mathscr{M}$ and the given ETL_{l+f} formula φ.

SEMANTIC BMC ENCODING FOR ETL$_{l+F}$

In this section, we will give a detailed description of the semantic BMC encoding for ETL$_{l+f}$. Firstly, we show how to extend the tableau construction of LTL [20] to that of ETL$_{l+f}$, and hence a product model can also be constructed. Subsequently, we interpret the fairness path finding problem (upon the product model) into SAT, and the size blow-up of this encoding is linear with the bound.

For the sake of convenience, in this section, we always assume that the given ETL$_{l+f}$ formulae have been normalized into NNF.

The Tableaux of ETL$_{l+f}$ Formulae

Given an ETL$_{l+f}$ formula φ, we first inductively define its elementary formula set el(φ) as follows.

- $\mathrm{el}(\top) = \mathrm{el}(\bot) = 0$.
- $\mathrm{el}(p) = \mathrm{el}(\neg p) = \{p\}$ for each $p \in AP$.
- $\mathrm{el}(\varphi_1 \wedge \varphi_2) = \mathrm{el}(\varphi_1 \vee \varphi_2) = \mathrm{el}(\varphi_1) \cup \mathrm{el}(\varphi_2)$.
- $\mathrm{el}(\mathrm{o}\varphi) = \mathrm{el}(\varphi) \cup \{\mathrm{o}\varphi\}$.
- If $\varphi = \mathscr{A}^q(\varphi_1,\ldots,\varphi_n)$ or $\varphi = \neg\mathscr{A}^q(\varphi_1,\ldots,\varphi_n)$ and the states set of $\mathscr{A}$ is Q, then

$$\mathrm{el}(\varphi) = \bigcup_{1\le k\le n} \mathrm{el}(\varphi_k) \cup \left\{\mathrm{o}\mathscr{A}^{q'}(\varphi_1,\ldots,\varphi_n) \mid q' \in Q\right\}. \tag{1}$$

Hence, if $\psi \in \mathrm{el}(\varphi)$, then ψ is either an atomic proposition or a formula rooted at the next operator.

Subsequently, we define the function sat, which maps each subformula ψ of φ to a set of members in $2^{\mathrm{el}(\varphi)}$. Inductively the following hold.

- $\mathrm{sat}(\top) = 2^{\mathrm{el}(\varphi)}$; $\mathrm{sat}(\bot) = 0$.
- $\mathrm{sat}(p) = \{\Gamma \subseteq \mathrm{el}(\varphi) \mid p \in \Gamma\}$ and $\mathrm{sat}(\neg p) = \{\Gamma \subseteq \mathrm{el}(\varphi) \mid p \notin \Gamma\}$.
- $\mathrm{sat}(\mathrm{o}\psi) = \{\Gamma \subseteq \mathrm{el}(\varphi) \mid \mathrm{I}\psi \in \Gamma\}$.
- $\mathrm{sat}(\varphi_1 \wedge \varphi_2) = \mathrm{sat}(\varphi 1) \cap \mathrm{sat}(\varphi_2)$ and $\mathrm{sat}(\varphi_1 \vee \varphi_2) = \mathrm{sat}(\varphi_1) \cup \mathrm{sat}(\varphi_2)$.
- Suppose that $\mathrm{A} = \langle\{a_1,\ldots,a_n\}, Q, \delta, q, F\rangle$.

(1) If $\mathscr{A}$ is a looping acceptance automaton or a finite acceptance automaton and $q \notin F$, then

$$\mathrm{sat}\left(\mathscr{A}^q(\varphi_1,\ldots,\varphi_n)\right) = \bigcup_{1\le k\le n}\left(\mathrm{sat}(\varphi_k)\cap\bigcup_{q'\in\delta(q,a_k)}\mathrm{sat}\left(\circ\mathscr{A}^{q'}(\varphi_1,\ldots,\varphi_n)\right)\right) \tag{2}$$

(2) If $\mathscr{A}$ is a finite acceptance automaton and $q \in F$, then $\mathrm{sat}(\mathscr{A}^q(\varphi_1,\ldots,\varphi_n)) = 2^{\mathrm{el}(\varphi)}$.

(vi) $\mathrm{sat}(\neg\mathscr{A}^q(\varphi_1,\ldots,\varphi_n)) = 2^{\mathrm{el}(\varphi)} \setminus \mathrm{sat}(\mathscr{A}^q(\varphi_1,\ldots,\varphi_n))$.

Recall the tableau construction for LTL [20], an "until subformula" would generate a fairness constraint to the tableau. Indeed, such a subformula corresponds to a "leastfixpoint subformula" if we translate the specification into a logic employing higher-order quantifiers, such as μ-calculus. Similarly, for ETL_{l+f}, the PF- and NL-subformulae also impose fairness constraints. For this reason, we need to define the following two auxiliary relations before giving the tableau construction.

For a PF-subformula $\psi = \mathscr{A}(\varphi_1,\ldots,\varphi_n)$ of φ, where $\mathscr{A} = \langle\{a_1,\ldots,a_n\}, Q, \delta, q, F\rangle$, we define a relation $\Delta^+_\psi \subseteq (2^{\mathrm{el}(\varphi)}\times 2Q)\times(2^{\mathrm{el}(\varphi)}\times 2Q)$ as follows: suppose that $\Gamma,\Gamma' \subseteq \mathrm{el}(\varphi)$ and $P, P' \subseteq Q$; then $((\Gamma,P),(\Gamma',P')) \in \Delta^+_\psi$ if and only if the following hold.

- When $\neq\emptyset$, then, for each $q \in P\backslash F$, there exists some $1\le k\le n$ such that $\Gamma \in \mathrm{sat}(\varphi_k)$ and $P' \cap \delta(q, ak) \neq \emptyset$.
- When P=0, then $q\in P'$ if and only if $\Gamma' \in \mathrm{sat}(\mathscr{A}^q(\varphi_1, \ldots,\varphi_n))$ for each $q\in Q$.

Likewise, for each NL-subformula $\psi = \neg\mathscr{A}(\varphi_1,\ldots,_n)$ of φ, we also define a relation $\Delta-\psi \subseteq (2^{\mathrm{el}(\varphi)}\times 2Q)\times(2^{\mathrm{el}(\varphi)}\times 2^Q)$. In detail, for any $\Gamma, \Gamma' \subseteq \mathrm{el}(\varphi)$ and $P, P' \subseteq Q$, we have $((\Gamma, P), (\Gamma', P')) \in \Delta^-_\psi$ if and only if the following hold.

- When $\neq\emptyset$, then, for each $q\in P$ and $1\le k\le n$, we have: $\Gamma \in \mathrm{sat}(\varphi k)$ implies $\delta(q, a_k)\subseteq P'$.
- When P=0, then $q\in P'$ if and only if $\Gamma' \notin \mathrm{sat}(\mathscr{A}^q(\varphi_1,\ldots,\varphi_n))$, for each $q\in Q$.

We now describe the tableau construction for φ. Suppose that $\psi_1,\ldots,\psi_m$ and $\neg\eta_1, \ldots, \neg\eta_n$ are, respectively, all the PFsubformulae and NL-subformulae occurring in φ then the tableau $\mathscr{T}_\varphi$ is such an LTS $\langle S_\varphi, \rho_\varphi, I_\varphi, L_\varphi, \mathscr{F}_\varphi\rangle$, where:

- $S\varphi$ consists of tuples like $\langle \Gamma; P_1,\ldots,P_m; R_1,\ldots,R_n\rangle$, where $\Gamma \subseteq \mathrm{el}(\varphi)$ and each P_i (resp., R_i) is a subset of ψ_i's (resp., η_i's) connective's state set.
- For two states $s = \langle \Gamma; P_1,\ldots,_m; R_1,\ldots,R_n\rangle$ and $s' = \langle \Gamma'; P'_1,\ldots,P'_m; R'_1,\ldots,R'_n\rangle$, $(s,s') \in \rho_\varphi$ if and only if the following three conditions hold.

(1) $\Gamma \in \mathrm{sat}(\mathrm{o}\psi)$ if and only if $\Gamma' \in \mathrm{sat}(\psi)$ for each $\mathrm{o}\psi \in \mathrm{el}(\varphi)$.

(2) $((\Gamma, P_i), (\Gamma', P'_i)) \in \Delta^+_{\psi_i}$ for each $1 \le i \le m$.

(3) $((\Gamma, R_j), (\Gamma', R'_j)) \in \Delta^-_{\neg\eta_j}$ for each $1 \le j \le n$.

(iii) $I_\varphi = \{\langle \Gamma; P_1,\ldots,_m; R_1,\ldots,R_n\rangle \in S_\varphi \mid \Gamma \in \mathrm{sat}(\varphi)\}$.

(iv) $L(\langle \Gamma; P_1,\ldots,P_m; R_1,\ldots,R_n\rangle) = \Gamma \cap AP$.

(v) $\mathscr{F}_\varphi = \{F_i^+ \mid 1 \le i \le m\} \cup \{F_j^- \mid 1 \le j \le n\}$, where

$$F_i^+ = \left\{\langle \Gamma; P_1,\ldots,P_m; R_1,\ldots,R_n\rangle \in S_\varphi \mid P_i = \emptyset\right\},$$

$$F_j^- = \left\{\langle \Gamma; P_1,\ldots,P_m; R_1,\ldots,R_n\rangle \in S_\varphi \mid R_j = \emptyset\right\}. \quad (3)$$

The below two theorems (Theorems 4 and 5) reveal the language property of ETL_{l+f} tableaux. To remove the lengthiness, we here just provide the proof sketches, and rigorous proofs of them are postponed to the appendices.

Theorem 4. For each $\pi \in (2^{AP})^\omega$, if $\pi \in \mathscr{L}(\mathscr{T}_\varphi)$, then $\pi \vDash \varphi$.

Proof (sketch). Just assume that $\sigma = s_0 s_1 \cdots \in S_\varphi^\omega$ is the corresponding fair path of $\mathscr{T}_\varphi$ such that $\pi = L_\varphi(\sigma)$, where $si = \langle \Gamma_i; P_{1,i},\ldots,P_{m,i}; R_{1,i},\ldots,R_{n,i}\rangle$. We may inductively prove the following claim.

"For each $\psi \in \mathrm{sub}(\varphi) \cup el(\varphi)$, we have: $\Gamma_i \in \mathrm{sat}(\psi)$ implies $\pi, i \vDash \psi$."

Because we require that $\Gamma_0 \in \mathrm{sat}(\varphi)$, hence we have $\pi, 0 \vDash \varphi$.

Theorem 5. For each $\pi \in (2^{AP})^\omega$, if $\pi \vDash \varphi$, then $\pi \in \mathscr{L}(\mathscr{T}_\varphi)$.

Proof (sketch). Suppose that $\pi \vDash \varphi$; to show $\pi \in \mathscr{L}(\mathscr{T}_\varphi)$, we need to first construct an infinite state sequence $\sigma = s_0 s_1 \cdots \in S_\varphi^\omega$ guided by π (the detailed construction is given in Section A.2), and then we will subsequently show that σ is a fair path of $\mathscr{T}_\varphi$ and $\pi = L_\varphi(\sigma)$.

The following theorem is immediate from Theorems 4 and 5.

Theorem 6. The model M violates the ETL_{l+f} property φ if and only if $\mathscr{L}(\mathscr{M}) \cap \mathscr{L}(\mathscr{T}_{\neg\varphi}) \neq \emptyset$, equivalently; there exists some fair path in $\mathscr{M} \times \mathscr{T}_{\neg\varphi}$.

Theorem 7. For an ETL_{l+f} formula φ, its tableau $\mathscr{T}_\varphi$ has at most $4^{|\mathrm{el}(\varphi)|}$ states.

Proof. Observe that a state should be of the form $\langle \Gamma;\ P_1,\ldots, P_m;\ R_1,\ldots,R_n\rangle$. For Γ, there are $2^{|\mathrm{el}(\varphi)|}$ possible choices. Suppose that $\psi_j = \mathscr{A}_j(\varphi_1,\ldots,\varphi_k)$ (resp., $\neg\eta_i = \neg\mathscr{A}_i(\varphi'_1,\ldots,\varphi'_r)$) and the state set of $\mathscr{A}_j$ is Q_j (resp., $\mathscr{A}_i$ with Q'_i). According to the construction, each $q \in Q_i$ (resp., $q \in Q'_i$) corresponds to a unique elementary formula $\circ\mathscr{A}_j^q(\varphi_1,\ldots,\varphi_k)$ (resp., $\circ\mathscr{A}_i^q(\varphi'_1,\ldots,\varphi'_r)$), and such a mapping is an injection. Hence we have

$$\left(\sum_{1\le j\le m} |Q_j| + \sum_{1\le i\le n} |Q'_i|\right) \le |\mathrm{el}(\varphi) \setminus AP|. \tag{4}$$

Note that $P_j \subseteq Q_j$ (resp., $R_i \subseteq Q'_i$), and hence we have $|S_\varphi| \le (2^{|\mathrm{el}(\varphi)|})^2 = 4^{|\mathrm{el}(\varphi|)}$.

The Linear Semantic Encoding

Practically, a model's state space is determined by the evaluation of a set of variables. Further, we may assume that each of them is a "Boolean variable" (which corresponds to a proposition belonging to AP), because every variable over finite domain could be encoded with several Boolean variables.

Let $\mathscr{C} = \langle S, \rho, I, L, \mathscr{F}\rangle$ be an arbitrary LTS, and we also assume that the corresponding variable set is $V = \{p_1, \ldots,_n\}$; then each state $s \in S$ uniquely corresponds to an assignment of such p_is.

If we use (p_i) to denote the value of p_i at s, then each subset $Z \subseteq S$ can be represented by a Boolean formula Φ_Z over V. In detail, it fulfills

$$s \in S \Longleftrightarrow s \Vdash \Phi_Z, \tag{5}$$

where $s \Vdash \Phi_Z$ means that Φ_Z is evaluated to be true if we assign each p_i with the value $s(p_i)$.

Let $V' = \{p'_1, \ldots, p'_n\}$, and each binary relation $\lambda \subseteq S \times S$ also has a Boolean representation Φ_λ over the variable set $V \cup V'$. That is,

$$(s_1, s_2) \in \lambda \Longleftrightarrow (s_1, s_2) \Vdash \Phi_\lambda, \tag{6}$$

where $(s_1, s_2) \Vdash \Phi_\lambda$ means that Φ_λ is evaluated to be true if we assign each p_i with $s_1(p_i)$ and assign each p'_i with $s_2(p_i)$.

Hence, all components of $\mathscr{M}$ can be encoded: I and ρ can be represented by two Boolean formulae Φ_I and Φ_ρ, respectively; we subsequently create a Boolean formula Φ_F for each $F \in \mathscr{F}$; note that the labeling function L is not concerned any longer, because the sates labeled with p can be captured by the Boolean formula p.

For example, from Theorem 7, we have that the symbolic representation of $\mathscr{T}_\varphi$ requires $2 \times |\mathrm{el}(\varphi) \setminus AP|$ new Boolean variables—because variables in $\mathrm{el}(\varphi) \cap AP$ can be shared with the encoding of the original model.

A canonical Boolean encoding of fair path existence detection upon LTSs is presented in [22]: given a model $\mathscr{C} = \langle S, \rho, I, L, \mathscr{F} \rangle$ and a bound $k \in \mathbb{N}$, one may use the formula

$$\Phi_I^{(0)} \wedge \bigwedge_{0 \le i < k} \Phi_\rho^{(i,i+1)} \wedge \bigvee_{0 \le \ell \le k} \left(\Phi_\rho^{(k,\ell)} \wedge \bigwedge_{F \in \mathscr{F}} \bigvee_{\ell \le j \le k} \Phi_F^{(j)} \right), \tag{7}$$

where $\Phi_I^{(j)}$ and $\Phi_F^{(j)}$ are, respectively, the Boolean formulae obtained from Φ_I and Φ_F by replacing each variable p with a new copy $p^{(j)}$, and $\Phi_\rho^{(i,j)}$ is obtained from Φ_ρ by replacing each p with $p(i)$ and replacing each p' with $p^{(j)}$.

It can be seen that this formula is satisfiable if and only if $\mathscr{C}$ involves a fair path of the form $s_0 s_1 \cdots s_{\ell-1}(s_\ell \cdots s_k)^\omega$ (call it is of the lasso shape). Since that $\mathscr{L}(\mathscr{C}) \neq \emptyset$ if and only if $\mathscr{C}$ contains some lasso fair path (note that from each fair path we may derive another fair path of lasso shape), hence we may convert the fair path detection into the satisfiability problem of the above Boolean formula.

However, a closer look shows that the size of such encoding is quadratic with the bound. To reduce the blow-up in size, we need to introduce the following new variables (the linearization can also be done with the syntactic fashion presented in [23, 24]. We would draw a comparison of these two approaches in Section 4.).

(1) For each $0 \leq \ell \leq k$, we introduce a new variable r_ℓ. Intuitively, r_ℓ indicates that s_ℓ is a successor of s_k.

(2) For each fairness constraint $F \in \mathscr{F}$ and each $0 \leq \ell \leq k$, we introduce a variable $f_F^{(\ell)}$, and this variable is evaluated to be true only if there is some $\Phi_F^{(j)}$ which is evaluated to true, where $\ell \leq j \leq k$.

And the new encoding (with the bound $k \in \mathbb{N}$) can be formulated as

$$
\begin{aligned}
\Psi_{\mathscr{C}}^{(k)} = {} & \Phi_I^{(0)} \wedge \bigwedge_{0 \leq i < k} \Phi_\rho^{(i,i+1)} \wedge \bigvee_{0 \leq \ell \leq k} r_\ell, \\
& \wedge \bigwedge_{0 \leq \ell \leq k} \left(r_\ell \longrightarrow \left(\Phi_\rho^{(k,\ell)} \wedge \bigwedge_{F \in \mathscr{F}} f_F^{(\ell)} \right) \right) \\
& \wedge \bigwedge_{F \in \mathscr{F}} \left(\bigwedge_{0 \leq i < k} \left(f_F^{(i)} \longrightarrow \left(f_F^{(i+1)} \vee \Phi_F^{(i)} \right) \right) \right. \\
& \left. \wedge \left(f_F^{(k)} \longrightarrow \Phi_F^{(k)} \right) \right).
\end{aligned}
\tag{8}
$$

Hence, both the number of variables and the size of this encoding are linear with k. Moreover, the following theorem guarantees the correctness of such encoding.

Theorem 8. $\mathscr{L}(\mathscr{C}) \neq \emptyset$ if and only if $\Psi_{\mathscr{C}}^{(k)}$ is satisfiable for some k.

Proof. We begin with the "if" direction: suppose that the variable set is $\{p_1,\ldots,p_m\}$; if there is some k such that $\Psi_{\mathscr{C}}^{k}$ is evaluated to 1 (i.e., true) under the assignment e, then we denote $s_i = (e(p_1^{(i)}), \ldots, e(p_m^{(i)}))$ for each $1 \leq i \leq m$. Hence, each s_i is a state of $\mathscr{C}$.

- Since the truth value of $\Phi_I^{(0)}$ is 1 under e, then we have $s_0 \Vdash \Phi_I^{(0)}$; this implies that $s_0 \in I$.
- For each $0 \leq i < k$, we have $(s_i, s_{i+1}) \Vdash \Phi_\rho^{(i,i+1)}$, and thus $(s_i, s_{i+1}) \in \rho$.
- Because we have the conjunct $\bigvee_{0 \leq \ell} r_\ell$, then there is some $0 \leq \ell \leq k$ such that $e(r_\ell)=1$. In the following, we fix this specific value ℓ for the discussion.
- According to the constraint $r_\ell \rightarrow (\Phi_\rho^{(k,\ell)} \wedge \bigwedge_{F \in \mathscr{F}} f_F^{(\ell)})$, we have the following.

(1) $e(\Phi_\rho^{(k,\ell)}) = 1$, which indicates that $(s_k, s_\ell) \in \rho$.

(2) For each $F \in \mathscr{F}$, we have $e(f_F^{(\ell)}) = 1$.

(v) For each fairness constraint $F \in \mathscr{F}$ and $0 \leq i \leq k$, we now inductively show that "$e(f_F^{(i)}) = 1$ implies $e(\Phi_F^{(j)}) = 1$ (alternatively, $s_j \in F$) for some $i \leq j \leq k$." First of all, it holds in the case of $i=k$, because we have the constraint $f^{(k)} \rightarrow \Phi_F^{(k)}$. In addition, the fact of "when $i=c$, it holds" can be immediately inferred from the hypothesis "when $i = c+1$, it holds," according to the conjunct $f_F^{(i)} \rightarrow (f_F^{(i+1)} \vee \Phi_F^{(i)})$. Since we have shown that $e(f_F^{(\ell)}) = 1$, we can conclude that there exists some $\ell \leq j \leq k$ such that $s_j \in F$.

The above shows that $s_0 s_1 \cdots s_{\ell-1}(s_\ell \cdots s_k)^\omega$ is a fair path of $\mathscr{C}$, and hence $\mathscr{L}(\mathscr{C}) \neq \emptyset$.

Conversely, for the "only if" direction, it suffices to find some k and some assignment e evaluating $\Psi_{\mathscr{C}}^{(k)}$ to be true. Since $\mathscr{L}(\mathscr{C}) \neq \emptyset$, there must exist some fair path of lasso shape in $\mathscr{C}$. Without loss of generality, assume that $\sigma = s_0 s_1 \cdots s_{\ell-1}(s_\ell \cdots s_k)^\omega$ is such a path; just let k be this value, and we now illustrate how the assignment e is constructed.

- For each variable pj and each $i \leq k$, let $e(p_j^{(i)}) = s(p_j)$. Since s_0 is an initial state, according to the definition, we have $e(\Phi_I^{(0)}) = 1$. Meanwhile, because each $(s_i, s_{i+1}) \in \rho$, we have that the conjunction $\bigwedge_{0 \leq i \leq k} \Phi_\rho^{(i,i+1)}$ is satisfied under e.
- For each $0 \leq i \leq k$, we let

$$e(r_i) = \begin{cases} 1, & i = \ell, \\ 0, & i \neq \ell. \end{cases} \tag{9}$$

Then it can be seen that $e(\bigvee_{0 \leq i \leq k} r_i) = 1$.

(iii) For each $F \in \mathscr{F}$ and each $0 \leq i \leq k$, we let

$$e\left(f_F^{(i)}\right) = \begin{cases} 1, & \text{if there is some } j \geq i \text{ such that } s_j \in F, \\ 0, & \text{otherwise.} \end{cases} \tag{10}$$

Then it can be directly checked that the conjunct

$$\bigwedge_{F \in \mathscr{F}} \left(\bigwedge_{0 \leq i < k} \left(f_F^{(i)} \longrightarrow \left(f_F^{(i+1)} \vee \Phi_F^{(i)}\right)\right) \wedge \left(f_F^{(k)} \longrightarrow \Phi_F^{(k)}\right) \right) \tag{11}$$

is evaluated to be true under e.

(iv) Since $(s_k, s_\ell) \in \rho$, we have $e(\Phi_\rho^{(k,\ell)}) = 1$. Also note that σ is a fair path; then for each $F \in \mathscr{F}$ there is some $s_j \in F$, where $\ell \leq j \leq k$. According to the previous definition, we can infer that $e(\bigwedge_{F \in \mathscr{F}} f_F^{(\ell)}) = 1$. Thus the conjunct

$$\bigwedge_{0 \leq i \leq k} \left(r_i \longrightarrow \left(\Phi_\rho^{(k,i)} \wedge \bigwedge_{F \in \mathscr{F}} f_F^{(i)} \right) \right) \tag{12}$$

is also satisfied under e (recall that we have assigned $e(r_i)=0$ in the case of $i \neq \ell$).

Thus, the formula $\Psi_{\mathscr{C}}^{(k)}$ is satisfiable.

For bounded model checking, an important issue is the completeness threshold, which is the specific value k such that we may declare

$\mathscr{L}(\mathscr{C}) = \emptyset$ in the case that $\Psi_{\mathscr{C}}^{(k)}$ is not satisfiable, and we denote it by $\mathrm{CT}(\mathscr{C})$ in this paper.

Since we need only to concern about fair paths of the form $\sigma_1(\sigma_2)^{\omega}$, as pointed in [22], a possible candidate for the completeness threshold $\mathrm{CT}(\mathscr{C})$ is

$$D^I(\mathscr{C}) + |\mathscr{F}| \times D(\mathscr{C}), \tag{13}$$

where D and D^I are, respectively, the diameter and the initialized diameter (cf. [22]). Since [22] just considers LTSs having only one fairness constraint, we here add the factor $|\mathscr{F}|$.

Observe that the part σ_2 must be enclosed in some SCC (i.e., strongly connected component) of $\mathscr{C}$, and we may replace $D(\mathscr{C})$ with $D(S)$, where S is the largest SCC that intersects all fairness constraints.Therefore, we may get a more compact upper bound of the completeness threshold.

Then, for a given LTS $\mathscr{M}$ and the given ETLl+f formula φ, since we have shown that $\mathscr{M} \nvDash \varphi$ if and only if $\mathscr{M} \times \mathrm{T}_{\neg\varphi}$ involves some fair path, we now just need to test if there is some k making $\Psi_{\mathscr{M}\times\mathscr{T}_{\neg\varphi}}^{(k)}$ satisfiable, where $k \leq \mathrm{CT}(\mathscr{M}\times\mathscr{T}_{\neg\varphi})$.

Remark 9. In the case that $\mathscr{C} = \mathscr{C}_1 \times \mathscr{C}_2$, where $\mathscr{C} = \langle S, \rho, I, L, \mathscr{F}\rangle$ and $\mathscr{C}_i = \langle S_i, \rho_i, I_i, L_i, \mathscr{F}_i\rangle$ for i = 1, 2, we have $\Phi_I = \Phi_{I_1} \wedge \Phi_{I_2}$, $\Phi_\rho = \Phi_{\rho_1} \wedge \Phi_{\rho_2}$, and $\Phi_{\mathscr{F}} = \Phi_{\mathscr{F}_1} \cup \Phi_{\mathscr{F}_1}$, and in addition, the variable set of $\mathscr{C}$ is just the union of the variable sets of $\mathscr{C}_1$ and $\mathscr{C}_2$.

Remark 10. Actually, for an ETL$_{l+f}$ formula φ, the symbolic representation of $\mathscr{T}_\varphi$ can be directly given without the detour of explicit construction of the LTS. Because, the relation sat could be inductively constructed if we introduce corresponding new variables in el(φ). Subsequently, encodings of $\Phi_{\Delta^+_\psi}$ or $\Phi_{\Delta^-_\eta}$ could be naturally obtained from the underlying Boolean variables corresponding to states of automata connectives. Hence, the Boolean representation of Φ_{ρ_φ} is obtained. And, encodings of other components are as routine.

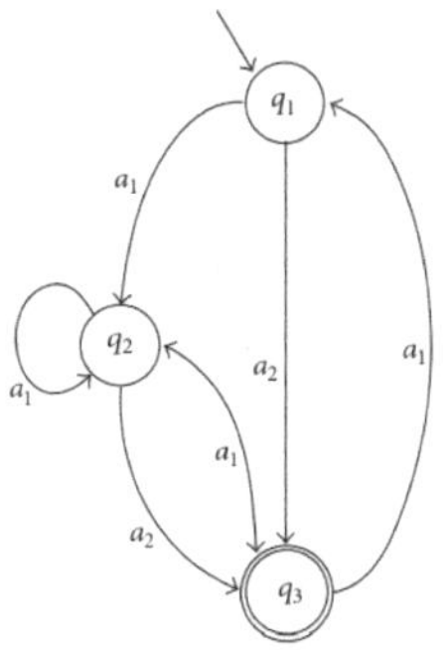


```
CONNECTIVE A (a1, a2) : FIN
STATES >q1, q2, q3<;
TRANSITIONS(q1)
case a1 : q2; a2 : q3; esac;
TRANSITIONS(q2)
case

   a1 : {q2, q3};
   a2 : q3;
esac;
TRANSITIONS(q3)
case a1 : {q1, q2}; esac;
```

Figure 1: An automata connective declaration in ENuSMV.

Table 1: Comparative results of BDD-based and BMC approaches

Cells	BDD			BMC				
	Time (s)	Memory (MB)	C.L.	Variables	Clauses	Time (s)	Memory (MB)	C.L.
5	170.56	68.32	78	3939	8126	13.50	34.98	39
6	513.06	209.17	84	4677	9620	29.14	42.37	39
7	2372.28	976.14	90	5415	11113	41.32	58.48	39
8	8074.05	1482.15	96	6154	12609	52.45	67.32	39
9	≥5 h	≥2 G	—	6892	14102	68.17	82.13	39
10	≥5 h	≥2 G	—	7630	15599	85.13	101.21	39

EXPERIMENTAL RESULTS

To justify our idea, we have integrated (the tool is available at https://sourceforge.net/projects/enusmv12/) the ETL_{l+f} BMC algorithm into ENuSMV (Ver. 1.2). This tool is completely compatible with NuSMV [25], and it allows end-users to customize new temporal connectives by defining automata.

For example, Figure 1 illustrates how to declare a finite acceptance automata connective (to define a looping acceptance automata connective, just replace the keyword FINwith LOOP), namely, A. Since it has three states q1, q2, and q3 (where q1 is the initial state and q3 is an accepting state), then A[q1], A[q2], and A[q3] are also connectives—for example, A[q2] just replaces the initial state with q2. Subsequently, one may define ETL_{l+f} specifications; for example,

```
ETLSPEC A(p,A[q2] (q,p))
```

is a proper declaration.

In this redistribution, both BDD-based and bounded model checkings for ETL_{l+f} are supported. To perform (semantic encoding based) BMC, we need to use the command option bmc tab.

We have conducted some experiments to test the correctness and efficiency of our algorithm. In this paper, we are especially concerned with the following issues.

(1) The comparison of BDD-based symbolic model checking and bounded model checking.

(2) The overhead contrast in verifications of ETL_f and ETL_{l+f} upon both star-free and nonstar-free properties.

(3) The comparison of performances with syntactic/semantic BMC of LTL and semantic BMC of ETL_{l+f}.

To compare the efficiencies between BDD-based MC and BMC, we chose the (distributed mutual exclusion) DME circuit as the model (which involves a buggy design), as described in [3]. It consists of n cells for n users that want to have exclusive access to a shared resource. We conducted the experiment by describing the liveness property that "a request for using the resource will eventually be acknowledged" (with ETL_{l+f} formula). The max bound are set to 100, and the comparative results are shown in Table 1, where "C.L." stands for the length of counterexample.

As a previous work, we have implemented the symbolic model checking algorithm for ETL_f in ENuSMV 1.0. To justify that in general ETL_{l+f} could be more effectively checked, we would make a comparison of BMC for ETL_f and ETL_{l+f}. To draw the comparison upon non-start-free regular properties, we use a "mod 2^n counter" as the model. The model consists of n "cells" bit 0,...,bit n-1. Each cell is a (mod 2) counter having an input carry in and an output signal carry out. These cells are connected in a serial manner; that is, bit 0's carry in is set to 1, and bit i's carry in is connected to bit i−1's carry out as described in Figure 2.

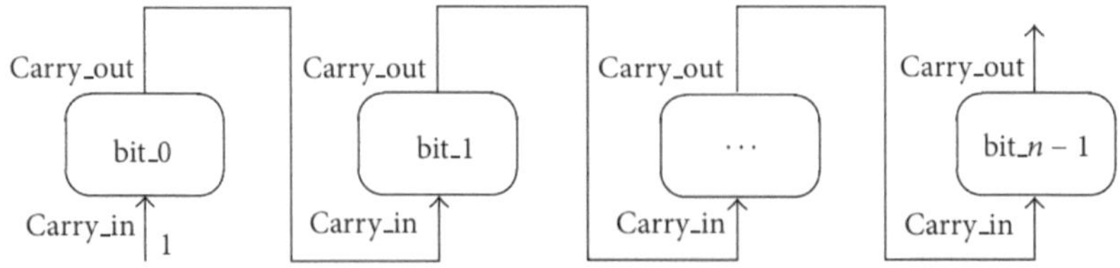

Figure 2: The circuit of MOD 2^n counter.

Table 2: Comparison of ETL_f and ETL_{l+f} with periodicity properties

Cells	ETL_f				ETL_{l+f}			
	Max bound	Memory (MB)	Variables	Clauses	Max bound	Memory (MB)	Variables	Clauses
10	11032	279.44	154461	1180431	13009	382.784	182140	1417990
11	5138	148.22	77084	631981	6037	173.59	90570	754461
12	3702	129.571	3702	518287	4316	143.802	69072	612888

We, respectively, describe the periodicity property that "bit 0 carries out at every even (except for 0) moment" with ETL_f and ETL_{l+f}. We set the time bound to 1 hour, and Table 2 provides the max bounds (together with related information) which can be handled by the SAT solver within the time bound. From it, we can see that a deeper search could be done when specifications are described with ETL_{l+f}.

To compare the overhead of ETL_f and ETL_{l+f} upon star-free properties, we would first use the DME model to check the safety property: "no two cells will be simultaneously acknowledged." The results are shown in Table 3, and the time bound is also set to 1 hour.

Table 3: Comparison of ETL_f and ETL_{l+f} with safety properties

Cells	ETL_f				ETL_{l+f}			
	Max bound	Memory (MB)	Variables	Clauses	Max-bound	Memory (MB)	Variables	Clauses
3	68	29.14	4002	8091	101	38.03	4958	9898
4	62	34.76	4864	9764	87	40.34	5712	11512
5	54	40.12	5170	10313	73	45.37	6072	12088

At the same time, we can also compare the verification performances of the DME model upon the aforementioned liveness property that "each request will be acknowledged in the further." Note that for this property, the verification could be accomplished within the given time bound, and a counterexample could be detected at the bound $k = 39$. The comparative results are given in Table 4.

Table 4: Comparison of ETL_f and ETL_{l+f} with liveness properties

Cells	ETL_f				ETL_{l+f}			
	Time (s)	Variables	Clauses	Memory (MB)	Time (s)	Variables	Clauses	Memory (MB)
3	25.09	2400	4796	29.66	13.78	2340	4747	26.34
4	31.42	3120	6200	38.61	20.06	3042	6115	31.38
5	42.56	3840	7604	44.28	24.70	3744	7843	36.12

The last group of experiments aims at comparing the efficiencies of (syntactic/semantic) LTL BMC and ETL_{l+f} BMC. First of all, for LTL BMC, we are also concerned with two types of encoding approaches.

(1) The Syntactic Approach. We here adopt the linear incremental syntactic encoding proposed in [26]—to the best of our

knowledge, this is the most effective syntactic encoding for full LTL.

(2) The Semantic Approach. ENuSMV 1.2 also supports semantic encoding for LTL—this is tailored from our linear encoding presented in Section 3.2.

We still use the DME circuit as the model and the liveness property as specification; Table 5 provides the experimental results on LTL BMC based on syntactic and semantic encodings. From that, we can see that, with semantic encoding, it tends to generate less clauses and tends to terminate earlier than that with the syntactic encoding, whereas the latter requires fewer variables.

Meanwhile, we can also make a comparison between Tables 4 and 5; we may find that the variable numbers of semantic ETL BMC and LTL BMC are almost at a fixed ratio—for this experiment, the ratio is 1.09 (approximately).

Table 5: Comparison of syntactic/semantic LTL BMC with liveness properties

Cells	Syntactic LTL BMC encoding				Semantic LTL BMC encoding			
	Time (s)	Variables	Clauses	Memory (MB)	Time (s)	Variables	Clauses	Memory (MB)
3	9.34	2052	5474	32.66	8.59	2128	4291	24.79
4	16.71	2736	6806	35.74	10.87	2812	5623	28.44
5	24.06	3420	8138	37.02	15.90	3469	6955	32.28

CONCLUDING REMARKS

The logic ETL_{l+f} is a variant of extended temporal logic, it employs both finite and looping acceptance automata connectives, and it can be considered a mixture of ETL_l and ETL_f. Thus, any omega-regular properties can be succinctly described with this kind of logic, particularly for safety and liveness properties.

We have presented the semantic bounded model checking algorithm for ETL_{l+f}. The central part of this approach is the tableau construction. Meanwhile, we also illustrate how to give a linear BMC encoding for it. To justify it, we have implemented the presented algorithm (in ENuSMV 1.2). Experimental results show that ETL_{l+f} could be more efficiently verified via BMC (in comparison to our previous implementation for ETL_{l+f}).

In this paper, verification of ETL_r, namely, extended temporal logic using Buchi (alternatively, repeating) automata ¨ as connectives, has not been studied. This is partly because of the inherited difficulties of Buchi complementation ¨ [27]. Indeed, we may mimic the ranking complementing technique of Buchi automata [¨ 28–30]. However, this would cause an asymptotically quadratic blow-up of variable number in building the tableaux. Hence, a further work is about to study the semantic BMC encodings of ETL_r.

Acknowledgments

The authors thank the anonymous reviewers for their helpful comments on a previous version of this paper. This work is supported by the NSFC in China under grant number 61103012, 61272335, 61133007, 91118007, 61120106006, the 863 Program 2011AA010106, 2012AA011201, and the Program for New Century Excellent Talents in University.

REFERENCES

1. R. E. Bryant, "Graph-based algorithms for Boolean function manipulation," *IEEE Transactions on Computers C*, vol. 35, no. 8, pp. 677–691, 1986.
2. K. L. McMillan, *Symbolic model checking, an approach to the state explosion problem [Ph.D. thesis]*, Carnegie Mellon University, Pittsburgh, Pa, USA; Kluwer Academic, Boston, Mass, USA, 1993.
3. A. Biere, A. Cimatti, E. M. Clarke, and Y. Zhu, "Symbolic model checking without BDDs," in *Proceedings of the 5th International Conference on Tools and Algorithms for the Construction and Analysis of Systems (TACAS '99)*, vol. 1579 of *Lecture Notes in Computer Science*, pp. 193–207, Springe, Berlin, Germany, 1999.
4. P. Wolper, "Temporal logic can be more expressive," *Information and Control*, vol. 56, no. 1-2, pp. 72–99, 1983.
5. A. Pnueli, "Linear and branching structures in the semantics and logics of reactive systems," in *International Colloquium on Automata, Language and Programming*, W. Brauer, Ed., vol. 194 of *Lecture Notes in Computer Science*, pp. 15–32, Springer, Berlin, Germany, 1985.
6. O. Lichtenstein, A. Pnueli, and L. Zuck, "The glory of the past," in *Proceedings of the Workshop on Logics of Programs*, vol. 193 of *Lecture Notes in Computer Science*, pp. 97–107, Springer, Brooklyn, NY, USA, 1985.
7. Accellera, "Accellera property languages reference manual," June 2004, http://www.eda.org/vfv/docs/PSL-v1.1.pdf.
8. B. Banieqbal and H. Barringer, "Temporal logic with fixed points," in *Temporal Logic in Specification*, vol. 398 of *Lecture Notes in Computer Science*, pp. 62–74, Springer, Berlin, Germany, 1987.
9. A. P. Sistla, M. Y. Vardi, and P. Wolper, "The complementation problem for Büchi automata with applications to temporal logic," *Theoretical Computer Science*, vol. 49, no. 2-3, pp. 217–237, 1987.
10. M. Leucker and C. Sanchez, "Regular linear temporal logic," in *Proceedings of the 4th International Conference on Theoretical Aspects of Computing*, vol. 4711 of *Lecture Notes in Computer Science*, pp. 291–305, Springer, Berlin, Germany, 2007.
11. R. Armoni, L. Fix, A. Flaisher et al., "The ForSpec temporal logic: a new temporal property-specification language," in *Proceedings of the International Conference on Tools and Algorithms for Construction*

and Analysis of Systems (TACAS '02), vol. 2280 of *Lecture Notes in Computer Science*, pp. 296–311, Springer, Berlin, Germany, 2002.

12. I. Beer, S. Ben-David, C. Eisner, D. Fisman, A. Gringauze, and Y. Rodeh, "The temporal logic sugar," in *Proceedings of the 13th International Conference on Computer Aided Verification*, G. Berry, H. Comon, and A. Frinkel, Eds., vol. 2102 of *Lecture Notes in Computer Science*, pp. 363–367, Springer, London, UK, 2001.
13. M. Y. Vardi and P. Wolper, "Reasoning about infinite computations," *Information and Computation*, vol. 115, no. 1, pp. 1–37, 1994.
14. J. R. Büchi, "On a decision method in restricted second order arithmetic," in *Proceedings of the International Congresses in Logic, Methodology and Philosophy of Science 1960*, pp. 1–12, Stanford University Press, Palo Alto, Calif, USA, 1962.
15. W. Liu, J. Wang, and Z. Wang, "Symbolic model checking of ETL," *Journal of Software*, vol. 20, no. 8, pp. 2015–2025, 2009.
16. W. Liu, J. Wang, H. Chen, X. Ma, and Z. Wang, "Symbolic model checking APSL," *Frontiers of Computer Science in China*, vol. 3, no. 1, pp. 130–141, 2009.
17. M. Jehle, J. Johannsen, M. Lange, and N. Rachinsky, "Bounded model checking for all regular properties," *Electronic Notes in Theoretical Computer Science*, vol. 144, no. 1, pp. 3–18, 2006.
18. A. Pnueli and A. Zaks, "PSL model checking and run-time verification via testers," *Formal Methods*, Springer, Berlin, Germany, vol. 4085, pp. 573–586, 2006.
19. A. Cimatti, M. Roveri, S. Semprini, and S. Tonetta, "From PSL to NBA: a modular symbolic encoding," in *Formal Methods in Computer Aided Design (FMCAD '06)*, Lecture Notes in Computer Science, pp. 125–133, Springer, 2006.
20. E. M. Clarke, O. Grumberg, and K. Hamaguchi, "Another look at LTL model checking," in *Computer Aided Verification, 6th International Conference (CAV '94)*, vol. 818 of *Lecture Notes in Computer Science*, pp. 415–427, Springer, Berlin, Germany, 1994.
21. A. Biere, K. Heljanko, T. Junttila, T. Latvala, and V. Schuppan, "Linear encodings of bounded LTL model checking," *Logical Methods in Computer Science*, vol. 2, no. 5, article 5, 2006.
22. E. Clarke, D. Kroening, J. Ouaknine, and O. Strichman, "Completeness and complexity of bounded model checking," in *Verification, Model*

Checking, and Abstract Interpretation (VMCAI '04), vol. 2937 of *Lecutre Notes in Computer Science*, pp. 85–96, Springer, Berlin, Germany, 2004.

23. T. Latvala, A. Biere, K. Heljanko, and T. Junttila, "Simple bounded LTL model checking," in *Formal Methods in Computer-Aided Design (FMCAD '04)*, A. Hu and A. Martin, Eds., vol. 3312 of *Lecture Notes in Computer Science*, pp. 186–200, Springer, Berlin, Germany, 2004.
24. T. Latvala, A. Biere, K. Heljanko, and T. Junttila, "Simple is better: efficient bounded model checking for past LTL," in *Verification, Model Checking, and Abstract Interpretation (VMCAI '05)*, vol. 3385 of *Lecture Notes in Computer Science*, pp. 380–395, Springer, Berlin, Germany, 2005.
25. R. Cavada, A. Cimatti, C. A. Jochim et al., "NuSMV 2. 5 user manual," April 2010, http://nusmv.fbk.eu/NuSMV/userman/v25/nusmv.pdf.
26. K. Heljanko, T. Junttila, and T. Latvala, "Incremental and complete bounded model checking for full PLTL," in *Proceedings of the 17th International Conference of Computer Aided Verification (CAV '05)*, K. Etessami and S. K. Rajamani, Eds., vol. 3576 of *Lecture Notes in Computer Science*, pp. 98–111, Springer, Berlin, Germany, 2005.
27. Q. Yan, "Lower bounds for complementation of ω-automata via the full automata technique," *Journal of Logical Methods in Computer Science*, vol. 4, no. 1, article 5, 2008.
28. O. Kupferman and M. Y. Vardi, "Weak alternating automata are not that weak," *ACM Transactions on Computational Logic*, vol. 2, no. 3, pp. 408–429, 2001.
29. E. Friedgut, O. Kupferman, and M. Y. Vardi, "Büchi complementation made tighter," in *Automated Technology for Verification and Analysis (ATVA '06)*, vol. 3299 of *Lecture Notes in Computer Science*, pp. 64–78, Springer, Berlin, Germany, 2004.
30. S. Schewe, "Büchi complementation made tight," in *STACS 2009: 26th International Symposium on Theoretical Aspects of Computer Science*, vol. 3, pp. 661–672, IBFI, 2009.

Chapter

AN AUTOMATA-BASED APPROACH TO PATTERN MATCHING

13

Ali Sever

Pfeiffer University, Misenheimer, USA

ABSTRACT

Due to its importance in security, syntax analysis has found usage in many high-level programming languages. The Lisp language has its share of operations for evaluating regular expressions, but native parsing of Lisp code in this way is unsupported. Matching on lists requires a significantly more complicated model, with a different programmatic approach than that of string matching. This work presents a new automata-based approach centered on a set of functions and macros for identifying sequences of Lisp S-expressions using finite tree automata. The objective is to test that a given

Citation: A. Sever, "An Automata-Based Approach to Pattern Matching," Intelligent Control and Automation, Vol. 4 No. 3, 2013, pp. 309-312. doi: 10.4236/ica.2013.43036.

list is an element of a given tree language. We use a macro that takes a grammar and generates a function that reads off the leaves of a tree and tries to parse them as a string in a context-free language. The experimental results indicate that this approach is a viable tool for parsing Lisp lists and expressions in the abstract interpretation framework.

Keywords: Computation and Automata Theory, Pattern Matching, Regular Languages

INTRODUCTION

There have been many studies on how to generate a tree parser over Lisp lists, so that the existence of a list as an element of a regular language can be determined through equality checks [1]. Matching on lists using Automatebased approach requires a significantly more complicated model, with a different programmatic approach than that of string matching which is studied in [2]. We present that one must abandon the use of regular expressions, for list pattern matching, in favor of tree parsers. Regular expressions are convenient in string matching because of the relative simplicity and predictability of string structures and their straightforward representation [3,4]. In the realm of trees, the closest we come to this is in comparisons at tree nodes. Rather, to conclude that a tree is a member of a regular tree language, we must represent it in a useful format for our purposes or else require each "cons" cell in the tree to have its own production.

Another requirement of this approach is to provide a certain malleability in the interface—the ability for a programmer to define his or her own transition rules (or rewrite rules) on-the-fly—in the make-tree-matcher **Figure 1** function call, for example. We made sure the code would run on Common Lisp as well as on different Lisp implementations. This required some bending of the constraints of Lisp, like redefinition of Boolean symbols out of utility, and working around the organization of nested lists.

```
(defparameter *falsehood* '*falsehood*)

(defun nonterminal? (x)
  (and (symbolp x)
       (let* ((str (symbol-name x))
              (fr (char str 0))
              (ls (char str (- (length str) 1))))
         (and
          (char= fr #\<)
          (char= ls #\>)))))


(defun yield (tree)
  (cond
   ((null tree) nil)
   ((atom tree) (list tree))
   (t (append (yield (car tree)) (yield (cdr tree))))))

;; Use a special symbol rather than nil for false.
;(defmacro n-and (&rest conjuncts)
; `(and ,@(mapcar #'(lambda (x) `(if (eq ,x *falsehood*) nil ,x))
conjuncts) t))

(defun n-and (&rest conjuncts)
 (if (eq (car conjuncts) *falsehood*)
   *falsehood*
   (if (eq (cdr conjuncts) nil)
     (car conjuncts)
     (apply #'n-and (cdr conjuncts)))))


;(defmacro n-or (&rest disjuncts)
; `(or ,@(mapcar #'(lambda (x) `(if (eq ,x *falsehood*) nil ,x)) dis-
juncts) nil))

(defmacro n-not-null (v)
 (if (null v)
   *falsehood*
   v))


(defun n-or (&rest disjuncts)
 (if (null disjuncts) *falsehood*
   (if (eq (car disjuncts) *falsehood*)
     (apply #'n-or (cdr disjuncts))
```

```
   (car disjuncts))))

(defmacro make-tree-matcher (name start &body rules)
  (labels
      ((gencall (right)
          (format t "generating expansion for ~A~%" right)
          `(if (or (null y) (eq *falsehood* y)) *falsehood*
             (let ((l (copy-list y)))
               (if (not (eq *falsehood* (n-and ,@(mapcar
#'(lambda (x)
                                    (if (nonter-
minal? x)
                                        `(if
(null l) *falsehood*
                                       (setf l (,x
l)))
                                        `(if
(null l) *falsehood*
                                       (if (eql
(quote ,x) (pop l))
                                            l

*falsehood*))))
                                 right))))
                  l
                  *falsehood*))))

      (genproduction (p)
                (let ((name (car p))
                      (rules (cdr p)))
                 (format t "Genproduction:~% name is:
~A~% rules is/are: ~A~%~%" name rules)
                `(,name (y)
                       (n-or ,@(mapcar #'gencall rules))))))
    (format t "Defining function named ~a~%" name)
    `(defun ,name (y)
    (labels ,(mapcar #'genproduction (car rules))
      (not (eq (,start (yield y)) *falsehood*))))))

(defmacro tst ()
 (make-tree-matcher plusone <plus>
  ((<plus> (+ <num>))
   (<num> (1 <num>) (1)))))

(defun ld () (load "tyield.lisp"))

(make-tree-matcher booleval <true>
  ((<true> (and <true> <true>)
         (not <false>)
         (or <false> <true>)
         (or <true> <false>)
         (or <true> <true>)
         (1))
   (<false> (and <false> <false>)
          (and <true> <false>)
          (and <false> <true>)
          (or <false> <false>)
          (not <true>)
          (0))))

(defun add-match-tracers ()
  (trace n-and)
  (trace n-or))
```

Figure 1. The make-tree-matcher function.

The inability to use regular expressions in a simple format was perhaps the biggest hurdle. We will use standard definitions, but we give some of these for convenience of the reader.

- **Analytic grammar**—A set of rules for parsing and returning of truth values confirming a string as either consistent or inconsistent with the rules of a formal language.
- **Context-free grammar**—A set of rules over an alphabet, defined by a quad-tuple $G = (V_t, V_n, P, S)$, where P is a set of production rules;
- V_n is a set of non-terminals;
- V_t is a set of terminals; and S is a starting non-terminal and an element of V_n.
- **Context-free language**—All strings which can be generated by a context-free grammar.
- **Finite automaton**—A set of states and transitions, commonly expressed in a flow diagram. Also called a finite state machine.
- **Finite state machine**: a model of computation composed of states, a transition function, and an input alphabet.
- **Transition function**: describes a condition that would need to be fulfilled to enable the transition.
- **input alphabet**: input recognized by the finite state machine
- **Formal grammar**—A description of a set of rules for a given alphabet over which a set of finite strings can be defined.
- **Kleene closure/Kleene star**—The set of all possible combinations of non-terminals of a regular language. Specifically, the superset of a set of strings containing the empty string ε and closed on the string concatenation function. Every string that is part of a regular language can be found in its Kleene expansion.
- **Parse tree**—A description of the syntax of a string within a formal grammar.
- **Parser**—A method or algorithm or its implementation which examines the application of a given string within an analytic grammar.
- **Regular tree language**—The set of trees accepted by a finite tree automaton.

- **Terminal**—A constant, indivisible value that cannot be further reduced to a more simplified form within its own grammar.
- **Tree automaton**—While finite automata typically act on strings, tree automata are used for tree expressions.
- **Yield**—The string pattern formed from a tree's leaves as encountered in an ordered traversal.

ANALYSIS

Finite tree automata are much more difficult to implement than finite automata, and regular tree languages do not have a nice compact notation like regular (string) languages do. Instead of reading only one next symbol, finite state machines that are used to recognize regular expressions can read any finite number of next symbols. Each of these next symbols can have any finite number of next states. Since parse trees are not generally unique, we do not know anything about the structure of the tree when we decide it's a member of the regular tree language by parsing its yield. We know it's a parse tree for a string in the grammar, but there can be more than one, and we don't know which it is.

Although our initial vision involved the usage of regular expressions, which would be appropriate for normal strings [5,6], the nested form of S-expressions required a more iterative and comprehensive method. Traversing a Lisp list involves the usage of a parse tree. Through this application, one can back trace the sequence of an expression through the parent nodes that generated each step [7].

Generalized nondeterministic finite automata can be defined as the 5-tuple (S, Σ, δ, s, a), where:

S = A complete set of states;

Σ = A finite alphabet;

δ = A transition function ($\delta: (S - \{a\}) \times (S - \{s\}) \rightarrow R$);

s = A start state; and a = A accept state;

where R is the set of all regular expressions over Σ ("Automata").

This can be similarly migrated to tree parsing. Recognizing trees as elements of a regular language only allows us to say that the input is part of the language, or set of all possible elements of the regular grammar. This is done through Boolean comparisons in a top-down tree automaton context. Such automata can be represented with the four-tuple (Q, F, Qf, Δ), where:

Q = A set of states;

F = A ranked alphabet;

Qf = A subset of terminal states in Q; and

Δ = A set of transition rules [8,9].

As our implementation uses a nondeterministic pushdown approach, we can only test for the presence of a list in a regular tree language. As the language can have any number of similarly organized tree structures, we cannot tell with our algorithm which one of them has been found—only that the pattern in question is present in the language.

Using a finite tree automaton to match a Lisp list would require every cons cell in the pattern to have its own production, all labeled "cons".

Theorem 1. A regular tree language is the set of parse trees for a context-free grammar [10].

We parse the yield of the tree to show that it's a member of the context-free language, and use the theorem 1 to conclude that it's a member of the regular tree language. This does not provide us any information about the structure of the original tree, but we do know whether it is a member of the tree language we defined. If it is a member, then we know that it matches the pattern we are looking for.

EXPERIMENTS

Parsing the string allowed us to test whether it was a member of a given string language. A language is a set of sentences (strings and trees are "sentences" in the sense that we use them in formal language theory). A contextfree language is the set of all strings that can be generated by its grammar. We also know because of the theorem that there is a regular tree language composed of all its parse trees.

We test whether that tree is in the regular tree language by testing whether the string is a parse tree of the context-free language.

Our tree pattern matching implementation allows for parsing of a tree structure as a string for subsequent comparison. By seeing that a given list is a parse tree accepted by a finite tree automaton, we know, by the general theorem which states "A regular tree language is the set of parse trees for a context-free grammar". That means the list in question is a member of the regular language tested by the automaton. The use of tree yield functions simplified this step. Conceptually, trees consist of parent nodes, where child nodes extend from the right leaf of each node, with the outermost levels on the left side of the graph. Our custom yield function generates usable strings from tree inputs. With the make-tree-matcher function, the yield of a given tree is produced, and recursive-descent parsing is performed on the value by the same function **Figure 1**.

We experimented and provided the results with the make-tree-matcher function in Figures 2 and 3. Notice that the BOOLEVALUATION function is the language of all true Boolean expressions without variables using efficient implementations of automata operations.

From the definition of a context-free language, we know that a regular tree language is the set of all parse trees of this language's grammar. Therefore, a tree is in the regular tree language if its yield is in the context-free language.

```
(make-tree-matcher boolevaluation <true>
  ((<true> (and <true> <true>)
          (not <false>)
          (or <false> <true>)
          (or <true> <false>)
          (or <true> <true>)
          (1))
   (<false> (and <false> <false>)
           (and <true> <false>)
           (and <false> <true>)
           (or <false> <false>)
           (not <true>)
           (0))))
```

Figure 2. The make-tree-matcher implementation of automata.

```
Defining function named BOOLEVALUATION
Genproduction:
   name is: <TRUE>
   rules is/are: ((AND <TRUE> <TRUE>) (NOT <FALSE>)
(OR <FALSE> <TRUE>) (OR <TRUE> <FALSE>) (OR
<TRUE> <TRUE>) (1))
generating expansion for (AND <TRUE> <TRUE>)
generating expansion for (NOT <FALSE>)
generating expansion for (OR <FALSE> <TRUE>)
generating expansion for (OR <TRUE> <FALSE>)
generating expansion for (OR <TRUE> <TRUE>)
generating expansion for (1)
Genproduction:
   name is: <FALSE>
   rules is/are: ((AND <FALSE> <FALSE>) (AND <TRUE>
<FALSE>) (AND <FALSE> <TRUE>) (OR <FALSE>
<FALSE>) (NOT <TRUE>) (0))
generating expansion for (AND <FALSE> <FALSE>)
```

```
generating expansion for (AND <TRUE> <FALSE>)
generating expansion for (AND <FALSE> <TRUE>)
generating expansion for (OR <FALSE> <FALSE>)
generating expansion for (NOT <TRUE>)
generating expansion for (0)
BOOLEVALUATION

This is the language of all true Boolean expressions (without
variables).
        > (boolevaluation '(or (not 1) (or 1 0)))
        T
        > (boolevaluation '(0))
        NIL
        > (boolevaluation '(1))
        T
        > (boolevaluation '(or 0 1))
        T
        > (boolevaluation '(not 0))
        T
        > (boolevaluation '(and 1 (not 1)))
        NIL
```

Figure 3. Sample run of "boolevaluation" function.

CONCLUSION

We proposed a symbolic approach for pattern matching on LISP programs. We use a symbolic automata representation and implement set of functions and macros for identifying sequences of Lisp S-expressions using finite Ali Sever (e.g., [9]). Therefore, to obtain these kind of particular but interesting results are of substantial and growing interest for many applied problems in symbolic computations [10].

REFERENCES

1. M. Sipser, "Theory of Computation," 3rd Edition, Course Technology, 2012.
2. M. Bojańczyk and T. Colcombet, "Tree-Walking Automata Cannot Be Determinized," Theoretical Computer Science, Vol. 350, No. 2-3, 2006, pp. 164-170. doi:10.1016/j.tcs.2005.10.031
3. H. Comon, M. Dauchet, R. Gilleron, C. Löding, F. Jacquemard, D. Lugiez, S. Tison and M. Tommasi, "Tree Automata Techniques and Applications," 2007.
4. H. Comon, M. Dauchet, R. Gilleron, C. Löding, F. Jacquemard, D. Lugiez, S. Tison and M. Tommasi, "Tree Automata Techniques and Applications II," 2007.
5. C.-H. Chen, "A Neural Network Arhitecture for Syntax Analysis," IEEE Transactions of Neural Networks, Vol. 10, No. 1, 1999, pp. 94-114. doi:10.1109/72.737497
6. J. Power, "Notes on Formal Language Theory and Parsing," National University of Ireland, Maynooth, Kildare, 2002.
7. I. Bagrak and O. Shivers, "trx: Regular-Tree Expressions, Now in Scheme," Scheme Workshop, September 2004.
8. F. Yu, et al., "Symbolic String Verification," SPIN'08 Proceedings of the 15th International Workshop on Model Checking Software, pp. 306-324.
9. A. Bouajjani, B. Johnson, M. Nilsson and T. Touili, "Regular Model Checking," Proceedings of the 12th International Conference on Computer Aided Verification, 2007, pp. 403-418.
10. L. Segoufin and V. Vianu, "Validating Streaming XML Documents," ACM, 2002, pp. 53-64.

SECTION 4 - FORMAL METHODS AND SEMANTICS IN DISTRIBUTED SOFTWARE

Chapter 14

BUILDING REQUIREMENTS SEMANTICS FOR NETWORKED SOFTWARE INTEROPERABILITY

Bin Wen, Keqing He, and Jian Wang

State Key Lab of Software Engineering, Wuhan University, Wuhan, China.

ABSTRACT

Naturally, like the web, integrated software systems in Internet will have to be distributed and heterogeneous. To im-prove the interoperability of services for SAAS, it is crucial to build requirements semantics that will cross the entire lifecycle of services especially on requirements stage. In this paper, a requirements semantics interoperability extend-ing approach called Connecting Ontologies (CO) that will act as semantics information carrier designing to facilitate the requirements identification and services

Citation: B. Wen, K. He and J. Wang, "Building Requirements Semantics for Networked Software Interoperability," Journal of Software Engineering and Applications, Vol. 3 No. 2, 2010, pp. 125-133. doi: 10.4236/jsea.2010.32016.

composition is proposed. Semantic measurement of Chinese scenario is explored. By adopting the approach, a series of tools support for transport domain are developed and applied based on CO and DPO (Domain Problem Ontology) to enforce requirements engineering of networked software efficiently.

Keywords: Networked Software, Requirements Semantics, Requirements Engineering, Connecting Ontologies

INTRODUCTION

Ideally, users can access services based on their requirements without regard to where the services are hosted or how they are delivered. Various computing paradigms have promised to deliver IT as services including grid computing, P2P computing, and more recently Cloud computing. The latter term denotes the infrastructure as "Cloud" from which businesses and users are able to access application from anywhere in the world on demand. Thus, the computing world is rapidly transforming towards developing software for millions of consume as a service, rather than to run on their individual computers [1].

The development of networked software has emerged varied forms and definitions. One is pervasive computing, such as grid computing, e-science, and transparent computing, which focus on resource sharing. Another category is cloud computing based on SAAS (software as a service) and related studies include SOA, Web Service, Semantic Web Service etc. SAAS and virtualization of hardware and software are two main features for Cloud computing. Networked software that this paper refers to [2] belongs to the second sort that is complex information system based on Internet towards service computing. Distribution, autonomy, opening and heterogeneity are its basic features and stakeholders to be faced having various sorts and interests. Typically, supporting diversified, personalized and dependable services to improve user QoE (Quality of Experience) is the highest goal.

Requirements engineering (RE) is crucial to the success of software engineering, especially for networked software, and considering issues mainly include dynamic elicitation and analysis, evolution modeling, requirements management and model verification of user requirements and so on. Requirements modeling methods mostly are classified as structural requirements modeling and object-oriented requirements modeling according to paradigm, and both of them can deal with functional and nonfunctional requirements analysis. Now the typical software RE approaches are goal-oriented, ontologyoriented, scenario-based, problem framework, prerequirements analysis based on domain modeling, document driving and aspect-oriented method [3].

The most widely significant approaches for networked software RE are goal-oriented and pre-requirements analysis based on domain ontology approach. Goal-oriented approach concentrates on analysis and modeling of early requirements so as to help developer understand the motivation and expectation for various roles, and involves the identification and analysis of functional and nonfunctional requirements goal. At present software RE is switching from object-oriented to goal-oriented [4,5], whereas goal-oriented approach has produced commercial products for tool supporting, for instance Cediti goal analyzer: Objectiver. Accordingly goal-oriented requirements analysis has become the hot spot of the studying of RE.

Virtually, pre-requirements analysis based on domain modeling [6,7] is the process of requirements analysis based on domain-level ontology knowledge. The issue of ODE method based ontology [8] only acquires domain conceptual knowledge especially, but it ignores the modeling for task and functional knowledge.

All the above-mentioned requirements modeling methods consider only for object-orient development. The applicability and feasibility of those approaches for service-oriented computing must be reconsidered. Regarding the features for service computing, role, goal, process and service, the four fundamental elements can be used to modeling for the users' truly intentions of networked software. A meta-modeling framework containing the four fundamental elements, namely RGPS [9], is presented for conducting synergy and ordered structure requirements specification from disordered requirements information. Furthermore, choosing ontology metamodeling [10] and encapsulating domain reusable core services asset, O-RGPS (Ontology-RGPS) meta-model proposal [2] is also put forward (see Figure 1).

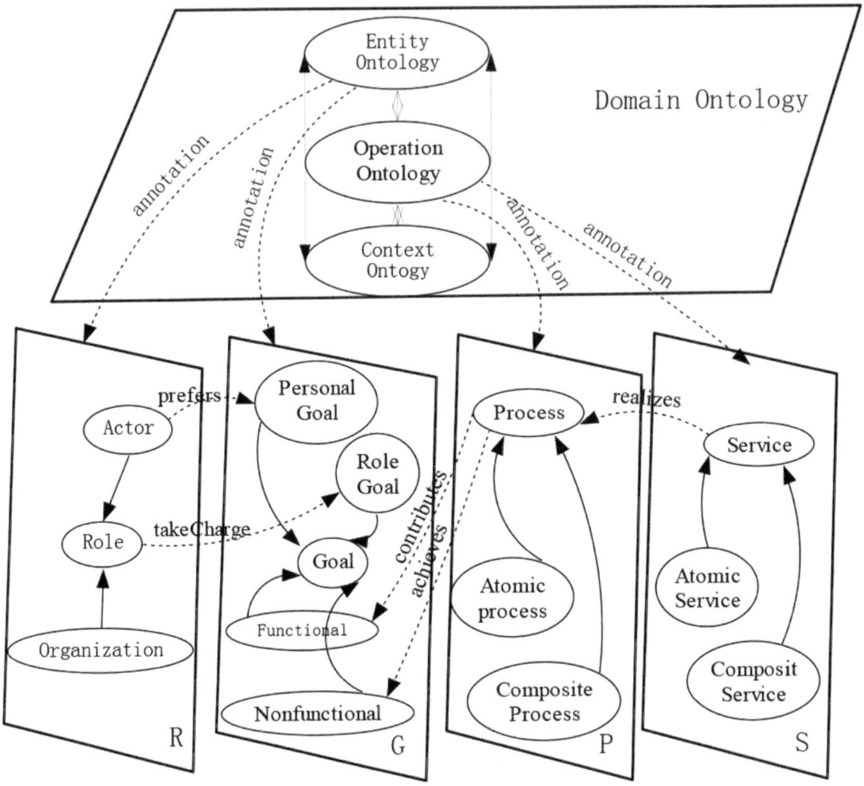

Figure 1. Domain asset customizing based-on O-RGPS

Based on O-RGPS requirements meta-model framework, user requirements can be described from different angle, level and granularity in order to form domain requirements asset and store as OWL for reuse.

Interaction and collaboration of networked software is a restricted semantic interoperable issue on essence. Then, how to constrain and extend the semantic interoperability in the process of self-organization and action emergence for the distributing services resource? How to categorize the structure of interoperability? How to satisfy stakeholders' requirements?

Regarding the above issues, this paper proposes an requirement semantic interoperable extending approach for networked software based on connecting ontologies (CO) and furnishes the unified and dynamic semantic information carrier for service aggregating and evolution modeling.

The rest of the paper is organized as follows: Section 2 explores software RE method based on domain ontology and requirements asset; furthermore, provides formal definition and aggregating method of connecting ontologies, and presents the related algorithm and integrating environment design for interoperable extending of networked software requirements semantics; Section 3 summarizes the related cutting-edge work in the research community; at the last, we conclude the paper and survey the future work.

CONNECTING ONTOLOGIES FOR NETWORKED SOFTWARE

Networked software system includes the overall architecture and goal software system that can embody dynamic property of the architecture. Goal software system is composed of services, whereas service resources distribute in network and are loosely coupled, dynamic binding and permit various levels of semantic interoperability.

Since service resources are dynamically distributed, for the sake of acquiring requirements knowledge from multi-domain service resources, disseminated ontology registry repositories in network require ontology encapsulation which is unified annotation of service with respect to requirements semantic. Ontology registry repositories will accord with ISO meta-model framework MFI (ISO/IEC SC32 19763) [11] that we participate. Requirements are gained by requirements acquiring & analysis (RAA) approach, and Requirements Sign Ontology (RSO, Definition 11, similar to process specification or workflow of application) is generated. Based on RSO, published ontologies of requirements semantic for available services are dynamic found and matched in network. Matched ontologies and RSO form ontologies group that is loosely coupled connected and dynamic generated, named Connecting Ontologies (CO). Stated in Figure 2, is requirements modeling approach for networked software based on CO. In ontology level, requirements semantic are dynamic acquired with semantic extending and matching. Furthermore, initial requirements model is generated by reusing multi-domain requirements asset. CO is the process of dynamic generating and continuous evolving, as stakeholders' requirements are uninterruptedly changed and loosely coupled for multi-domain requirements asset.

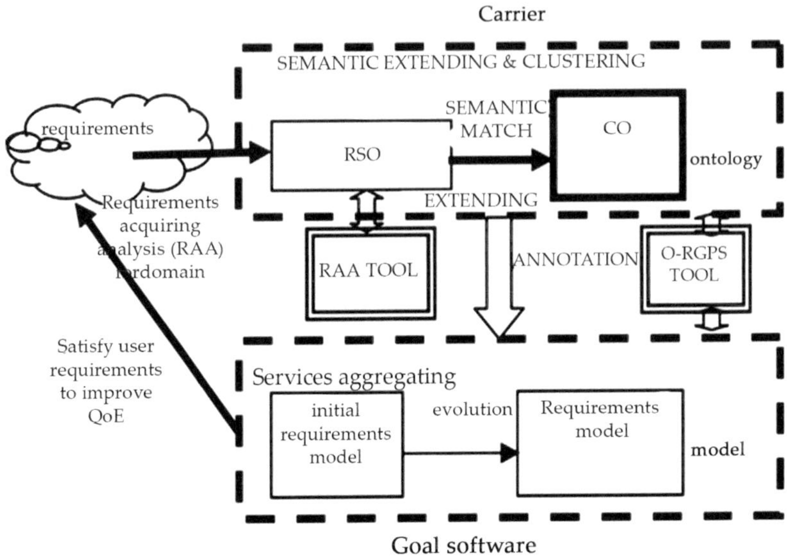

Figure 2. Requirements modeling for networked software based-on connecting ontologies.

Domain Ontology Based on Description Logic

In the line of computer, ontology is explicit representation and description of conceptualization objects. Ontology can also be used for software RE as requirements representation and carrier. At the same time, since reusability of broad-spectrum ontology is relatively hard, the principal application direction of ontology for software requirements should be domain-oriented and problem-oriented.

Firstly, this section gives the definition of domain ontology based on description logic and other related definitions. Next section will apply these definitions. Then requirements elicitation based on domain ontology and requirements asset is designed and implemented.

Definition 1 (Domain Ontology based on description logic). Domain Ontology is expressed as DO=<D, C, T, A, LH>, where D represents domain; C represents a set of concepts; T represents TBox; A represents ABox; and LH represents lattice hierarchy of concepts.

Definition 2 (Relation Triple among concepts). For domain ontology DO=<D, C, T, A, LH> with $c_p, c_q \in C$ and relation $r(c_p, c_q) \in T$, if

c_p and c_q satisfy (1) $c_p \sqsubseteq c_q$ (2). $c_p \sqsubseteq \exists r.c_q$ (3). $c_p \sqsubseteq \forall r.c_q$, and $\sqsubseteq$ is concepts inclusion relation, then $\chi =< c_p, r, c_q >$ represents relation triple between c_p and c_q.

Definition 3 (Semantic Association). For two relations $\chi_1 =< c_{p1}, r_1, c_{q1} >$ and $\chi_2 =< c_{p2}, r_2, c_{q2} >$, $\chi_1 \Box \chi_2$ denotes semantic association between χ_1 and χ_2 where $\chi_1.c_{q1} = \chi_2.c_{p2}$.

Definition 4 (semantic association path). For a set of relations $X = \{\chi_i \mid \chi_i =< c_{pi}, r_i, c_{qi} >, i = 1, 2, \ldots, n\}$ and relation triples $\chi_s =< c_{ps}, r_s, c_{qs} >$, $\chi_d =< c_{pd}, r_d, c_{qd} >$, $\chi_s, \chi_d \in X$, if $\exists m \le n-2$, for $\forall \chi_{i+j} \in X, j = 1, 2, \ldots, m$, then DO have a semantic association path in X from χ_s to χ_d where semantic associations $\chi_s \Box \chi_{i+1}, \chi_{i+1} \Box \chi_{i+2}$ and $\chi_{i+m} \Box \chi_d$ exist, namely semantic association path between concept c_{qs} and c_{qd}.

Definition 5 (concept semantic depth, Depth). Apart form the class for itself, the meaning of ontology concept is also described by the associated classes, namely concept semantic depth. To calculate semantic depth, let the Depth of ontology root concept is zero, if the Depth of concept c, Depth(c), is I, then the Depth of its father concept (if existed) is I-1 and the Depth of its child concept (if existed) is I+1.

Connecting Ontologies

Connecting ontologies based on semantic matching of multi-domain requirements asset only utilize local or part of ontologies registry repositories for services. Modularization is an important technique of ontology reuse for services. Different researchers have different definitions or designations including segment, module, view or subontology etc. The paper adopts sub-ontology [12] notion. Some definitions and algorithms are presented as follows.

Definition 6 (sub-ontology). For domain ontology DO=<D,C,T,A,LH>, a sub-ontology sub-Onto consists of 5 elements $<C_{sub}, T_{sub}, A_{sub}, LH_{sub}, I>$, where C_{sub} represents the set of sub-Onto concepts which denotes the context of sub-ontology; $|C_{sub}| < |C|$; there exist semantic association or semantic association path in C_{sub}; $T_{sub} \subseteq T$, $A_{sub} \subseteq A$ represent sub-Onto's local knowledge base for T_{sub}, A_{sub}; LH_{sub} represents lattice hierarchy of concepts; and I represents index pointer towards DO. If sub-ontology=DO or sub-ontology have nondeterministic domain, then I is nil.

Definition 7 (sub-ontology space in same source). For DO=<D,C,T,A,LH>, sub-ontology space in same source Space represents $\{<\text{sub-Onto}_b, B, DO|\forall\ \text{sub-Onto}_b.I=DO, B \in \text{Index}>\}$

Algorithm: Sub-Ontology Extracting Algorithm

For DO=<D,C,T,A,LH>, <CON,n,DO> is the input of sub-ontology extracting, where CON={con_1, con_2, …, con_k} represents a set of concepts which will be matched; DO represents father ontology; n represents the depth of travel. Based on [12], we can get sub-ontology extracting algorithm. The outcome of the algorithm is a sub-ontology sub-Onto.

Sub-ontology extracting algorithm can be seen from Algorithm 1.

Algorithm 1 sub-ontology extracting algorithm.

```
INPUT: <CON, n, DO>
OUTPUT: sub-Onto
i<-1
Repeat
    get i-th concept coni from CON
    remove coni from CON
    if exist a concept c satisfy concept coni semantic
similarity match constraint in DO
   then
      do breadth-first traversal from c through rela-
tion in DO
       if another concept c' in DO is reached in
traversal
      then
      add the triples along the traversed path from c
to c' into Tsub
      add c to a set Csub
      add c' to a set Csub
      remove c' from CON
      end if
         stop when traversal up to a depth of n;
end if
i<- i+1
Until CON is empty or i>=k
m<- |Csub|
If m<k then return nil
end if
get the set of propertys Tsub from DO for all the
concepts in CON
get the set of individuals Asub from DO for all the
concepts in CON
        get        the        extracting
```

Attentively, semantic similarity matching can be described in details: for any two concepts C1 and C2, assuming string S1 and S2 is the name of C1 and C2 respectively. Firstly, lexical analysis that preposition, conjunction, pronoun and interjection are cancelled is carried out for two strings, whereas continuous and meaning words are reserved. Strings S1 and S2 will be transferred to $<S1_{w1},\ldots,S1_{wn}>$ and $<S2_{w1},\ldots,S2_{wm}>$. For any words $S1_{wi} \in \langle S1_{w1},\ldots,S1_{wn}\rangle$ and $S2_{wj} \in \langle S2_{w1},\ldots,S2_{wm}\rangle$, we can calculate two words' similarity similarityScore($S1_{wi}$, $S2_{wj}$)=wst.lookup($S1_{wi}$,$S2_{wj}$). This similarity is acquired by looking up similarity table which is generated by experts in matching computing by using words association tool (such as WordNet) in advance. If n<=m, then for $S1_{wi}$, we can find $S2_{wj}$ in accordance with maximum similarity, namely matchscore($S1_{wi}$, $S2_{wj}$)=similarityScore($S1_{wi}$, $S2_{wj}$). Finally similarity between two concepts is matchscore(C1,C2)=Sum(matchscore($S1_{wi}$, $S2_{wj}$))/n.

Algorithm: Sub-Ontology Merging Algorithm

For a set of sub-ontology, onto-set consists of {$SubOnto_1$, $Sub\text{-}Onto_2$, …, $Sub\text{-}Onto_n$}, $n \geq 2$, and the outcome of the algorithm generates a sub-ontology Onto= Merge(onto-set).

Sub-ontology merging algorithm can be seen in Algorithm 2 in details.

Algorithm 2 sub-ontology merging algorithm.
INPUT:onto-set={Sub-Onto1, Sub-Onto2,….., Sub-Onton},n >=2
OUTPUT: Merge(onto-set)

```
add all concepts in Sub-Onto1.Csub from onto_set to a new set  C
add all items in Sub-Onto1.Tsub from onto_set to a new set  T
add all items in Sub-Onto1.Asub from onto_set to a new set  A
i<- 2;
repeat
   get i-th Sub_Onto  Sub-Ontoi from  onto_set
   for each concept CK in Sub-Ontoi.Csub
      if Ck not in C then
         add Ck to  C
      end if
   end for
   for each item Mj in Sub-Ontoi.Tsub
      if Mj not in T then
         add Mj to T
      end if
   end for
   for each item Aq in Sub-Ontoi.Asub
      if Aq not in A then
         add  Aq to A
      end if
   end for
   i<- i+1
until i>=n
get the corresponding Onto = <C,T,A,LH,nil>
```

Definition 8 (maximum self-contained sub-ontology on concepts). For a set of concepts C which will be matched and a sub-ontology extracting algorithm, the last sub-ontology represents maximum self-contained subontology on concepts C, where the set of concepts in the extracted sub-ontology unable to increase along with addition of travel depths to cease the extracting process.

Definition 9 (domain requirements ontology). For convenient requirements acquisition and matching, domain requirements ontology is a special DO which only have two concepts with semantic depth Depth=1 in the sets of concepts: Operation denotes requirements verb concept and Entity denotes requirements noun concepts. Maximum self-contained sub-ontology of the set of operation is called operation ontology and maximum selfcontained sub-ontology of the set of entity is called entity ontology for domain requirements ontology.

Definition 10 (Domain Problem Ontology, DPO). Domain Problem Ontology (DPO) represents as Merge ($\cup_{asseti} \in_{RGPS}$(Extracting (P, Dep,asseti),index$_i$)), where P represents a set of problem's concepts; Dep refers to travel depth; RGPS represents domain-customized asset based RGPS; indexi represents source ontology index with respect to matched problem concepts of RGPS asset.

Note that the Problem is a specific application context, for example travel is a Problem for traffic domain.

Definition 11 (Requirements sign ontology, RSO). Requirements sign ontology RSO consists of 3 elements <DSorl, Concept, Control>, where DSorl represents input in domain requirements service language; C represents the set of extracting concepts from DSorl; Concept ⊇ DPO.C; Control represents control structure among matched service ontologies mainly including sequence, choice, split-union, any order, cycle.

CO are a sub-ontologies set with different sources in which involve dynamic finding and matching ontologies of published services, and RSO serves as mediator and conducts the process of generating CO for service-oriented requirements.

Definition 12 (connecting ontologies, CO).Connecting ontologies (CO) consists of <RSO, DPO, Mapping-OntoSet>, where RSO represents requirements sign ontology; DPO represents problem-oriented domain problem ontology; Mapping-Onto-Set represents matched sub- ontology set of different source.

Based on sub-ontology extracting algorithm and the direction of RSO, requirements semantic of CO firstly execute the matching for DPO. The rest of unabsorbed parts by DPO for CO run ontologies finding and matching from multi-domain services in network to satisfy requirements semantic for stakeholders. General speaking, the matched ontologies always denote some subontologies of ontologies with respect to multi-domain services, and they are semantically matching with RSO, namely O_i (i=1~n). Then, as seen in Figure 3, connecting sub-ontology O_0 of DPO and sub-ontologies O_i of ontologies for multi-domain services according to RSO that acts as the center will dynamically generate CO. Accordingly, dynamically generated CO not only contain O_0 which is domain-oriented and tightly couple with DPO, but also do it include some services ontologies O_i for different domain i and loosely coupling with RSO. A few of unmatched services based CO will be solved by customizing manufacture.

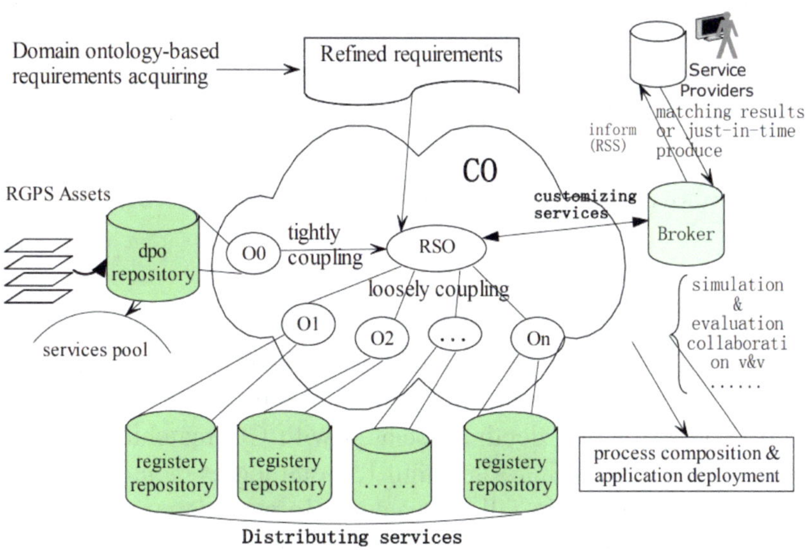

Figure 3. Connecting ontologies.

Domain Problem Ontology

According to Definition 10, Domain Problem Ontology (DPO) is really a composite sub-ontology in terms of problem by extracting from Domain Ontology and RGPS requirements assets that express as OWL format. DPO is very important in the creating process of CO and acts as problem vision for CO. Creating CO firstly need adopting and matching with DPO, so the quality of DPO is crucial for the success of appropriate and preferred match regarding the contract ontology (i.e. CO) of all circles for software web clustering.

We believe that: 1) semantic distance is only necessary and fundamental measure method for semantic interoperability capability; 2) for semantic interoperability measurement, semantic distance is not sufficient condition; 3) not only do semantic interoperability capability relate to similarity but also tightly associate with the contracted standard (i.e. CO) for both sides and really CO is sufficient condition for interoperability.

Generating DPO can adopt two fashions: semi-automated method directed by domain experts and fully automated method. We have realized the first fashion in our domain modeling tool designing to acquiring RGPS assets and automated fashion is now designing and optimizing. For automated fashion, we considered problem as follows: 1) the relation between DPO extracting depth (traverse depth) and CO matching degree with RSO; 2) the relation between DPO extracting depth (traverse depth) and extracting time cost.

For the above issues, we work out an experiment for evaluating these relations.

Experiment Design

Regarding low-scale Transport ontology (concepts number below 200) and OWL formatted R, G and P, experiment will evaluate the capability between DPO extracting depth associated with CO matching degree and time spending. Firstly, using Algorithm 1, 4 ontologies including Transport ontology, R, G and P [9], will be executed in accordance with the word "travel" and its synonym and outcome will be merged to generate DPO by Algorithm 2. RSO can be obtained by requirements acquiring tool [13] that we have implemented. Matching degree is manually achieved by domain experts between RSO and DPO.

Result Evaluation and Discussion

In the simulate experiment, the initial value of DPO extracting depth is 1. Through changeable extracting depth, we can get different matching degree and time cost for different depth value in order to analysis the influence of depth for entire CO generating process. Figure 4 is the result for different depth value.

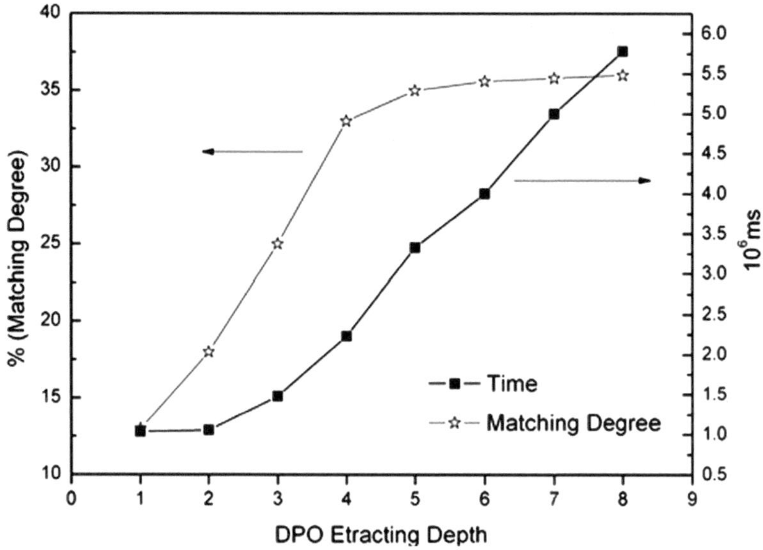

Figure 4. DPO extracting simulation result.

According to the result, higher depth value will have higher matching degree with RSO. When DPO extracting depth is higher, the scale of DPO sub-ontology is also biggish correspondingly. Considering the principle of space locality, the reuse probability of DPO will evidently increase to enhance the matching degree with RSO. But higher depth value will lead to more time spending for creating DPO. At the same time, matching degree do not obviously enhance when the depth value increase from 6 to 8. It shows that only increasing depth value is not always efficient for improving matching degree. Since adopting sound depth value is very important for DPO to optimize the matching performance. The time cost of the experiment is higher than large-scale single ontology extracting in [12] because the experiment adds the spending of merging process.

The drawback of this experiment is low-scale original ontology, so future work will execute on large-scale ontologies to obtain valuable result for real-world.

Interoperability Extending Integrating Environment for Requirements Semantic Based CO

Regarding travel problem in urban traffic domain, simulation tests for acquiring requirements semantic based on CO [14] have shown that the semantic interoperability extending approach provides semantic information carrier for networked software and furnishes semantic goal for on-demand service aggregating. But now both RSO perfection and CO dynamic generating mainly rely on manually participating and customizing by requirements analyzers frequently, and quantitative measurement is absent for denoting semantic distance and interoperability level. Farther studies are listed as follows: 1) interoperability extending integrating environment for requirements semantic; 2) measurement system for requirements semantic interoperability.

Requirements Semantics Distance for Chinese Context

Now, software requirements semantics mainly adopts ontology encapsulation style, and requirements matching will reduce to similarity comparing among entities. Basic elements of entity include concept, relation and instance. Main measurement feature of concept are: concept name (no semantics, only consider linguistic and literal similarity, such as some distance formula [15]), concept semantics similarity, concept structure. Main measurement feature in relation involve property name, domain and range. Instance is auxiliary measurement for concept.

Semantics distance refers to a measurement of semantics similarity or association between two semantic entities. Semantic entities involving this paper are key words of documents. In general, semantics distance is a real number in $[0,\infty)$. Semantics distance has tight association with word similarity. Between two words, the bigger semantics distance is, the lower semantics similarity is and vice versa. They can be built a simple correspondence that need satisfy some conditions as follows: 1) similarity is 1 when semantics distance is 0 between two words; 2) similarity is 0 when semantics distance is infinity between two words; 3) between two words, the bigger semantics distance is, the lower semantics similarity is (monotony descend).

For two words w_1 and w_2, similarity expressed as $Sim(w_1,w_2)$, semantics distance is $Dis(w_1,w_2)$, then one can define a simple transfer relation that satisfy the above conditions:

$$Sim(w_1,w_2)=\frac{\alpha}{Dis(w_1,w_2)+\alpha} \quad (1)$$

α is a adjustable parameter that embody the words' distance value when similarity is 0.5. In the most cases, directly computing the words' similarity is difficult, so distance measurement can be calculated in advance and then transfer the similarity for words.

In general, thesaurus is the basis of the semantics distance measurement throughout computing MSCA (the Most Specific Common Abstraction) to acquire. To calculate semantics distance, one must use a comprehensive and exact structural semantic resource repository. Hownet (http://www.keenage.com) that involves more complete semantics knowledge content and is referred in some Chinese information processing is suitable for this studying.

Hownet includes two main definitions: concept and sememe. Concept is a description for vocabulary's semantics and every word can be expressed several concepts. Concept applies a knowledge representation language that uses sememe as vocabulary to describe.

Differentiated from the other thesaurus (e.g. Wordnet), Hownet don't reduce concept to a tree-like hierarchical architecture and that try to depict every concept using a series of sememes. Hownet adopts 1500 sememes which are divided into some categories as follows:

1) Event; 2) entity; 3) attribute; 4) aValue; 5) quantity; 6) qValue; 7) SecondaryFeature; 8) syntax; 9) EventRole; 10) EventFeatures.

For these sememes, they can be reduced to 3 groups: group 1 is called basic sememe to describe semantics feature for single concept containing sememes from category 1 to category 7; syntactic sememe only include category 8 to describe syntactic feature for words; group 3 contain category 9 and 10 called relation sememe to denote relation between concepts (similar to lattice relation from lattice syntax).

Semantics distance $d_1(p_1,p_2)$ between two sememes p_1 and p_2 is the path length from p_1 to p_2 in the sememe hierarchy structure.

For concept S_1 and S_2 which they have only one sememe in Hownet, semantics distance $d_1(S_1,S_2)$ is called the first basic sememe; except from the first basic sememe expression, for concept S_1 and S_2 which their semantics in Hownet is a set of basic sememes, $d_2(S_1,S_2)$ is defined as this part's semantic distance.

Corresponding to relation sememe description, its value is a feature structure. Considering every feature for the feature structure, its attribute is a relation sememe and its value is a basic sememe or a concrete word. This part of semantics distance for two concept S_1 and S_2 denote as $d_3(S_1,S_2)$.

For every feature of the above feature structure, if its value is a set in which the element of the set is a basic sememe or a concrete word, $d_4(S_1,S_2)$ can be designed to describe the part of relation signal sememe's semantics distance for concept S_1 and S_2.

Naturally, for the first basic sememe $d_1(S_1,S_2)$, $S_1(S_2)$ have a element-sememe $p_1(p_2)$ in Hownet, then $d_1(S_1,S_2) = d_1(p_1,p_2)$.

For the other basic sememes, if S_1 includes m sememes, S_2 includes n sememes, then

$$D_2(S_1,S_2)=\begin{cases} avg[d(p_{1i},p_{2j})], m>0, n>0 \\ |m-n|, else \end{cases} \tag{2}$$

where p_{1i} is the sememe of S_1, p_{2j} is the sememe of S_2. The following is a java program for calculating relation sememe:

```
private double disMap(
      Map<String, List<String> map1,
      Map<String, List<String> map2)
  {
      if (map1.isEmpty() || map2.isEmpty())
      {return
      Math.abs(map1.keySet().size() - map2.keySet().size());
      }
      double min = DEFAULT_DISTANCE;
      for (String key : map1.keySet( )) {
        if (map2.containsKey(key)) {
        List<String> list1 = map1.get(key);
        List<String> list2 = map2.get(key);
        double sim = disPriList(list1, list2);
        if (sim < min) {
          min = sim;
      }}}
      return min;
  }
```

Similarly, we can also get the java program for calculating relation signal sememe's semantics distance.

Considering the above-mentioned factors, for two concepts S_1 and S_2, semantics distance is defined as [15]:

$$d(S_1,S_2)=\sum_{i=1}^{4}\beta_i \square d_i(S_1,S_2) \tag{3}$$

where β_i ($1 \le i \le 4$) is adjustable parameter and $\beta_1 + \beta_2 + \beta_3 + \beta_4 = 1$, $\beta_1 \ge \beta_2 \ge \beta_3 \ge \beta_4$; if $d_i = 0$, then β_i will assign other item proportionally. The act for global similarity from d_1 to d_4 is descending order. Since the first basic sememe expression reflects the main feature for concept, its weigh value should be defined comparatively bigger and larger than 0.5 usually.

Based on semantics distance between Chinese concepts, we can calculate semantics distance between two sentences w_1 and w_2 for Chinese SORL [16], where w_1 contains m concepts (S_{11}, …, S_{1m}), w_2 has n concepts (S_{21}, …, S_{2n}).

If w_1 is context-unaware and S_{1i} is unknown, then Dis(w_1, w_2)=min Dis(S_{1i}, S_{2j}), $1 \le i \le m$, $1 \le j \le n$.

If w_1 is context-aware and S_{1i} is definite, then Dis(w_1,w_2)=min Dis(S_{1i}, S_{2j}), $1 \le j \le n$.

Similarity measurement between two ontologies will be calculated based on the above parts according to weight value synthetically. The relation between ontology similarity measurement and connecting ontologies can be induced as follows: firstly the extracting operation for ontologies is processed to adopt limited candidate ontologies; then calculating ontology similarity among ontologies will be run in order to choose the most similar ontologies for matching.

On the basis of studying in this section, we have designed Chinese semantics distance measurer and matcher for software requirements semantics matching measurement on connecting ontologies to build a measurement ground for connecting ontologies generating.

Integrating Environment

This section presents the design of interoperability extending integrating environment for requirements semantic based CO in Figure 5. Applying sub-ontology extracting algorithm, DPO can be generated from requirements asset that has been produced by domain modeling tool in the phase of requirements elicitation. DPO and domain requirements asset together become reusable asset for requirements acquiring and modeling tool.

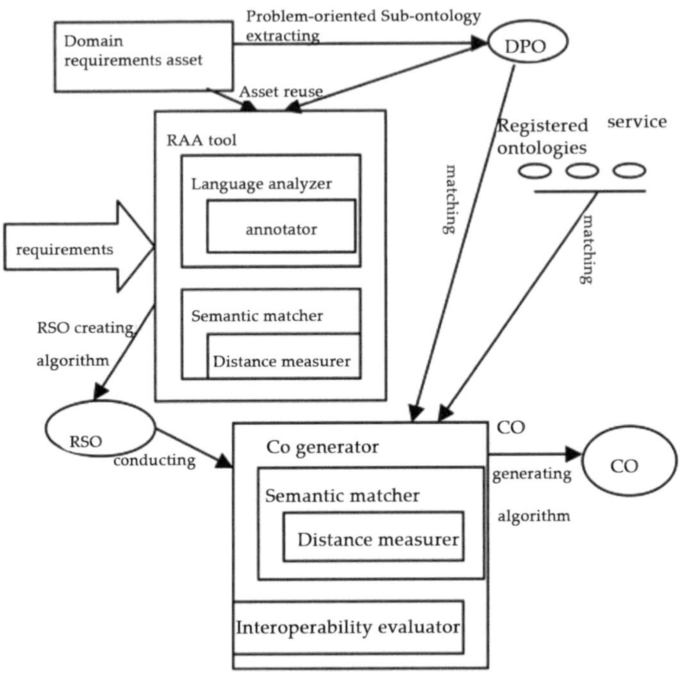

Figure 5. Requirements semantics interoperability extending integrating environment based-on CO.

Within the requirements acquiring and modeling tool, semantic matcher, which can execute matching operation with semantic distance measurement tool to achieve the matching for role, goal, and process of requirements asset, will be added. Main functions of semantic distance measurer include: measure semantic distance between two concepts; measure semantic distance between two ontologies; measure semantic distance between two services. Existing basis is: 1) thesaurus: WordNet (English), HowNet (Chinese); 2) similarity calculating based on two thesaurus.

Adopted approach is: calculating two concepts similarity from words similarity; calculating ontologies similarity based on concepts similarity; calculating services similarity based on concepts similarity.

To generate CO, the function of CO generator is driven and conducted by the control structure of RSO, and it will use semantic matcher and interoperability level evaluator. It can automatically complete the task for looking up reusable resources with CO generating algorithm purposed in the above part to the more extent.

After received CO, interoperability level evaluator, which will evaluate semantic interoperability level, able to decide the preference grade for candidate services and forecast the QoE of users.

We have designed and implemented a series of tools for supporting service identifying and composition based on CO and DPO. Relative prototype and validation of the proposed approach have also partly achieved. Experiment has demonstrated that the proposed approach is useful for service finding and integrating. The snapshot of primary tools and Prototype system for context of traffic travel problem domain can see from Figure 6.

Figure 6. Prototype context and tools of traffic travel problem domain.

RELATED WORK

Application of ontology in RE starts from domain engineering. As reusable core resources in product line, domain requirements [17] mainly solve requirements modeling issue for component-oriented software system.

Dr. Jerome Euzenat from INRIA Grenoble RhoneAlpes in France has studied semantic interoperability issues based ontology mapping [18,19] and acts as principal in NeOn project of EU FP6 plan. In June 2008, In formatics of EU startup semantic interoperability central plan for Europe and set up first session in Brussels aiming at realizing semantic data interoperability for E-government in Europe. Open source SILIME project of MITSemantic Interoperability of Metadata and Information in unLike Environments attempts to semantic interoperability for data resources (such as data library).

The studying of connecting ontologies is new direction in the world. Initial investigation studies original domain-level ontology for heterogeneity and explores how to create new ontology for covering original ontology with collaboration and consistence, and also containing ontology grouping technology (for example ontology mapping, ontology aligning, ontology merging etc.). In 2007, the paper by Shuaib Karim [20] presented a CO application framework that need not cover original ontology and focus on studying transfer principle and intermediate concept among original ontologies. Cregan Anne [21] proposes to build semantic interoperability by CO and gives some CO examples of gene ontology in 2008. However, in Cregan Anne's paper, connecting manners of CO, incentive of connecting, method and critical content of building semantic interoperability are absent. We also notice that Linked Open Data [22] initiative has become the existing foundation for federal Web of Data.

Now, together with CO and RE, the investigation of requirements semantic interoperability extending for networked software with respect to service-oriented computing just begins to proceed, and a great deal of theoretical and technological issues will require to solve.

CONCLUSIONS

This paper explores ontology-based RE, for interoperability extending of requirements semantics; we present CO approach to improve requirements modeling under the condition of distributed services aggregation with loosely coupling and different domain. Some formal definition and generating algorithm of CO are given. With the novel approach, a integrating environment and measurement system based on CO is designed and implemented.

Further work can be classified as follows: studying partial meaning of semantic interoperability for networked software requirements; build CO based on Linked Open Data infrastructure; empirical testing for integrating environment with multi-domain, such as financial risk assessment, environment protection and so on.

ACKNOWLEDGMENTS

This research has been partly supported by the National Basic Research Program of China (Grant No. 2007CB310801) and the National Natural Science Foundation of China under Grant No.60970017 and 60903034.

REFERENCES

1. Rajkumar Buyyaa, Chee Shin Yeoa, Srikumar Venu- gopala, James Broberg, and Ivona Brandic, "Cloud com-puting and emerging it platforms: Vision, hype, and real-ity for delivering computing as the 5th utility," Future Generation Computer Systems, Vol. 25, No. 6, pp. 599–616, June 2009.
2. K. Q. He, R. Peng, W. Liu, et al. "Networked Software," Science Press, Beijing, 2008.
3. Z. Jin, L. Liu, and Y. Jin, "Software Requirements Engi-neering: Principles and Method," Science Press, Beijing, 2008.
4. J. Mylopoulos, L. Chung, and E. Yu, "From object-oriented to goal-oriented requirements analysis," Communications of ACM, Vol. 42, No. 1, pp. 31–37, January 1999.
5. A. V. Lamsweerde and E. Letier, "From object orienta-tion to goal orientation: A paradigm shift for requirements engineering," Radical Innovations of Software and Sys-tem Engineering in the Future, pp. 325–340, 2004.
6. R. Q. Lu, Z. Jin, and G. Chen, "Ontology-oriented re-quirements analysis," Journal of Software, Vol. 11, No. 8, pp. 1009–1017, August 2000.
7. Z. Jin, "Ontology-based requirements elicitation," Chi-nese Journal of Computers, Vol. 23, No. 5, pp. 486–492, May 2000.
8. R. A. Falbo, G. Guizzardi, and K. C. Duarte, "An onto-logical approach to domain engineering," In Proceedings of the International Conference on Software Engineering and Knowledge Engineering (SEKE02), Ischia, Italy, pp. 351–358, 2002.
9. J. Wang, K. He, P. Gong, et al. "RGPS: A unified re-quirements meta-modeling frame for networked soft-ware," In Proceedings of Third International Workshop on Advances and Applications of Problem Frames (IWAAPF'08) at 30th International Conference on Soft-ware Engineering (ICSE'08), Leipzig, Germany, pp. 29–35, May 2008.
10. K. Q. He, F. He, and B. Li, "Research on service oriented ontology meta modeling theory and methodology," Chi-nese Journal of Computers, Vol. 28, No. 4, pp. 524–533, April 2005.
11. K. Q. He, Y. F. He, and C. Wang, "International standard: Information technology-metamodel framework for inter-operability (mfi)-3: metamodel for ontology registration," (ISO/IEC19763–3), online at:

http://www.iso.org/iso/iso_ catlogue/catalogue_tc/catalogue_detail.htm?csnumber=3863 7. ISO, 2007.

12. Y. Mao, Z. Wu, and H. Chen, "Sub-ontology based re-source management for web-based e-learning," doi: http://doi.ieeecomputersocirty.org/10.1109/TKDE.2008.127, 2008
13. B. Hu, K. Q. He, H. F. Chen, and J. Wang, "Require-ments driven web service composition based on RGPS domain assets: Approach and realization," Journal of Chi-nese Computer System, Vol. 30, No. 5, pp. 859–862, May 2009.
14. K. Q. He, "Semantic interoperability refining and clus-tering theory and its application in on demand service ag-gregation," Science in China, F: Information Science (un-published).
15. L. Lin, "Text clustering research based on semantic dis-tance," Master's thesis, Xiamen University, April 2007.
16. W. Liu, "Research on services-oriented software re-quirements elicitation and analysis," PhD thesis, Wuhan University, June 2008.
17. M. Mikyeong and Y. Keunhyuk, "An approach to devel-oping domain requirements as a core asset based on commonality and variability analysis in a product line," IEEE Software Engineering (unpublished), Vol. 31, No. 7, pp. 551–569, July 2005.
18. J. Euzenat, "An api for ontology alignment," In Proceed-ings of 3rd International Semantic Web Conference (ISWC), Hiroshima, Japan, Lecture Notes in Computer Science, Vol. 3298, pp. 698–712, 2004.
19. J. Euzenat and P. Shvaiko, "Ontology matching springer," Heidelberg, Germany, 2007.
20. Shuaib Karim, Khalid Latif1, and A. Min Tjoa1, "Pro-viding universal accessibility using connecting ontologies: A holistic approach," Lecture Notes in Computer Science 4556, pringer-Verlag, Berlin Heidelberg, Vol. 3, pp. 637–646, S 2007.
21. Cregan Anne, "W3c semantic web ontology languages: Owl and rdf tutorial," Technical Report, ISO/IEC JTC1 SC32 11th Open Forum on Metadata Registries, Sydney, Australia, May 2008. Tutorial.ppt.
22. C. Bizer, T. Heath, and T. Berners-Lee, "Linked data: The story so far," International Journal on Semantic Web and Information Systems, Vol. 5, No. 3, pp. 1–22, 2009.

Chapter

FORMAL SEMANTICS OF OWL-S WITH REWRITE LOGIC

15

Ning Huang[1], Xiao Juan Wang[1], and Camilo Rocha[2]

[1]Beihang University, Beijing, China
[2]University of Illinois at Champaign Urbana, USA

ABSTRACT

SOA is built upon and evolving from older concepts of distributed computing and modular programming, OWL-S plays a key role in describing behaviors of web services, which are the essential of the SOA software. Although OWL-S has given semantics to concepts by ontology technology, it gives no semantics to control-flow and data-flow. This paper presents a formal semantics framework for OWL-S sub-set, including its abstraction, syntax,

Citation: N. Huang, X. Wang and C. Rocha, "Formal Semantics of OWL-S with Rewrite Logic," Journal of Software Engineering and Applications, Vol. 2 No. 1, 2009, pp. 25-33. doi: 10.4236/jsea.2009.21004.

static and dynamic seman-tics by rewrite logic. Details of a consistent transformation from OWL-S SOS of control-flow to corresponding rules and equations, and dataflow semantics including "Precondition", "Result" and "Binding" etc. are explained. This paper provides a possibility for formal verification and reliability evaluation of software based on SOA.

Keywords: SOA, Web Services, OWL-S, Formal Semantics, Rewrite Logic, Consistent Transformation, Reliability Evaluation

INTRODUCTION

Service Oriented Architecture (SOA) adopts the Web services standards and technologies and is rapidly becoming a standard approach for enterprise information systems. We believe that there will be a heavy demand of reliability evaluation for the SOA software in the early development phase.

Web services are the essential of the SOA software, because SOA offers one such architecture, it unifies business processes by structuring large applications as an ad hoc collection of smaller modules called "services", Services-orientation aims at a loose coupling of services with operating systems, programming languages and other technologies which underlie applications. There are many higher level standards such as BPEL [1], WSCI, BPML, DAML-S (the predecessor of OWL-S), etc. OWL-S (Web Ontology Language for services) is a well-established language for the description of Web services based on ontology, it has been recommended by Web-ontology Working Group at the World Wide Web Consortium [2].

In order to undergo more accurate reliability evaluation and prediction of software based on SOA in the early phase, we should give software architecture the description semantics for gaining more components information, according to this motivation, we plan to use OWL-S to describe the software architecture, and then use formal method to construct a well-defined mathematical model of the system described by OWL-S, once we have this formal model, we will compute the reliability of the software based on the formal model. Here, as a meta-language of rewrite logic, Maude has been chosen as a language to give OWL-S formal semantics in a framework in static and dynamic aspects. With this framework in hand, an OWL-S specification can be transformed into an easy-understand formal one only concerning syntax, and this transformation makes a critical contribution for the formal verification and reliability evaluation of OWL-S model using the Maude language.

The rest of this paper is organized as follows. We will give an overview of related works in Section 2, the background of our method will be presented in Section 3 by an overview of OWL-S, rewrite logic and Maude, section 4 presents the abstraction of the model to introduce what we should abstract from the model. How to give OWL-S model the formal semantics using Maude in static and dynamic aspects will be presented respectively in Section 5 and Section 6. Finally, we will give a conclusion in Section 7 and an acknowledgement in Section 8.

RELATED WORKS

Most of previous related works focus on model checking instead of providing a precise formal semantics for the specification. For example, [3] converts OWL-S Process Model using a C-Like code specification language and then uses BLAST to validate it by the test cases automatically generated in the model checking process. [4] proposes a Petri Net-based operational semantics, which models the control flow of DAML-S (the former of OWL-S) Process Model, but lack of dataflow. Based on the work of [4,5] extends it to translate OWL-S Process Model using Promela, a SPIN specification language, and uses SPIN to do model checking. [6] presents a formal denotational model of OWL-S using the Object-Z(OZ) specification language, but it focuses on the formal model for the syntax and static semantics of OWL-S, and does not discuss the dynamic semantics.

These researches give good examples of formal specification of OWL-S and applications, but there are some problems:

- Some semantics is undefined: Because OWL-S is not a traditional language, so some properties can't be expressed directly. Some papers didn't claim this problem explicitly, but some do so, for example, in [5], "Precondition" and "Effect" can't be expressed in Promela and so ignored. In this paper, these properties are declared as important ones for processes and defined clearly.
- Transforming consistency: The structural operational semantics (SOS) of OWL-S hasn't been mapped to the specification language directly in related researches. This gives less consistency assurance for the transformation. But for rewrite logic, the mapping is rather clear. On the other hand, the later transformation for OWL-S specification only concerns syntax.

- Lack of analysis of dataflow: Among the above researches, only [5] explicitly stated the dataflow, it simplifies "Input" and "Output" as "Integer" and connects them to "channel". On the other hand, rewrite logic used in this paper is much more appropriate and efficient for properties deal not only with causality of events but also with data types or recursive constructs.

[6] and [7] use an algebra specification language Z. But the former focuses on the analysis of ontology, including some properties verification and reasoning without analysis of control flow and dataflow. The latter gives a good explanation for static semantics of OWL-S but without dynamic semantics.

BACKGROUND

Introduction of OWL-S

OWL-S is an OWL-based (Recommendation produced by the Web-Ontology Working Group at the World Wide Web Consortium) Web service ontology, which supplies Web service providers with a core set of constructs for describing the properties and capabilities of their Web services in unambiguous, computer interpretable form. OWL-S markup of Web services will facilitate the automation of Web service tasks, including automated Web service discovery, execution, composition and interoperation.

It is the first well-researched Web Services Ontology, which has numerous users from industry and academe, and is still undergoing. Details of the latest version of OWL-S submission document can be referred to [7].

Rewrite Logic and Maude

Rewriting logic is a computational logic that can be efficiently implemented and that has good properties as a general and flexible logical and semantic framework, in which a wide range of logics and models of computation can be faithfully represented [8].

Definition 1: A rewrite theory R is a triple R = (∑, E, R), with:

- (∑,E) a membership equational theory, and
- R a set of labeled rewrite rules of the form: " l :t → t' ⇐ cond", with "l" as a label, t, t′ ∈ $T_{\Sigma}(X)_{k}$ for some kind k, and "cond" is a condition (involving the same variables X).

In general, a rule in rewrite logic is like:

$$l : t \rightarrow t' \Leftarrow \left(\underset{i}{\Lambda} u_i = u'_i \right) \wedge \left(\underset{j}{\Lambda} v_j : s_j \right) \wedge \left(\underset{k}{\Lambda} w_k \rightarrow w'_k \right)$$

Maude is a formal programming language based on the mathematical theory of rewriting logic. With Maude system's support, this kind of language specifications can be executed and model can be checked, what is more, its mathematic semantics can be obtained. A program in Maude is just a rewrite logic theory, and Maude offers a comprehensive toolkit for the analysis of specifications, such as LTL model checker, Inductive Theorem Prover (ITP), Maude Termination Tool, Church Rosser Checker, Coherence Checker, etc.

In Maude, object-oriented systems are specified by object-oriented modules, defined by the keyword "omod ... endom", in which classes and subclasses are declared. A class declaration has the form class C|a_1: S_1, ..., a_n: S_n ,where C is the name of the class, the a_i are attribute identifiers, and the S_i are the sorts of the corresponding attributes. An object of a class C is defined as:< O : C | a1 : v1, ... , a_n : v_n >, where O is the object's name, and the v_i are the corresponding values of object's attributes, for i=1 . . . n. Objects can interact in a number of different ways such as messages passing, messages between objects can be defined by the keyword "msg". Details of Maude can be referred to [9].

The main reasons why we choose rewrite logic to give the OWL-S specification the formal semantics are listed as follows, and more detail advantages can be referred to [10]:

- Consistency in transforming: Rewrite logic is a flexible and expressive one that unifies algebraic denotational semantics and structural operational semantics (SOS) in a novel way, which can be seamlessly transformed from OWL-S SOS into rewrite rules/ equations [8], and suitable for describing data types and their relationships. On the other hand, the latter transforming from an OWL-S specification model into an algebraic one only concerns syntax.
- Suitable for many logic formula expressions in OWL-S: In [11], it is argued that rewriting logic is suitable both as a logical framework in which many other logics can be represented, and as a semantic framework.

- Efficiency implementation: Maude is a high performance rewriting logic implementations [12]. It is demonstrated that the performance of the Maude model checker "is comparable to that of current explicit-state model checkers" such as SPIN [11].
- Analyzing tools: It has been well established now that a rewrite logic specification can be benefited for comprehensive tool-supported formal verification [13].

ABSTRACTION OF THE MODEL

There are three methods to transform an OWL-S specification into a rewrite logic one:

- Translate every lines in OWL-S specification into corresponding rewrite logic ones directly. The translating is the same as giving semantics to OWL-S specification.
- Give OWL-S (the language) rewrite logic semantics for every parts of its syntax. Then the OWL-S specification itself is a rewrite logic one with semantics. None transformation is needed.
- Abstract the main parts of OWL-S (the language), define syntax in rewrite logic for the sub-set and give semantics for the syntax. And then translate the OWL-S specification into a rewrite logic one by abstracting it into the syntax in rewrite logic.

Method (1) is direct, but difficult to ensure the consistency. Method (2) is a complete one. For example, a service is modeled by a process in OWL-S, some tags such as < *process: CompositeProcess rdf:ID="CP"*>, <*process: hasInput rdf:resource="#inputownright1"*/> are used to give a detailed perspective within a composite web service. All tags should have semantics (here the tags are "*process: CompositeProcess rdf: ID*" and "*process: hasInput rdf: resource*").

In this paper, method (3) is accepted. One reason to do so is that abstraction can reduce the complexity of analyzing a model; another reason is that we can go to the most difficult and challenge problems quickly.

In the following section, we will explain what has been abstracted from OWL-S, including control flow and data flow. This abstraction becomes a sub-set of OWL-S.

Parameters and Expressions

Parameters are the basis of representing expressions, conditions, formulas and the state of an execution. In OWL-S, parameters are distinguished as "ProcessVar", "Variables" and "ResultVar", etc. They can even be identified as variables in SWRL. Our abstraction in this paper doesn't distinguish these, but refer them all as parameters.

Expressions can be treated as literals in OWL-S, either string literals or XML literals. The later case is used for languages whose standard encoding is in XML, such as SWRL or RDF. In this paper, expressions are separated into Arithmetic and Boolean expressions.

Precondition

If a process's precondition is false, the consequences of performing or initiating the process are undefined. Otherwise, the result described in OWL-S for the process will affect its "world".

Input

Inputs specify the information that the process requires for its execution. It is not contradictive with the definition of messages between web services, because a message can bundle as many inputs as required, and the bundling is specified by the grounding of the process model.

Result and Output

The performance of a process may result in changes of the state of the world (effects), and the acquisition of information by the client agent performing it (returned to it as outputs). In OWL-S, the term "Result" is used to refer to a coupled output and effect. Having declared a result, a process model can then describe it in terms of four properties, in which, the "inCondition" property specifies the condition under which this result occurs, the "withOutput" and "hasEffect" properties then state what ensures when the condition is true. The "hasResultVar" property declares variables that are bound in the "inCondition".

Precondition and Result are represented as logical formulas in OWL-S, but when they are abstracted, Boolean expression and assignment are used separately in this paper.

Process

A Web service is regarded as a process. There are three different processes: Atomic process corresponds to the actions that a service can perform by engaging it in a single interaction; composite process corresponds to actions that require multi-step protocols and/or multiple services actions; finally, simple process provides an abstraction mechanism to provide multiple views of the same process. We focus on atomic process and composite process here.

Control structure

Composite processes are decomposable into other (non-composite or composite) processes; their decomposition can be specified by using eight control structures provided for web services, including Sequence, Split, Split-Join, Choice, Any-Order, If-Then-Else, Repeat-Until, and Repeat-While.

Dataflow and Variables Binding

When defining processes using OWL-S, there are many conditions where the input to one process component is obtained as one of the outputs of a preceding step. This is one kind of data flow from one step of a process to another.

A Binding represents a flow of data to a variable, and it has two properties: "toVar", the name of the variable, and "valueSpecifier", a description of the value to receive. There are four different kinds of valueSpecifier for Bindings: valueSource, valueType, valueData, and valueFunction. The widely used one "valueSource" is addressed in this paper.

The information listed above gives an overview of how web services are bound together with control structures and dataflows.

Syntax and Static Semantics in Maude

According to the method (3) described above, we now need to define how to express the information abstracted in Section 3 in rewrite logic, namely, syntax of the sub-set in Maude. Because of space limited, we only explain parts of it:

Parameters and Expressions

To express them, several rewrite logic modules have been defined. They are NAME, EXP and BEXP.

To specify process variables we define a module named "NAME", in which "op_._: Oid Varname-> Name "is defined to be the form "process. var" as a variable name, while "Oid" is the name of a process, which has been regarded as an object identification. And a "NameList" is used to be a list of variables.

The value of a variable is stored in a "Location" which is indicated by an integer. When we bind a location with a variable name, the variable get the value stored in that location.

Arithmetic expressions (sort name is "Exp") and Boolean expressions (sort name is "BExp") are defined separately in module EXP and BEXP, which gives a description of how to use variable names to describe expressions.

IOPR (Input/Output/Precondition/Result), Data Flow and Variable Bindings

"Input" and "Output" of a process are defined as "NameLists" which are attributions of a process.

In OWL-S, "Precondition" and "Effects" are represented as logical formulas by other languages such as SWRL. Here we first simplify Precondition as "BExp" to be an attribution of a process class.

"Result" is more complicated. After separate "Output" as an attribution of a process, "Result" combines a list of "Effect", while every Effect is simplified as a conditional assignment here. The definition in Maude is " op_<-_if_: Name Exp BExp -> Effect."

As discussed above, there are four types of binding "valueSpecifier". Here we defined binding as "op fromto : Name Name -> Binding " to specify "valueSource" in module WSTYPE. With this definition, dataflow in a composite web service is created.

Processes and Control Structures

Atomic and composite web services are defined as two classes with different attributions. In order to distinguish definitions of "ControlConstructList" and "ControlConstructBag" for control structure, "OList" is defined to represent the object list which should be executed in order, and "OBag" to represent there is no order for the objects.

It seems very hard to express that a web service set can be executed in any order. But benefited with Maude operator attribution "comm", we can get this with definition "op_#_: OBag OBag -> OBag [ctor assoc comm id: noo]". "comm" attribution means that this "op" is with commutative property, which makes the objects in this "bag" ignore the order unlike it is in "OList".

After defining two sorts as follows:

subsort Qid < Oid < Block < BlockList.

subsort Qid < Oid < Block < BlockBag.

We define a nested control structure. For example, "sequence" as "op sequence: BlockList->Block [ctor]" and "split" as "op split : BlockBag -> Block [ctor] ". This separates "Block" into three cases:

- Only a process.
- A group of processes within one control structure (we refer it as a control block).
- A group of processes and control blocks within one control structure.

Obviously, the (3) is a nested control structure. If the group is order sensitive, it is a "BlockList", otherwise, it is a "BlockBag".

Syntax of atomic web service: A class "Atomws" is defined in Definition 2. When an instance of atomic web service is created, it should be declared as an object of class "Atomws".

Definition 2: *class Atomws | initialized : Bool,*
father : Oid, input : NameList,
output : NameList, precondition : BExp,
result : Effect .

Syntax of composite web service: And a class "Compositews" is defined in Definition 3. We have explained "IOPR", "Result", "Precondition" and "Binding" above. Other attributions are: "initialized" to represent whether this instance object (composite web service) of the class has been initialized with actual values of its "IOPR", "Result", "Precondition", "Binding" and control structures. "father" denotes which composite web service (instance) it belongs to. "struc" is the control structure with "BlockList" and "BlockBag" as its subsort. Other attributions are defined to be used when the composite one is executed, especially for the nested control structures.

Definition 3: *class Compositews | initialized : Bool,*
input : NameList, output : NameList,
precondition : BExp, result : Effect,
binding : Binding, father : Oid,
struc : CStruc, nest : BlockList, wait : OList,
blockwait : NList, waitbag : OBag .

When an instance of composite web service is created, it should be declared as an object of class "Compositews". And prepare an initial equation for itself (how to define an initial equation is ignored here).

DYNAMIC SEMANTICS IN MAUDE

Auxiliary Modules

When "Precondition" of a process is true, it can be initialized and executed. It affects the "world" by various "Effect". So we need to define what the "world" will be for a web service. Here a module of "SUPERSTATE" is extended with "CONFIGURATION" which already defines as a "soup" of floating objects and messages in Maude.

A "Superstate" is the "world" of a process which defined as "op_|_: State Configuration -> Superstate". "State" is a group of variables with corresponding locations, and locations with corresponding values. A message is defined as "msg call: Oid Oid -> Msg" for a composite web service to trigger its sub-process to execute. Another is defined as "msg tellfinish: Oid Oid -> Msg" to tell its father that it has finished execution.

In module "SUPERSTATE", assignment, evaluation of an arithmetic expression and a Boolean expression are defined, which gives semantics to how these syntax can be executed to affect the "world" of a process.

An operator "k" is defined as "op k: Configuration -> Configuration" to indicate that one web service is ready to be executed. Two operators "val" and "bval" are defined to evaluate expression and Boolean expression values in a state.

A sort "NList" (natural number list) is also defined in "NLIST" module to give semantics of executions of a nested control structure, with the help of the four attributions: nest, wait, blockwait, and waitbag.

Dynamic Semantics

In this section, we first analyze how executions of web services can affect their "world", and then by giving out the SOS for control structure, explain the corresponding rewrite logic rules or equations.

Execution of a Service:

Execution of an Atomic One

As defined in OWL-S, atomic processes have no sub-processes and execute in a single step only if the service is concerned and its precondition is true. The execution gives result to its "world" by "Effect". The main parts for its execution semantics (Figure 1) have been chosen to be explained as below.

Equation (1) asks atomic web service "ws" do initialization if its precondition "Cd" is true and hasn't been initialized before. Initialization is designed as an equation in a module of an instance of "Atomws". It prepares an initial state for this web service.

Equation (2) explains that when an atomic web service "ws" gets a message from its father "F", it is the same meaning that it will be executed after initialized.

```
(1) ceq   < ws : Atomws | initialized : init-state, father : F,
      precondition : Cd >   call(F, ws)
     = ( initial( < ws : Atomws | initialized : true, father : F
     > )  call(F, ws) ) if ( not init-state) and bval(Cd, st1) .
(2) eq st1 | (conf < ws : Atomws | initialized : true , father :
     F> call(F, ws)) = st1 |(conf k( < ws : Atomws | father :
     F > )) .
(3) eq   st1 | (conf k(< ws : Atomws | father : F, result : (x1
                    <- exp1 if cond) Eff >))
     = st1 | (conf k(< ws : Atomws | father : F, result : (x1 <-
     exp1 if bval(cond, st1)) Eff >)) .
(4) rl [ execute ] :
     (mem(x1, location1) loc(location1, y1) st1) | (conf k(<
     ws : Atomws | father : F , result : (x1 <- y) Eff >))
     => (mem(x1, location1) loc(location1, y) st1) | (conf k(<
     ws : Atomws | father : F , result : Eff >)) .
(5) rl [finish] :   st1 |(conf k(< ws : Atomws | father : F,
     initialized : init-state , result : noEff >))
     => st1 | (conf < ws : Atomws | father: F , initialized:
     false, result: noEff > tellfinish(F, ws)).
```

Figure 1. Semantics of execution atomic web service.

Equation (3) explains that how to execute a condition inside an "Effect". Of course there are rules that explain how to evaluate expression inside an "Effect" (ignored here).

The forth rule (4) simulates state changes by one "Effect". And rule (5) ensure that only after all the "Effect" of this web service has been executed it tells its father it has finished, and prepares a same instance waiting for its "Precondition" to be true to be initialized to execute again.

Execution of a composite process

Composite web service changes its "world" by executing its sub-processes according to its control structures.

Different from atomic web service, before a composite one is going to be executed, it should prepare "binding" information. A rule below is used to explain how to do that. After that, "sourcedata" should be defined to affect the "world" by other rules.

```
rl  st1 | (conf call(F, ws) < ws : Compositews | initialized :
true, father : F, binding : fromto(v1 , v2) BD > )
=> (sourcedata(v1, v2, st1) st1) | (conf < ws : Compositews |
father : F, binding : BD >   call(F, ws) ) .
```

After that, composite web service will be executed when its precondition is true like the atomic one. The difference is that the sub-processes grouped in control structures should be executed according to semantics of control. How to execute these structures will be explained below.

Sequence

The SOS of "sequence" and the corresponding rewrite logic rule are showed in Figure 2. "BLK" is a "Block" and "BL" is a "BlockList". Obviously, attribution "nest" here is used to separate the "BlockList", leaves the first one in "struc" to be executed first.

The question is how to ensure the first control block be executed firstly? Especially when it is a nested control structure-because when the most inner to be executed, the decomposing difference (order sensitive of BlockList and opposite BlockBag) should be recorded.

As discussed above in Section 4, a "Block" has three cases. But all of them should ensure that this "Block" be executed before "BL" is going to be executed for "sequence". To ensure this, two attributions "wait" and "blockwait" are defined. "wait" is a "OList (object list)" to ensure that only after the list of objects are all finished that this service "ws" can be executed. "blockwait" is defined as "NList (nature number list)". When an order sensitive control block (here is sequence) is separated, it definitely asks the "Block" left in "struc" (here is BLK) should be finished before other "Block" left in "nest" are going to be executed. So we add natural number "1" into the "NList" (here is BW). Otherwise, "0" is added for no order sensitive one, such as for control structure "Split". And "2" will be added when the outer control structure asks order but this one doesn't.

$$\frac{\langle BLK, \sigma \rangle \rightarrow \sigma'' \langle BL, \sigma'' \rangle \rightarrow \sigma'}{\langle sequence\ (BLK;\ BL)\ \sigma \rangle \rightarrow \sigma'}$$

```
rl k( < ws : Compositews | father : F, struc : sequence(BLK ;
BL), nest : BL1, wait : nilo, blockwait : BW > )
=><ws:Compositews|father:F, struc:BLK, nest : sequence(BL) ;
BL1, wait: nilo, blockwait : (1 BW) >    call(F, ws) .
```

Figure 2. Semantics of execution sequence.

After separating a control structure, there are three cases waited to be explained of how to execute BLK.

Case 1: BLK is a process

In this case, if it is a composite one, it can be separated recursively. If it is an atomic one (here is "A"), different rules should be matched according to "blockwait" (here is BW). The value of head of "BW" is "1" means "A" should be completely finished before "ws" continues, showed as Figure 3 rule (1). So "A" is put into "wait", and "1" is added to "BW" of "ws" to indicate that this an order sensitive block. And then two messages are released to trigger "A" and "ws". Of course, there should be a corresponding rule when "A" completes its execution (Figure 3 rule (2)).

If head(BW)==0 and sum(BW) > 0 (rule (3)), then "2" is added to "BW" and "A" is added to "waitbag" (here is OO), this indicates that "A" need not to be executed firstly in this level of the control structure, but it need to be so in outer control structure. The corresponding rule to release the blocking is showed as rule (4).

Similarly if head (BW) == 0 and sum(BW) == 0, "0" is added to "BW" to indicate there is no need for "A" to be executed firstly.

(1) crl : k(< ws : Compositews | father : F, struc : A ,
wait :nilo, blockwait : BW >)
=> < ws : Compositews | father : F, struc : empty,
wait : A ,
blockwait: (1 BW) > call(ws, A) call(F, ws)
if head(BW) == 1 .
(2) crl k(< ws : Compositews | father : F, struc : CS1,
nest :BL, wait : A ~ L, blockwait : BW >) tellfin-
ish(ws, A)
=> (call(F, ws) < ws : Compositews | father : F,
struc : CS1,
nest : BL , wait : L, blockwait : BW >) if head(BW)
== 1 .
(3) crl :k(< ws : Compositews | father : F, struc : A ,
wait : nilo, blockwait : BW, waitbag : OO >)
=> call(ws, A) call(F, ws) < ws : Compositews | father :
F, struc : empty, wait : nilo, blockwait : (2 BW), wait-
bag : A # OO >
if head(BW) == 0 and sum(BW) > 0 .
(4) crl k(< ws : Compositews | father : F, struc : CS1,
nest :BL, waitbag : A # OO, blockwait : BW >) tell-
finish(ws, A)
=> < ws : Compositews | father : F, struc : CS1, nest :
BL , waitbag : OO, blockwait : BW > call(F, ws)
if head(BW) == 2 or sum(BW) > 0 .

Figure 3. Semantics of execution nested control Block.

Case 2 and Case 3: are similar with case1, but with recursive definition. Because of space limited, we will not discuss them here

Repeat-while

"Repeat-While" tests the condition, exits if it is false and does the operation if the condition is true, then loops. Its SOS and corresponding rewrite rule are showed in Figure 4.

$$\frac{\langle bexp1, \sigma \rangle \rightarrow false}{\langle \text{repeat-while } bexp1\ CS1, \sigma \rangle \rightarrow \sigma}$$

$$\frac{\langle bexp1, \sigma \rangle \rightarrow true \qquad \langle CS1, \sigma \rangle \rightarrow \sigma'' \ \langle \text{repeat-while } bexp1\ CS1, \sigma'' \rangle \rightarrow \sigma'}{\langle \text{repeat-while } bexp1\ CS1, \sigma \rangle \rightarrow \sigma'}$$

crl k(< ws : Compositews | father : F, struc : repeat CS1
while bexp1, nest : BL, wait : nilo >) =>
if bexp1
then < ws : Compositews | father : F, struc : CS1, nest :
(repeat CS1 while bexp1) ; BL, wait : nilo > call(F, ws)
else < ws : Compositews | father : F, struc : empty,
nest : BL, wait : nilo > call(F, ws)
fi.

Figure 4. Semantics of execution Repeat-while.

Actually, for the control structure itself, it is not complex. The rule in Figure 4 just gives semantics of how this structure can be executed. The difficult is how "Block" can be executed inside it.

For example, when there is simple composite web service which contents only one atomic web service "A" within its repeat-while, say (*k(< ws : Compositews | father: F, struc: repeat 'A while bexp1, nest: BL, wait: nilo >*)), how about the execution of "A"?

As discussed for Definition 2 and 3, before this composite "ws" enters its execution "k" state, it should prepare an atomic instance 'A. If the precondition of 'A is true, it can be initialized and go into "k" execution state. This may affect the "world" of "ws" according to the group rules and equations in Figure 1. After 'A has finished its execution, rule (5) in Figure 1 prepares another instance 'A waiting for its "precondition" again. If "bexp1" decides to execute 'A again, execution can be continue, and the "Result" may turn "bexp1" to be true by affecting the "world" of "ws".

Repeat-until

"Repeat-until" does the operation, tests for the condition, exits if the condition is true, and otherwise loops. Its SOS and corresponding rewrite rule are showed in Figure 5.

$$\frac{\langle CS1, \sigma\rangle \rightarrow \sigma' \quad \langle bexp1, \sigma'\rangle \rightarrow true}{\langle repeat\text{-}until\ bexp1\ CS1, \sigma\rangle \rightarrow \sigma'}$$

$$\frac{\langle bexp1, \sigma''\rangle \rightarrow false \qquad \langle CS1, \sigma\rangle \rightarrow \sigma'' \ \langle repeat\text{-}until\ bexp1\ CS1, \sigma''\rangle \rightarrow \sigma'}{\langle repeat\text{-}until\ bexp1\ CS1, \sigma\rangle \rightarrow \sigma'}$$

eq k(< ws : Compositews | father : F, struc : repeat CS1 until bexp1, nest : BL, wait : nilo >) =
k(< ws : Compositews | father : F, struc : CS1, nest : (repeat CS1 while not bexp1) ; BL, wait : nilo > call(F, ws)).

Figure 5. Semantics of execution Repeat-until.

Actually, Repeat-While may never act, whereas Repeat-Until always acts at least once. Other executions are the same.

Split

The components of a "Split" process are a bag of process components to be executed concurrently. Split completes as soon as all of its component processes have been scheduled for execution. The rule below creates "Block" without order (because of split definition). At last these "Block"s produce atomic web services and messages into the "world" of "ws". Benefitted with objects concurrent execution in Maude, all web services that meet its precondition can be executed concurrently.

```
rl    k( < ws : Compositews |   father : F, struc : split(BLK @
BB), nest : BL, wait : nilo, blockwait : BW > )
=> < ws : Compositews | father : F, struc : BLK, nest :
split(BB) ; BL , wait : nilo, blockwait : (0 BW) >        call(F,
ws).
```

Split-join

Here the process consists of concurrent execution of a bunch of process components with barrier synchronization. That is, "Split-Join" completes when all of its components processes have completed. To do this, a special object named "split-join" is defined, and then control structure "split-join(BB)" is equal to "sequence (split(BB) ; 'split-join)".

Choice

"Choice" calls for the execution of a single control construct from a given bag of control constructs. Any of the given control constructs may be chosen for execution. As discussed above, any "Block" inside the control bag may be chosen to match BLK@BB, because of commutative property. This gives a choice to the control bag. And then "0" is added to "BW" means there is no need waiting for this "BLK".

```
rl   k( < ws : Compositews | father : F, struc : choice(BLK @
BB), nest : BL, wait : nilo, blockwait : BW >)
=> < ws : Compositews | father : F, struc : BLK, nest : BL,
wait: nilo, blockwait : (0 BW) > call(F, ws).
```

Anyorder

"Anyorder" allows the process components (specified as a bag) to be executed in some unspecified order but not concurrently. Execution and completion of all components is required. The execution of processes in an Any-Order construct cannot overlap, i.e. atomic processes cannot be executed concurrently and composite processes cannot be interleaved. All components must be executed. As with Split+Join, completion of all components is required.

```
rl    k( < ws : Compositews |  father : F, struc : anyor-
der(BLK @ BB), nest : BL, wait : nilo, waitbag: OO,
blockwait : BW > )
=><ws:Compositews | father:F, struc: BLK, nest : anyor-
der(BB) ; BL, wait: nilo, waitbag: OO,  blockwait : (1 BW) >
call(F, ws).
```

If-then-else

The "If-Then-Else" class is a control construct that has a property ifCondition, then and else holding different aspects of the If-Then-Else. Its semantics is intended to be "Test If-condition; if True do Then, if False do Else". Its SOS and corresponding rewrite rule are showed in Figure 6.

$$\frac{\langle bexp1, \sigma \rangle \rightarrow true \quad \langle CS1, \sigma \rangle \rightarrow \sigma'}{\langle if\ bexp1\ then\ CS1\ else\ CS2, \sigma \rangle \rightarrow \sigma'}$$

```
 rl k( < ws : Compositews | father : F, struc : if bexp1 then CS1
else CS2, nest : BL, wait : nilo > )      =>
if bexp1
           then < ws : Compositews | father : F, struc : CS1, nest :
BL, wait : nilo > call(F, ws)
           else < ws : Compositews | father : F, struc : CS2, nest :
BL, wait : nilo > call(F, ws)
       fi.
```

Figure 6. Execution of if-then-else.

As discussed above, the rewrite logic rules are obviously consistent with the definition of SOS benefited from the great expressing capability of rewrite logic.

CASE STUDY

Through the modules discussed above, we get a "semantics-OWL-S.maude" rewrite logic theory for semantics of the sub-set OWL-S. With this theory in hand, a software requirement or design in OWL-S can be abstracted into a rewrite logic theory with the syntax described above by extending this frame. Different from other translating methods directly mapping an OWL-S model into another specification language, this way avoids explaining the semantics in an actual model, translating work only concerns syntax mapping while semantics have been given in "semantics-OWL-S. maude".

For verifying the rewrite logic theory, we give an example to translate the process model to a Maude program, and undergo simple verifications [14] on it.

This example presented by OWL-S in Figure 7 is a web service based on Amazon E-Commerce Web Services. The process is to search books on Amazon by inputted keyword and create a cart with selected items, it is composed of four atomic processes through a sequence control construct.

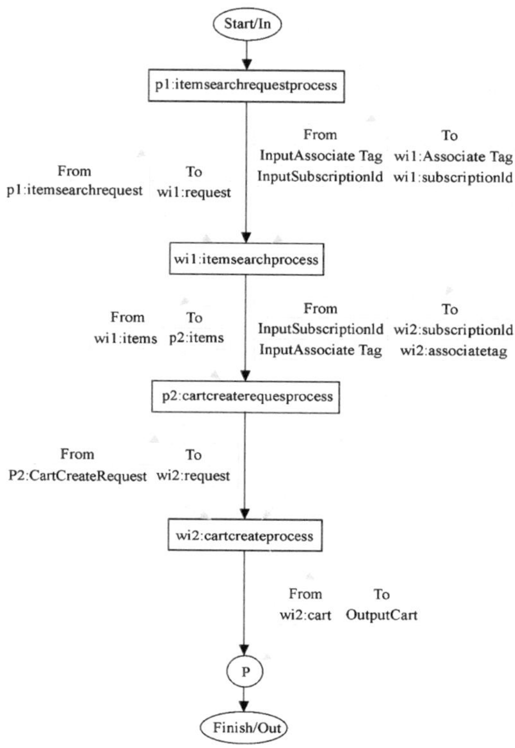

Figure 7. Structure of the web service process.

Through the rewrite theory discussed above, we get a complete Maude program. Here we only display the main parts of the Maude program. Figure 8 is the initializing equation for the third atomic process "cartCreateRequest Process". In this module, we should build attributes to express process's

IOPRs first. Two inputs and one output are translated name by name. Precondition and Effect in Result are translated to SWRL expression.

```
vars ws F : Oid .
var state : State .
var attrs : AttributeSet .
var conf : Configuration .
var cartCreateRequestProcess : cartCreateRequestProcess' .

eq state | conf initial(<ws : cartCreateRequestProcess |
father : F,attrs>)call(F,ws)
=state | conf<ws : cartCreateRequestProcess |
initialized : true,father : F,input : ws.'index || ws.'items,
output : ws.'cartCreateRequest,
precondition : #not(swrlb:length(ws.'items) #< (ws.'index)),
result:noEff>call(F,ws) .
```

Figure 8. Initial equation of atomic process.

The main process of this example is a composite process, and the translated Maude code of initializing equation of it has been shown in Figure 9.

```
eq initial( < WS : MainProcess | father : F, attrs> )
    = < WS : MainProcess | initialized : true, father : F, input : ws .
    'InputAssociateTag || (ws . 'InputAssociateTag || (ws . 'InputSub-
    scriptionId,output : ws . 'OutputCart, precondition : true, result :
     noEff, binding : fromto( 'Perform_Search . 'SubscriptionId, ws .
                                              'InputSubscriptionId)
  fromto('Perform_Search . 'AssociateTag, ws .'InputAssociateTag)
  fromto('Perform_Search . 'Request, ws . 'ItemSearchRequest)
  fromto('Perform_Search . 'Request, 'Perform_SearchReq . 'Item-
   SearchRequest)
  fromto('Perform_CreateReq . 'items, 'Perform_Search . 'Items)
  fromto('Perform_Create . 'SubscriptionId, ws . 'InputSubscrip-
   tionId)
  fromto('Perform_Create . 'AssociateTag, ws . 'InputAssociateTag)
  fromto('Perform_Create . 'Request, 'Perform_CreateReq . 'Cart-
   CreateRequest) ,
struc : sequence('Perform_SearchReq ; 'Perform_Search; 'Per-
   form_CreateReq; 'Perform_Create),
nest : null, wait : nilo, blockwait : niln, waitbag : noo >
  < 'Perform_SearchReq : itemSearchRequestProcess | initialized :
   false, father : ws,   precondition : true >
  < 'Perform_Search : ItemSearchProcess | initialized : false, fa-
   ther : ws, precondition : ture>
  < 'Perform_CreateReq : ItemSearchProcess | initialized : false,
   father : ws,   precondition : swrlb:length( 'Perform_CreateReq .
   'items) #>= ('Perform_CreateReq .'index) >
  < 'Perform_Create : CartCreateProcess | initialized : false, fa-
   ther : ws, precondition :   true>.
```

Figure 9. Initializing equation of composite process.

Load the Maude program, and then execute the process by the command "rew execute-aws", we can also search executing path using command "search execute-aws =>! S: State|C: Configuration tell finish (', 'mainProcess)." If input a right number less than or equal to the length of 'items' for 'index', the input of third atomic process "cart CreateRequestProcess", Maude will display the result in Figure 10. In other cases the result is like Figure 11.

```
omod EXEC-MAINPROCESS is
 pr MAINPROCESS .

 op MainProcess : -> MainProcess' .
 op execute-aws : -> Superstate .

 eq exec-MainProcess = loc(1, nilv) loc(2, 'SubscriptionId')
loc(3, 'AssociateTag') loc(4, nilv) loc(5, Keywords-1) loc(6,
SearchIndex-1) loc(7, nilv) loc(8, nilv) loc(9, nilv) loc(10,
index-1) loc(11, nilv) loc(12, nilv) loc(13, index-1) loc(14,
nilv) loc(15, nilv) loc(16, index-1) loc(17, nilv) loc(18,nilv)
mem('mainProcess . 'InputSubscriptionId, 2)
mem('MainProcess . 'InputAssociateTag, 3)
mem('Perform_SearchReq . 'ItemSearchRequest, 4)
mem('Perform_Req . 'Keywords, 5)
mem('Perform_SearchReq . 'SearchIndex, 6)
mem('Perform_Search . 'Items, 7)
mem('Perform_Search .'OperationRequest, 8)
mem('Perform_Search . 'Shared, 13)
mem('Perform_Search . 'Validate, 14) mem('Perform_Search .
  'XMLEscaping, 15) mem('Perform_CreateReq . 'CartCre-
    ateRequest, 9) mem('Perform_CreateReq . 'index, 10)
  mem('Perform_Create . 'Cart, 11) mem('Perform_Create .
  'OperationRequest, 12) mem('Perform_Create . 'Shared, 16)
mem('Perform_Create . 'Validate, 17) mem('Perform_Create .
'XMLEscaping, 18) | < 'MainProcess : MainProcess | initial-
        ized : false, father : ', precondition : true >
  call(' . 'MainProcess).
endom
```

Figure 10. Executing environment module.

Figure 11. Process can finish.

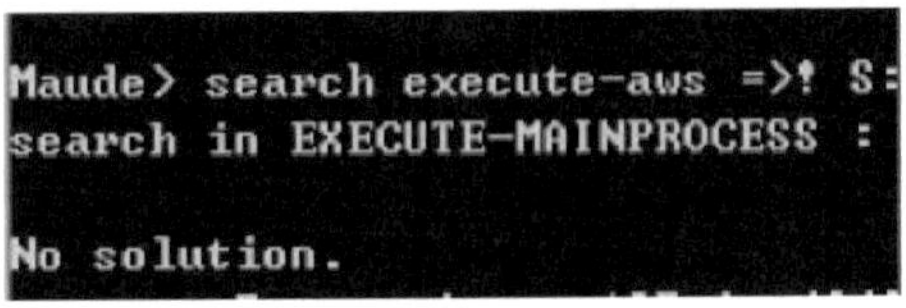

Figure 12. Process can not finish.

We have done more works concerning this framework, because of space limitation, details are ignored:

- Test framework: although the directly mapping from OWL-S SOS to rewrite logic gives the consistency, some web services have been constructed to test the eight control structures and nested ones, including the execution of atomic web services. The results are the same as expected.
- Model checking and analysis: several cases are constructed including "philosopher dining" which not only concern control flow but also get a deadlock because of data sharing in the dataflow; and "online shopping" which concerns an error in dataflow. These errors can be found by the Maude analysis tools.

CONCLUSIONS

This paper gives a formal semantics for OWL-S sub-set by rewrite logic, including abstraction, syntax, static and dynamic semantics. Compared with related researches, the contribution of this paper gives a translation consistency and benefited with formal specification, dataflow can be analyzed deeply, which makes formal verification, and reliability evaluation of software based on SOA possible.

The undergoing future works include: "Precondition" and "Effect" in SWRL format; WSDL and grounding information; and a more complex application analysis.

ACKNOWLEDGEMENT

This work has been greatly helped by Prof. Meseguer. Thanks to Michael Katelman, Feng Chen and Joe Hendrix in Formal Systems Lab of UIUC, and all the developers of the shared software.

REFERENCES

1. X. Fu, T. Bultan, and J. W. Su, "Analysis of interacting bpel web services," in Proceedings of the 13th Interna-tional Conference on World Wide Web, New York, NY,USA, pp. 621-630, May 2004.
2. D. Martin, M. Burstein, J. Hobbs, O. Lassila, D. McDer-mott, S. McIlraith, S. Narayanan, M. Paolucci, B. Parsia, T. R. Payne, E. Sirin, N. Srinivasan, and K. Sycara, "OWL-S: Semantic markup for web services," Technical Report UNSPECIFIED, Member Submission, W3C, http://www.w3.org/Submission/ OWL-S/, 2004.
3. H. Huang, W. T. Tsai, R. Paul, and Y. N. Chen, "Auto-mated model checking and testing for composite web ser- vices," in Proceedings Eighth IEEE International Sympo sium on Object-Oriented Real-Time Distributed Computing, Washington, DC, USA, pp. 300-307, May 2005.
4. S. Narayanan and S. A. McIlraith, "Simulation, verifi-cation and automated composition of web services," in Proceedings 11th International Conference on World Wide Web, Honolulu, Hawaii, USA, pp. 77-88, May 2002.
5. A. Ankolekar, M. Paolucci, and K. Sycara, "Spinning the OWL-S process model-toward the verification of the OWL-S process models," in Proceedings International Semantic Web Conference 2004 Workshop on Semantic Web Services: Preparing to Meet the World of Business Applications, Hiroshima, Japan, 2004.
6. H. H. Wang, A. Saleh, T. Payne, and N. Gibbins, "Formal specification of OWL-S with Object-Z: The static aspect," in Proceedings IEEE/WIC/ACM International Conference on Web Intelligence, Washington, DC, USA, pp. 431-434, November 2007.
7. J. S. Dong, C. H. Lee, Y. F. Li, and H. Wang, "Verifying DAML+OIL and beyond in Z/EVES," in Proceedings the 26th International Conference on Software Engineering, Washington, DC, USA, pp. 201-210, May 2004.
8. T. F. Serbanuta, F. Rosu, and J. Meseguer, "A rewriting logic approach to operational semantics (Extended Ab-stract)," Electronic Notes Theoretical Computer Science, Vol. 192, No. 1, pp. 125-141, October 2007.
9. H. Huang and R. A. Mason, "Model checking technolo-gies for web services," in Proceedings the 4th IEEE Workshop on Software

Technologies for Future Embed-ded and Ubiquitous Systems, and the Second International Workshop on Collaborative Computing, Integration, and Assurance, Wanshington, DC, USA, pp. 217-224, April 2006.

10. A. Verdejo and N. Marti-Oliet, "Executable structural operational semantics in maude," Journal of Logic and Algebraic Programming, Vol. 67, No. 1-No. 2, pp. 226-293, April-May 2006.
11. M. Birna van Riemsdijk, Frank S. de Boer, M. Dastani, and John-Jules Meyer, "Prototyping 3APL in the maude term rewriting language," in Proceedings of the fifth in-ternational joint conference on Autonomous agents and multiagent systems, Hakodate, Hokkaido, Japan, pp. 1279-1281, May 2006.
12. M. Clavel, F. Duran, S. Eker, P. Lincoln, N. M. Oliet, J. Meseguer, and C. Talcott, "All about maude-A high-per-formance logical framework," Springer-Verlag New York, Inc., 2007.
13. J. Meseguer and G. Rou, "The rewriting logic semantics project," Theoretical Computer Science, Vol. 373, No. 3, pp. 213-237, April 2007.
14. M. Clavel, F. Duran, etc., "Maude mannual," Department of Computer Science University of Illinois at Urbana- Champaign, 2007, http://maude.cs.uiuc.edu

Chapter 16

WEB SEMANTIC AND ONTOLOGY

Elodie Marie Gontier

Professor of French and History, Paris, France

ABSTRACT

Ontologies have become a popular research topic in many communities. In fact, ontology is a main component of this research; therefore, the definition, structure and the main operations and applications of ontology are provided. Web content consists mainly of distributed hypertext and hypermedia, and is accessed via a combination of keyword based search and link navigation. Hence, the ontology can provide a common vocabulary, and a grammar for publishing data, and can supply a semantic description of data which can be used to preserve the ontologies and keep them ready for inference. This

Citation: Gontier, E. (2015), "Web Semantic and Ontology". Advances in Internet of Things, 5, 15-20. doi: 10.4236/ait.2015.52003.

paper provides basic concepts of semantic web, and defines the structure and the main applications of ontology.

Keywords: Ontology, Semantic Web, Language OWL

WHAT DO WE REPRESENT IN AN ONTOLOGY?

In the context of Semantic Web, ontologies describe domain theories for the explicit representation of the semantics of the data. In other words, ontology should be seen as a right answer to provide a formal conceptualization. Indeed, ontology must translate an explicit consensus and develop a certain level of division. It has two essential aspects to allow the operation of the resources of web by various applications or software agents. The ontologies serve then:

- For the vocabulary, the structuring and the operation of metadatas;
- As representation pivot for the integration of springs of heterogeneous data;
- To describe the web departments, and generally, everywhere it is going to be necessary to press software modules on semantic representations requiring certain consensus.

Ontology (def. 1): all the objects recognized as existing in the domain. To build an ontology, it is also to decide on the way of being and to exist objects. To continue towards a definition of ontology, it seems to us essential to remind that the works on the ontologies are developed in an IT context that is the case, for instance, for Engineering of knowledge, Artificial intelligence or, more specifically here, the context of Semantics Web where the final goal is to specify an IT artefact. In this context, the ontology becomes then a model of the existing objects which makes a reference to it through concepts of the domain.

The developments are a free performance of the reasons adduced for the works of Guarino and Giaretta [1] . They aim at progressing towards a definition reporting an evolutionary process of construction.

Ontology (def. 2): an ontology involves or includes a certain worldview compared with a given domain. This sight is often conceived as a set of concept—e.g. entities, attributes, and process—their definitions and their interrelations. We call it a "conceptualization". An ontology can take various forms, but it will include inevitably a vocabulary of terms and specification of their meaning. So, it is a specification partially reporting a conceptualization. This second definition proposes another point of view compared with the first one, coherent with her but more precise, in terms of specification and compared with web operation.

Ontology is good at conceptualization, like said Thomas Gruber "it's an explicit specification of conceptualization":

- Afterward, it must be used in an IT artefact, but we have to specify it more later. Ontology will also have to be a logical theory for which we shall specify the manipulated vocabulary;
- Finally, the conceptualization is sometimes specified in a very precise way. That's why a logical theory cannot always report it in an exact way: she can accept the interpretative wealth of the domain conceptualized in an ontology and make it thus only partially. This gap between the conceptualization and the formal specification is described by Guarino as the ontological commitment which the designer has to accept in the passage of the one to the other one.

The ontology is a theory on the representation of the knowledge. As indicated in 2000, the ontology "defines the kinds of things that exist in the application domain". It is by this theory that it unifies in the domain of the computing. The ontology "is a formal, explicit specification of a shared conceptualization" (Gruber, 1993). For the IT specialist of semantic web, the ontology is a consensual model, because the conceptualization is shared and brings then to build a linguistic specification with the vocabulary RDF/RDFS and the language OWL. In the semiotic perspective, conceptualization according to Gruber relates to the domain of the speech because it is the abstraction. The domain of the speech takes place of the referent. In semiotics, we shall say that the ontology symbolizes the conceptualization, the terms, the notions and the relations which are conceptualized.

THE WEB ONTOLOGY LANGUAGE OWL

The rapid evolution of semantic web ontology languages was enabled by learning from the experiences in developing existing knowledge representation formalisms and database conceptual models, and by inheriting and extending some of their useful features. In particular, the semantic web significantly improves visibility and extensibility aspects of knowledge sharing in comparison with the previous approaches [2] . Its

URI-based vocabulary and XML-based grammar are key enablers to web scale knowledge management and sharing.

One of the strong results of semantic web on the ontologies is the normalization of their expression. This point, essential if we want that the ontologies can be shared, exactly seems to find a solution in the context of semantic web: the definition of the language OWL (Web Ontologies Language) at various levels of complexity (capacity of complexity of the descriptions versus calculability) is the best example.

Although already recognisable as an ontology language, the capabilities of RDF are rather limited: they do not, for example, include the ability to describe cardinality constraints (such as Hogwarts Students having at most one pet), a feature found in most conceptual modelling languages, or to describe even a simple conjunction of classes.

The need for a more expressive ontology language was widely recognised within the nascent semantic web research community, and resulted in several proposals for "web ontology languages", including SHOE, OIL and DAML + OIL. The architecture of the web depends on agreed standards and, recognising that an ontology language standard would be a prerequisite for the development of the semantic web, the World Wide Web Consortium (W3C) set up a standardisation working group to develop a standard for a web ontology language. The result of this activity was the OWL ontology language standard [3] . OWL exploited the earlier work on OIL and DAML + OIL, and also tightened the integration of these languages with RDF. The integration of OWL with RDF includes the provision of a RDF based syntax. This has the advantage of making OWL ontologies directly accessible to web based applications, but the syntax is rather verbose and not easy to read. For example, the description of the above mentioned class of Student Wizards would be written in RDF/XML as:

```
<owl:Class>
<owl:intersectionOf
rdf:parseType="Collection">
<owl:Class rdf:about="#Student"/>
<owl:Class rdf:about="#Wizard"/>
</owl:intersectionOf>
</owl:Class>
```

In the remainder of this paper, I will instead use an informal \human readable" syntax based on the one used in the Protege 4 ontology development tool [4] . A key feature of OWL is its basis in Description Logics (DLs), a family of logic-based knowledge representation formalisms that are descendants of Semantic Networks and KLONE, but that have a formal semantics based on rstorder logic [5] . These formalisms all adopt an objectoriented model, similar to the one used by Plato and Aristotle, in which the domain is described in terms of individuals, concepts (called classes in RDF), and roles (called properties in RDF). Individuals, e.g., "Hedwig", are the basic elements of the domain; concepts, e.g., "Owl", describe sets of individuals having similar characteristics; and roles, e.g., "hasPet", describe relationships between pairs of individuals, such as "HarryPotter hasPet Hedwig".

In order to avoid confusion, I will keep to the already introduced RDF terminology and from now on refer to these basic language components as individuals, classes and properties. As well as atomic class names such as Wizard and Owl, DLs also allow for class descriptions to be composed from atomic classes and properties. A given DL is characterised by the set of constructors provided for building class descriptions. OWL is based on a very expressive DL called SHOIN (D) a sort of acronym derived from the various features of the language [6] . The class constructors available in OWL include the Booleans and, or and not, which in OWL are called intersectionOf, unionOf and complement Of, as well as restricted forms of existential and universal quantication, which in OWL are called, respectively, "some Values From" and "all Values From" restrictions. OWL also allows for properties to be declared to be transitive| if has Ancestor is a transitive property, then Enoch has Ancestor Cain and Cain has Ancestor Eve implies that Enoch has Ancestor Eve. The S in SHOIN (D) stands for this basic set of features.

In OWL, some values from restrictions are used to describe classes whose instances are related, via a given property, to instances of some other class. For example, Wizard and hasPet some Owl describes those Wizards having pet Owls. Note that such a description is itself a class, the instances of which are just those individuals that satisfy the description; in this case, those individuals that are instances of Wizard and that are related via the hasPet property to an individual that is an instance of Owl. If an individual is asserted to be a member of this class, then we know that they must have a pet Owl, although we may not be able to identify the Owl in question, i.e., some values from restrictions specify the existence of a relationship. In contrast, all values from restrictions constrain the possible objects of a given property and are typically used as a kind of localised range restriction.

For example, we might want to state that Hogwarts students can have only Owls, Cats or Toads as pets without placing a global range restriction on the hasPet property (because other kinds of pet may be possible in general). We can do this in OWL as follows:

Wizard and hasPet some Owl describes those Wizards having pet Owls. Note that such a description is itself a class, the instances of which are just those individuals that satisfy the description; in this case, those individuals that are instances of Wizard and that are related via the hasPet property to an individual that is an instance of Owl. If an individual is asserted to be a member of this class, then we know that they must have a pet Owl, although we may not be able to identify the Owl in question, i.e., some values from restrictions specify the existence of a relationship. In contrast, all values from restrictions constrain the possible objects of a given property and are typically used as a kind of localised range restriction. For example, we might want to state that Hogwarts students can have only Owls, Cats or Toads as pets without placing a global range restriction on the has Pet property (because other kinds of pet may be possible in general). We can do this in OWL as follows:

Class: HogwartsStudent

SubClassOf: hasPet only (Owl or Cat or Toad)

In addition to the above mentioned features, OWL also allows for property hierarchies (the H in SHOIN (D)), extensionally denied classes using the one of constructor (O), inverse properties using the inverse of property constructor (I), cardinality restrictions using the minCardinality, maxCardinality and cardinality constructors (N), and the use of XML Schema datatypes and values (D) [7] . For example, we could additionally state that the instances of Hogwarts House are exactly Gryndor, Slytherin, Ravenclaw and Huepu, that Hogwarts students have an email address (which is a string) and at most one pet, that isPetOf is the inverse of hasPet and that a Phoenix can only be the pet of a Wizard:

Class: HogwartsHouse

EquivalentTo: {Gryffindor, Slytherin

Ravenclaw, Hufflepuff}

Class: HogwartsStudent

SubClassOf: hasEmail some string

SubClassOf: hasPet max 1

ObjectProperty: hasPet

Inverses: isPetOf

Class: Phoenix

SubClassOf: isPetOf only Wizard

An OWL ontology consists of a set of axioms. As in RDF, subClassOf and subPropertyOf axioms can be used to dene a hierarchy of classes and properties. In OWL, an equivalent Class axiom can also be used as an abbreviation for a symmetrical pair of subClassOf axioms. An equivalentClass axiom can be thought of as an "if and only if" condition: given the axiom C equivalentClass D, then an individual is an instance of C if and only if it is an instance of D. Combining subClassOf and equivalentClass axioms with class descriptions allows for easy extension of the vocabulary by introducing new names as abbreviations for descriptions. For example, the following axiom:

Class: HogwartsStudent

EquivalentTo: Student and attendsSchool

value Hogwarts

introduces the class name HogwartsStudent, and asserts that its instances are just those Students that attend Hogwarts. Axioms can also be used to state that a set of classes is disjoint, and to describe additional characteristics of properties: as well as being Transitive, a property can be Symmetric, Functional or Inverse Functional. For example, the axioms:

DisjointClasses: Owl Cat Toad

Property: isPetOf

Characteristics: Functional

state that Owl, Cat and Toad are disjoint (i.e., that they have no instances in common), and that isPetOf is Functional (i.e., pets can have at most one owner). The above mentioned axioms describe constraints on the structure of the domain, and play a similar role to the conceptual schema in a database setting; in DLs such a set of axioms is called a TBox (Terminology Box). OWL also allows for axioms asserting facts about some concrete situation, similar to data in a database setting; in DLs such a set of axioms is called an ABox (Assertion Box). These might, for example, include the facts:

Individual: HarryPotter

Types: HogwartsStudent

Individual: Fawkes

Types: Phoenix

Facts: isPetOf Dumbledore

Basic facts (i.e., those using only atomic classes) correspond directly to RDF triples|the above facts, for example, correspond to the following triples:

HarryPotter rdf:type, HogwartsStudent

Fawkes rdf:type Phoenix

Fawkes isPetOf Dumbledore

The term ontology is often used to refer just to a conceptual schema or TBox, but in OWL an ontology can consist of a mixture of both TBox and ABox axioms; in DLs, this combination is known as a Knowledge Base. Description Logics are fully edged logics and so have a formal semantics. DLs can, in fact, be seen as decidable subsets of rst-order logic, with individuals being equivalent to constants, concepts to unary predicates and roles to binary predicates. As well as giving a precise and unambiguous meaning to descriptions of the domain, this also allows for the development of reasoning algorithms that can provide correct answers to arbitrarily complex queries about the domain. An important aspect of DL research has been the design of such algorithms, and their implementation in (highly optimised) reasoning systems that can be used by applications to help them "understand" the knowledge captured in a DL based ontology.

ONTOLOGY LANGUAGE PROCESSORS

As we can see, ontologies are like taxonomies but with more semantic relationships between concepts and attributes; they also contain strict rules used to represent concepts and relationships. An ontology is a hierarchically structured set of terms for describing a domain that can be used as a skeletal foundation for a knowledge base. According to this definition, the same ontology can be used for building several knowledge bases.

Indeed, an ontology construct conveys descriptive semantics, and its actionable semantics is enforced by inference. Hence, effective tools, such as parsers, validators, and inference engines, are needed to fulfill the inferenceablity objective:

1. **OWLJessKB** is the descendent of DAMLJessKB and is based on the Jess Rete inference engine [7] .
2. **Java Theorem Prover (JTP)** developed at Stanford university [8] supports both forward and backward chaining inference using RDF/RDFS and OWL semantics.
3. **Jena** (http://jena.sourceforge.net/), developed at HP Labs at Bristol, is a popular open-source project. It provides sound and almost complete (except for blank node types) inference support for RDFS. Current version of Jena also partially supports OWL inference and allows users to create customized rule engines [9]
4. **F-OWL** developed at UMBC, is an inference engine which is based on Flora-218 [10] .
5. **FaCT ++** uses the established FaCT algorithms, but with a different internal architecture. Additionally, FaCT ++ is implementated using C ++ in order to create a more efficient software tool, and to maximise portability [11] .
6. **Racer** (https://www.ifis.uni-luebeck.de/index.php?id=385) is a description logic based reasoner. It supports inference over RDFS/DAML/OWL ontologies through rules explicitly specified by the user [12] .
7. **Pellet** (http://www.w3.org/2004/04/13-swdd/SwoopDevDay04.pdf), developed at the University of Maryland, is a "hybrid" DL reasoner that can deal both TBox reasoning as well as non-empty ABox reasoning [13] . It is used as the underlying OWL reasoner for SWOOP ontology editor [14] and provides in-depth ontology consis- tency analysis.
8. **TRIPLE** developed by Sintek and Decker into Proceedings of the 1st International Semantic Web Con- ference [15] , is a Horn Logic based reasoning engine (and a language) and uses many features from F-logic. Unlike F-logic, it does not have fixed semantics for classes and objects. This reasoner can be used by translating the Description Logics based OWL into a language (named TRIPLE) handled by the reasoner. Extensions of Description Logics that cannot be handled by Horn logic can be supported by incorporating other reasoners, such as FaCT, to create a hybrid reasoning system.
9. **SweetRules** (http://sweetrules.projects.semwebcentral.org/) is a rule toolkit for RuleML. RuleML is a highly expressive language based on courteous logic programs, and provides additional built-in semantics to OWL, including prioritized conflict handling and pro-

cedural attachments. The SweetRules engine also provides semantics preserving translation between a various other rule languages and ontologies (implicit axioms).

The semantics conveyed by ontologies can be as simple as a database schema or as complex as the back- ground knowledge in a knowledge base. By using ontologies in the semantic web, users can leverage the ad- vantages of the following two features:

- Data are published using common vocabulary and grammar;
- The semantic description of data is preserved in ontologies and ready for inference.

Ontology transformation [16] is the process used to develop a new ontology to cope with new requirements made by an existing one for a new purpose, by using a transformation function t. In this operation, many changes are possible, including changes in the semantics of the ontology and changes in the representation formalism. Ontology Translation is the function of translating the representation formalism of an ontology while keeping the same semantic. In other words, it is the process of change or modification of the structure of an ontology in order to make it suitable for purposes other than the original one.

There are two types of translation. The first is translation from one formal language to another, for example from RDFS to OWL, called syntactic translation. The second is translation of vocabularies, called semantic translation [17] . The translation problem arises when two Web-based agents attempt to exchange information, describing it using different ontologies. The goal of an ontology is to achieve a common and shared knowledge that can be transmitted between people and between application systems. Thus, ontologies play an important role in achieving interoperability across organizations and on the semantic web, because they aim to capture domain knowledge and their role is to create semantics explicitly in a generic way, providing the basis for agreement within a domain. Thus, ontologies have become a popular research topic in many communities. In fact, ontology is a main component of this research; therefore, the definition, structure and the main operations and applications of ontology are provided.

CONCLUSION

Elodie Marie Gontier Ontologies play an important role in achieving interoperability across organizations and on the semantic web, because they aim to capture domain knowledge and their role is to create semantics explicitly in a generic way, providing the basis for agreement within a domain. In other words, the current web is transformed from being machine-readable to machine-understandable. So, ontology is a key technique with which to annotate semantics and provide a common, comprehensible foundation for resources on the semantic web.

REFERENCES

1. Guarino, N. and Giaretta, P. (1995) Ontologies and Knowledge Bases. In: Towards Very Large Knowledge Bases, IOS Press, Amsterdam, 1-2.
2. Web Ontology Language (OWL) Offers Additional Knowledge Base Oriented Ontology Constructs and Axioms. http://www.w3.org/2002/Talks/04-sweb/slide12-0.html
3. Ian Horrocks, Ontologies and the Semantic Web, Oxford University Computing Laboratory.
4. http://protege.stanford.edu/
5. Baader, F., Calvanese, D., McGuinness, D., Nardi, D. and Patel-Schneider, P.F., Eds. (2003) The Description Logic Handbook: Theory, Implementation and Applications. Cambridge University Press, Cambridge.
6. Horrocks, I. and Sattler, U. (2007) A Tableau Decision Procedure for SHOIQ. Journal of Automated Reasoning, 39, 249-276.
7. Joseph, K. and William, R. (2003) DAMLJessKB: A Tool for Reasoning with the Semantic Web. IEEE Intelligent Systems, 18, 74-77.
8. Joseph, K. And William, R. (2003) DAMLJessKB: A Tool for Reasoning with the Semantic Web. IEEE Intelligent Systems, 18, 74-77.
9. Richard, F., Jessica, J. and Gleb, F. (2003) JTP: A System Architecture and Component Library for Hybrid Reasoning. Stanford University, Stanford.
10. Carroll, J.J, Ian, D., Chris, D., Dave, R., Andy, S. and Kevin, W. (2004) Jena: Implementing the Semantic Web Recommendations. Proceedings of the 13th International World Wide Web Conference on Alternate Track Papers & Posters, 2004, 74-83. ISBN:1-58113-912-8.
11. Zou, Y.Y., Finin, T. and Chen, H. (2004) F-OWL: An Inference Engine for the Semantic Web. Formal Approaches to Agent-Based Systems. Vol. 3228 of Lecture Notes in Computer Science. Springer-Verlag, Berlin. Proceedings of the Third International Workshop (FAABS), 16-18 April 2004.
12. Dmitry, T. and Ian, H. (2003) Implementing New Reasoner with Datatypes Support. Wonder Web: Ontology In- frastructure for the Semantic Web Deliverable.

13. Ian, H. (1998) The FaCT System. Automated Reasoning with Analytic Tableaux and Related Methods. International Conference Tableaux-98, Springer Verlag, Berlin, 307-312.
14. Evren, S. and Bijan, P. (2004) Pellet: An OWL DL Reasoner. In: Description Logics, CEUR-WS.org, 9.
15. Aditya, K., Bijan, P. and James, H. (2005) A Tool for Working with Web Ontologies. International Journal on Semantic Web and Information Systems, 1, 4.
16. Michael, S. and Stefan, D. (2002) TRIPLE—A Query, Inference, and Transformation Language for the Semantic Web. Proceedings of the 1st International Semantic Web Conference (ISWC-02), Springer-Verlag, Berlin, 364-378.
17. Chalupsky, H. (2000) OntoMorph: A Translation System for Symbolic Knowledge. Proceedings of KR, Morgan Kaufmann Publishers, San Francisco, 471-482.

Chapter 17

WEB SERVICES CONVERSATION ADAPTATION USING CONDITIONAL SUBSTITUTION SEMANTICS OF APPLICATION DOMAIN CONCEPTS

Islam Elgedawy

Computer Engineering Department, Middle East Technical University, Northern Cyprus Campus, Guzelyurt, Mersin 10, Turkey

ABSTRACT

Internet of Services (IoS) vision allows users to allocate and consume different web services on the fly without any prior knowledge regarding the chosen services. Such chosen services should automatically interact with one another in a transparent manner to accomplish the required users' goals.

Citation: Elgedawy Islam, "Web Services Conversation Adaptation Using Conditional Substitution Semantics of Application Domain Concepts", International Scholarly Research Notices, volume 2013, article ID 408267, https://doi.org/10.1155/2013/408267.

As services are chosen on the fly, service conversations are not necessarily compatible due to incompatibilities between services signatures and/or conversation protocols, creating obstacles for realizing the IoS vision. One approach for overcoming this problem is to use conversation adapters. However, such conversion adapters must be automatically created on the fly as chosen services are only known at run time. Existing approaches for automatic adapter generation are syntactic and very limited; hence they cannot be adopted in such dynamic environments. To overcome such limitation, this paper proposes a novel approach for automatic adapter generation that uses conditional substitution semantics between application domain concepts and operations to automatically generate the adapter conversion functions. Such conditional substitution semantics are captured using a concepts substitutability enhanced graph required to be part of application domain ontologies. Experiments results show that the proposed approach provides more accurate conversation adaptation results when compared against existing syntactic adapter generation approaches.

INTRODUCTION

Internet of Services (IoS) vision enables users (i.e. people, businesses, and systems) to allocate and consume the required computing services whenever and wherever they want in a context-aware seamless transparent manner. Hence, chosen services automatically interact with one another in a transparent manner to accomplish the required users' goals. Middleware software plays an essential role in supporting such interactions, as it hides services heterogeneity and ensures their interoperability. Middleware enables services to locate one another without a priori knowledge of their existences and enables them to interact with one another even though they are running on different devices and platforms [1]. Services interactions are conducted via exchanging messages. A conversation message indicates the operation to be performed by the service receiving the message. A sequence of messages exchanged between services to achieve a common goal constitutes what is known by a conversation pattern. A set of conversation patterns is referred to as a service conversation. However, services may use different concepts, vocabularies, and semantics to generate their conversation messages, raising the possibility for having conversation incompatibilities. Such incompatibilities must be automatically resolved in order to enable services conversations on the fly. This should be handled by a conversation adapter created on the fly by the middleware, please refer to Section 2.2 for more information about conversation adapters.

In general, in order to create a conversation adapter, first we have to identify the possible conversation incompatibilities and then try to resolve the incompatibilities using the available conversation semantics, which are constituted from service semantics (such as service external behavior, encapsulated business logic, and adopted vocabulary) and application domain semantics (such as concepts relations and domain rules). If solutions are found, the adapter can be created; otherwise the conversations are labelled as unadaptable, and the corresponding services cannot work together. Hence, we argue that in order to automatically generate conversation adapters the following prerequisites must be fulfilled.

- First, we require substitution semantics of application domain concepts and operations to be captured in application domain ontologies in a context-sensitive manner, as such semantics differ from one context to another in the same application domain, for example, the concepts Hotel and Resort could be substitutable in some contexts and not substitutable in others. Hence, capturing substitution semantics and its corresponding conversion semantics in a finite context-sensitive manner is mandatory to guarantee the adapter functional correctness, as these conversion semantics provide the basic building blocks for generating converters needed for building the required adapters.
- Second, we require services descriptions to provide details about the supported conversation patterns (that is the exchanged messages sequences), such as the conversation context, the supported operations, and the supported invocation sequences. Such information must be captured in a machine-understandable format and must be based on the adopted application-domain ontology vocabulary.
- Finally, as different conversation patterns could be used to accomplish the same business objective, different types of mappings between the conversation patterns operations must be automatically determined (whether it is many-to-many, or one-to-one, and etc). Such operations mappings are essential for determining the required adapter structure.

Unfortunately, existing approaches for adapter generation (such as the ones discussed in [2–9]) do not fulfill the mentioned prerequisites; hence they are strictly limited and cannot be adopted in dynamic environments implied by the IoS vision. More details and discussion about these approaches are given in the related work section (Section 3).

To overcome the limitations of the existing adaptation approaches, this paper proposes a novel approach for automatic adapter generation that is able to fulfill the above prerequisites by adopting and integrating different solutions from our previous research endeavors discussed in [10–16]. The proposed approach successfully adapts both signature and protocol conversation incompatibilities in a context-sensitive manner. First, we adopt the metaontology proposed in [13, 14, 16] to capture the conversion semantics between application domain concepts in a context-sensitive manner using the Concepts Substitutability Enhanced Graph (CSEG) (details are given in Section 4). Second, we adopt the G^+ model [10, 16] to semantically capture the supported service conversation patterns using concepts and operations defined in CSEG (details are given in Section 5). Third, we adopt the context matching approach proposed in [12] to match conversation contexts and adopt a Sequence Mediation Procedure (SMP) proposed in [10, 15] to mediate between different exchanged messages sequences (details are given in Section 2). Fourth, the proposed approach generates the conversation patterns from the services G^+ model then matches these patterns using context matching and SMP procedures to find the operations mappings, which determine the required adapter structure, and then generates converters between different operations using the concepts substitution semantics captured in the CSEG. Finally, it builds the required adapter from the generated converters between conversation operations (i.e. messages). Each couple of conversation patterns should have their own corresponding adapter. Experiments results show that the proposed approach provides more accurate conversation adaptation results when compared against existing syntactic adapter generation approaches. We believe the proposed automated approach helps in improving business agility and responsiveness and of course establishes a solid step towards achieving the IoS vision.

Contributions Summary

We summarize paper contributions as follows.

- We propose a novel approach for automatic service conversation adapter generation that uses conditional substitution semantics between application domain concepts and operations in order to resolve conversation conflicts on the fly, in a context-aware manner.

- We propose to use a complex graph data structure, known as the Concepts Substitutability Enhanced Graph (CSEG), which is able to capture the aggregate concept substitution semantics of application domain concepts in a context-sensitive manner. We believe CSEG should be the metaontology for every application domain.
- We propose a new way for representing a service behavior state that helps us to improve the matching accuracy and propose a new behavior matching procedure known as Sequence Mediator Procedure (SMP) that can match states in a many-to-many fashion.
- We propose an approach for service operation signature adaptation using CSEG semantics.
- We propose an approach for service conversation adaptation using CSEG semantics and SMP.

The rest of the paper is organized as follows. Section 2 provides some background regarding service conversation management, conversation adaptation, application domain representation, and concepts substitutability graph. Section 3 provides the related work discussions in the areas of conversation adaptation and ontology mapping. Section 4 provides an overview on the adopted metaontology and its evolution. Section 5 proposes the adopted conversation model and describes how to extract the corresponding behavior model. Section 6 proposes the adopted approach for signature adaptation, while Section 7 proposes the adopted approach for conversation protocol adaptation. Section 8 proposes the adopted approach and algorithms for automatic adapter generation. Section 9 shows the various verification experiment and depicts results. Finally, Section 11 concludes the paper and discusses future work.

BACKGROUND

This section provides some basic principles regarding conversation management and application domain representation needed to understand the proposed automatic adapter generation approach.

Service Conversation Management

Basically, we require each web service to have two types of interfaces: a functional interface and a management interface [17, 18]. The functional interface provides the operations others services can invoke to accomplish different business objectives, while the management interface provides operations to other services in order to get some information about the service internal state that enables other services to synchronize their conversations with the service, such as operations for conversation start and end, supported conversation patterns, and supported application domain ontologies. However, to generate the right adapters, we need first to know which conversation patterns will be used before the conversation started. Therefore, we require from the consuming service to specify which conversation pattern it will use and which conversation pattern required from the consumed service before starting the conversation. This information is provided to the consuming service via the matchmaker or the service discovery agent, as depicted in Figure 1. Figure 1 indicates consuming services call conversation adapters to accomplish the required operations, and in turn the adapter invokes the corresponding operations from the service providers side and does the suitable conversions between exchanged messages.

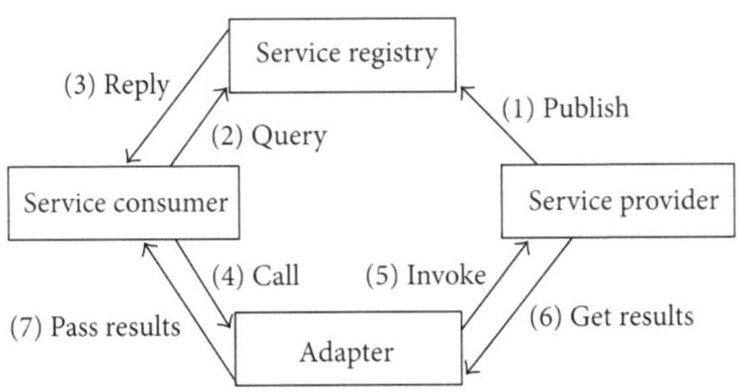

Figure 1. Service-Oriented Architecture with Adapters.

Once the consuming service knows which conversation patterns are required, it needs to communicate this information with the consumed service in order to build the suitable adapter. This could be achieved by invoking the conversation management operations defined in the management interface of the consumed service, as depicted in Figure 2. The figure indicates that the consuming service calls the consumed service management interface to specify the required conversation pattern; once it gets the confirmation, it starts the conversation and performs the conversation interactions via the conversation adapter. The adapter in turn will invoke the needed operations from the functional interface of the consumed service. This forms what we define as the *conversation management architecture*, in which each service is capable of monitoring and controling its conversations and can synchronize with other services via management interfaces.

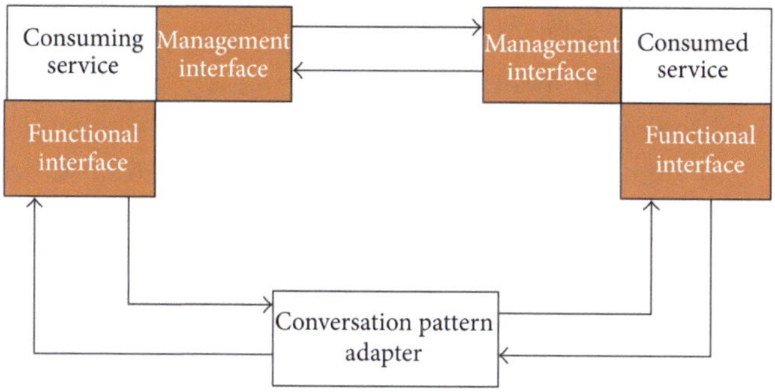

Figure 2. Conversation Management Architecture.

Specifying the required conversation patterns in advance has another benefit that services could determine the correctness of the interactions during the conversation and that if a service received any operation invocation request not in the specified conversation pattern or even in the wrong order, it could reject the request and reset the conversation. It is important to note that management interfaces should be created according to a common standard such as Web Services Choreography Interface (WSCI) [19].

Conversation Adaptation

Conversations incompatibilities are classified into signature incompatibilities and protocol incompatibilities [20, 21]. Signature incompatibilities arise when the operation to be performed by the receiving service is either not supported or not described using the required messaging schema (such as using a different number of input and output concepts, different concepts names, and different concepts types). On the other hand, protocol incompatibilities arise when interacting services expect a different ordering for the exchanged message sequences. An example for signature incompatibility occurs when one service needs to perform an online payment via operation PayOnline that has input concepts CreditCard, Amount, and Currency, and the output concept Receipt. The CreditCard concept contains the card information such as card holder name, card number, card type, and card expiration date while the Receipt concept contains the successful transaction number. Continuing with our example, another service performs online payment by invoking operation PaymentRequest that has one input concept Payment (which contains all the payment details) and one output concept Confirmation, which contains the transaction number. With purely syntactical matching between operations signatures, the first service cannot invoke the second service in spite of its ability to perform the required payment operation. An example for protocol incompatibility occurs when one service needs to perform a purchase operation and expects to send a message containing user details first and then another message containing purchase-order details, while the other interacting service is receiving the purchase-order details first and then the user details. One well-known approach for handling conversation incompatibilities is through the use of conversation adapters [2–6, 11]. A conversation adapter is the intermediate component between the interacting services that facilitates service conversations by converting the exchanged messages into messages "understandable" by the interacting services, as indicated in Figure 3.

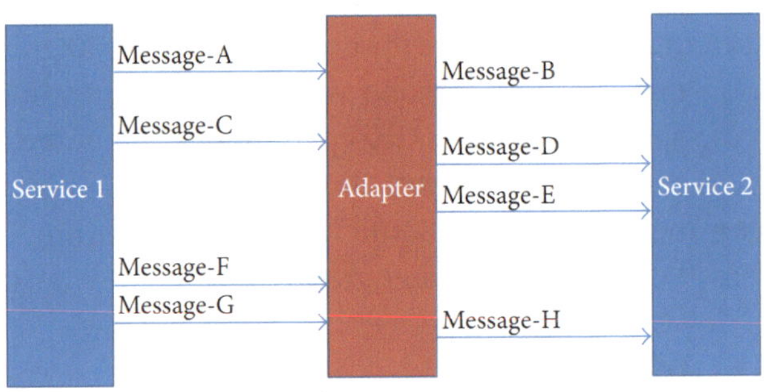

Figure 3. Conversation customization via an adapter.

Figure 3 shows an example of interactions between two incompatible services via a conversation adapter. Figure 3 shows that mapping between messages could be of different types (i.e. one-to-one, one-to-many, many-to-one, and many-to-many). For example, the adapter converts Message-A into Message-B, while it converts Message-C into the sequence consisting of Message-D and Message-E, and finally it converts the sequence consisting of Message-F and Message-G into Message-H. In other words, the adapter performs conversation customization and translation. The adapter can convert one message into another message or into another sequence of messages. It can also convert a sequence of messages into a single message or into another sequence of messages. Creating conversation adapters manually is a very time-consuming and costly process, especially when business frequently changes its consumed services, as in the IoS vision. This creates a need for automatic generation for web services conversations adapters, in order to increase business agility and responsiveness, as most of the business services will be discovered on the fly.

Automatic conversation adaptation is a very challenging task, as it requires understanding of many types of semantics including user semantics, service semantics, and application-domain semantics. All of these types of semantics should be captured in a machine-understandable format so that the middleware can use them to generate the required conversation adapters. One way of capturing different types of semantics in a machine-understandable format is using ontologies. Ontologies represent the semantic web architecture layer concerned with domain conceptualization. They are created to provide a common shared understanding of a given application domain that can be communicated across people, applications, and systems [10]. Ontologies play a very important role in automatic adapter generation, as they provide the common reference for resolving any appearing semantic conflicts. Therefore, we argue that the adopted application domain ontologies must be rich enough to capture different types of semantics in order to be able to resolve different conversation conflicts in a context-aware semantic manner. We argued in our previous work [10, 12, 13] that ontologies defined as a taxonomy style are not rich enough to capture complex types of semantics; hence more complex ontology models must be adopted. Therefore, we proposed in [10, 13, 14] to capture relationships between application domain concepts as a multidimensional hypergraph rather than a simple taxonomy; more details will be given in Section 4.

Application Domain Representation

Business systems use application domain ontologies in their modelling and design in order to standardize their models and to facilitate systems interaction, integration, evolution, and development. This is because application domain ontologies provide a common shared understanding of application domains that can be communicated across people, applications, and systems. Ontologies represent the semantic web architecture layer concerned with domain conceptualization; hence application domain ontology should include descriptions of the domain entities and their semantics, as well as specify any attributes of domain entities and their corresponding values. An ontology can range from a simple taxonomy to a thesaurus (words and synonyms), to a conceptual model (where more complex relations are defined), or to a logical theory (where formal axioms, rules, theorems, and theories are defined) [22, 23].

It is important to note the difference between application-domain ontologies and service modelling ontologies. In Application-domain ontologies the vocabulary are needed for describing the domain concepts, operations, rules, and so forth. Application-domain ontologies could be represented by existing semantic web standards, such as Web Ontology Language (OWL 2.0) [24]. On the other hand, service modelling ontologies provide constructs to build the service model in a machine-understandable format; such constructs are based on the vocabulary provided by the adopted application domain ontologies. Web Services Modelling Ontology (WSMO) [25], Web Ontology Language for Services (OWL-S) [26], and Semantic Annotations for Web Services Description Language (SAWSDL) [27] are examples of existing service modelling ontologies. The conversation modelling problem has attracted many research efforts in the areas of SOC and agent communication (such as in [28–30]). Additionally, there are some industrial standards for representing service conversations, such as Web Services Choreography Interface (WSCI) [19] for modelling service choreography and Web Services Business Process execution Language (WS-BPEL) [31] for modelling service orchestration. In this paper, we preferred to conceptually describe our conversation and applications-domain models without being restricted to any existing standards. However, any existing standard that is sufficiently rich to capture the information explained below will be suitable to represent our models. In general, there are two approaches that can be adopted for application domain conceptualization: the single-ontology approach and the multiple-ontology approach. The single-ontology approach requires every application domain to be described using only one single ontology, and everyone in the world has to follow this ontology. The multiple-ontology approach allows the application domain to be described by different ontologies such that everyone can use a different preferred ontology. As we can see both approaches have serious practicality concerns if adopted, the single-ontology approach requires reaching world consensus for every application domain conceptualization, which is far from feasible. On the other hand, the multiple-ontology approach requires determining the mappings between different ontologies in order to be able to resolve any appearing incompatibilities, which is not feasible approach when the number of ontologies describing a given application domain is big. Ontologies incompatibilities result due to many reasons. For example, two different concepts could be used to describe the same entity, or the same concept could be used to represent different entities. An entity could appear as an attribute in a given ontology and appear as a concept in other ontology, and so forth [32].

The ontology mapping process is very complex, and it requires identification of semantically related entities and then resolving their appearing differences. We argued before in [10] that any appearing conflicts should be resolved according to the defined semantics of involved application domains as well as the semantics of the involved usage contexts. When services in the same domain adopt different ontologies, ontology mapping becomes crucial for resolving conversation incompatibilities. To maintain the flexibility of application domain representation without complicating the ontology-mapping process, we propose to adopt a metaontology approach, which is a compromise between the consensus and multiple-ontology approaches, as depicted in Figure 4. Figure 4 shows the difference between the single-ontology, multiple-ontology, and metaontology approaches. Adopting a metaontology approach for application domain conceptualization provides users with the flexibility to use multiple ontologies exactly as in the multiple-ontology approach, but it requires ontology designers to follow a common structure indicating the entities and the types of semantics to be captured, which indeed simplifies the ontology mapping process. Furthermore, having a common structure ensures that all application domain ontologies capture the same types of semantics; hence we can systematically resolve any appearing conflicts; more details are given in Section 4.

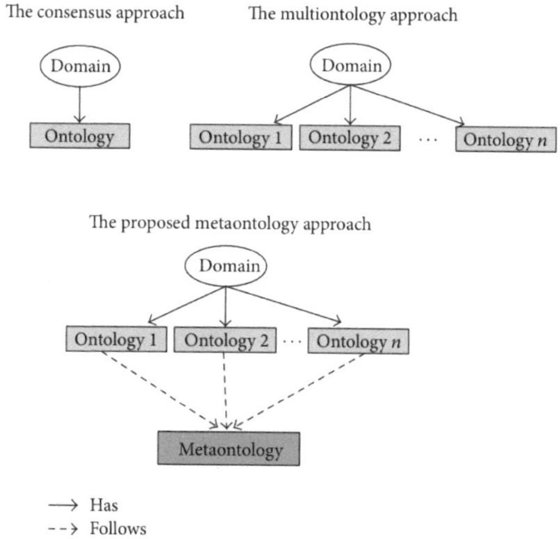

Figure 4. Approaches for application domain conceptualization.

Concepts Substitutability Graph (CSG)

As we indicated before that we adopt a metaontology approach for describing application domain ontologies. Following the separation of concerns design principle, we argue that the metaontology should consist of two layers: a schematic layer and a semantic layer [10, 13, 14]. The schematic layer defines which application domain entities need to be captured in the ontology, which will be used to define the systems models and their interaction messages. The semantic layer defines which entities semantics need to be captured in the ontology.

In the metaontology schematic layer, we propose to capture the application domain concepts and operations. *An application domain concept* is represented as a set of features defined in an attribute-value format. *An application domain operation* is represented as a set of features defined in an attribute-value format. In addition it has a set of input concepts, a set of output concepts, a set of preconditions and a set of postconditions. The preconditions are over the input concepts and must be satisfied before the operation invocation. The postconditions are over the output concepts and guaranteed to be satisfied after the operation finishes its execution.

A *conversation message* is basically represented by an application domain operation. A sequence of conversation messages constitutes a *conversation pattern*, which describes an interaction scenario supported by the service. Each conversation pattern has a corresponding *conversation context* that is represented as a set of preconditions and a set of postconditions. The context preconditions are the conditions that must be satisfied in order to be able to use the conversation pattern, while the context postconditions are the conditions guaranteed to be satisfied after the conversation pattern finishes its execution. A set of conversation patterns constitutes the *service conversation model*. In general, service conversation models are not necessarily linear. However, linear models (in which interactions are described as a sequence of operations) could be extracted from the nonlinear models (in which interactions are described as a graph of operations) by tracing all possible paths in the nonlinear model. During runtime, having linear conversation patterns provides faster performance than subgraph matching approaches, as graph paths are analyzed and enumerated (which could be performed offline) only once when a service is published and not repeated every time a matching process is needed as in subgraph matching approaches; additional details about this approach may be found in [10].

In our previous work [10, 12, 15, 16], we argued that concept substitutability should be used for concept matching that our approach maps a concept A to a concept B only if the concept A can substitute the concept B in the involved context without violating any conditions in the involved context or any rule defined in the application domain ontology. Matching concepts based on their conditional substitutability is not a straightforward process due to many reasons. First, there exist different types of mappings between concepts such as one-to-one, one-to-many, many-to-one, and many-to-many mappings, which require taking concept aggregation into consideration. For example, the Address concept could be substituted by a composite concept constituted from the Country, State, City, and Street concepts, as long as the usage context allows such substitution. Second, concept substitution semantics could vary according to the logic of the involved application domain operation; hence substitution semantics should be captured for each operation separately. Third, concept substitutability should be determined in a context-sensitive manner and not via generic schematic relations in order to be able to check if such concept substitution violates the usage context or not. In order to fulfill these requirements and capture the concept conditional substitution semantics in a machine-understandable format, we propose to use a complex graph data structure, known as the Concepts Substitutability Enhanced Graph (CSEG), which is able to capture the aggregate concept substitution semantics in a context-sensitive manner with respect to every application domain operation. Hence, we propose the metaontology semantic layer to include CSEG as one of its basic constructs.

CSEG extends the Concept Substitutability Graph (CSG) previously proposed in [10], which captures only the bilateral conditional substitution semantics between concepts. CSEG captures both bilateral as well as aggregate conditional substitution semantics of application domain concepts. Hence, we first summarize CSG graph depicted in Figure 5 and then discuss CSEG in more details. Figure 5 indicates that CSG consists of segments, where each segment captures the substitution semantics between application domain concepts with respect to a given application domain operation. For every pair of concepts the following are defined: substitutable attributes and their substitution constraints, conversion functions, and operator mapping matrices. The substitution context is represented by a set of substitution constraints that must be satisfied during substitution in order to have a valid substitution. A CSG captures the concepts functional substitution semantics at the scope level (a scope is defined by a combination of concept C_i and attribute $attr_k$ with the form $C_i.attr_k$), and not at the concept level only.

This is needed because attributes with similar names could have different semantics when they are used to describe different concepts.

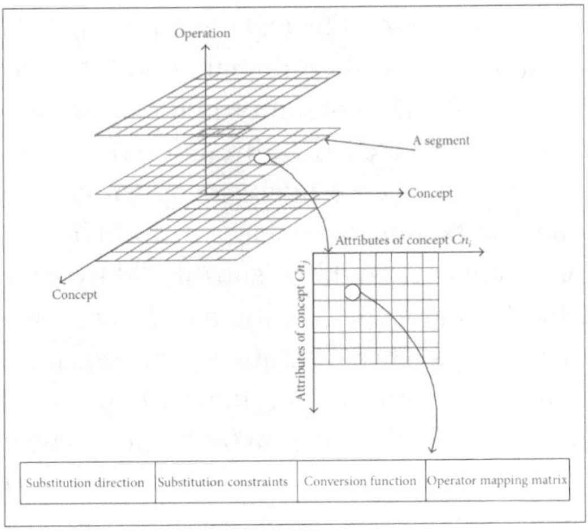

Figure 5. Concepts substitutability graph.

The proposed concept matching approach maps a concept A to a concept B only if the concept A can substitute the concept B in the involved context without violating any conditions in the involved context or any rule defined in the application domain ontology. This is done by defining the conditional substitution semantics of application domain concepts in application domain ontologies and then using such conditional semantics to resolve appearing incompatibilities by checking if the conditions representing the involved context satisfy the required substitution conditions between concepts before performing any concepts substitutions. In other words, concept mapping is conditional and not generic that concept mapping will be only valid in the contexts satisfying the required substitution conditions. Table 1 shows an example of a segment of a CSG in the logistics application domain that corresponds to the *CargoTransportation* operation. A row represents an edge in a segment in the substitutability graph. For example, the first row indicates the existence of an edge in the CSG going from the scope Cargo.Det (the cargo details) into the scope Freight.Det (the freight details). This edge has also the corresponding substitution constraint as well as conversion function. Substitutability semantics defined in CSG can be seen as conditional conversion semantics, as it allows conversion only when the substitution constraints are valid. Also it provides the details of how to perform such conversion via conversion functions and operator mapping matrices.

Table 1. A Part of CSG segment for CargoTransportation operation, adapted from [10].

From scope	To scope	Conversion function	Substitution constraints
Cargo.Det	Freight.Det	Freight.Det = Cargo.Det	
Freight.Det	Cargo.Det	Cargo.Det = Freight.Det	
Credit.Period	Payment.Type	IF (Credit.Period > 0) THEN Payment.Type = Credit ELSE Payment.Type = Cash END IF	Credit.Period ≥ 0
Payment.Type	Credit.Period	IF (Payment.Type = Credit) THEN Credit.Period ∈ {15, 30, 45, 60} ELSE Credit.Period = 0 END IF	Payment.Type ∈ {Credit, Cash}

CSG managed to provide a conditional ontology mapping approach that is able to resolve appearing concepts incompatibilities in a context-sensitive manner (more details will be given later in Section 4.1). Unfortunately, this approach cannot resolve cases requiring concept aggregation, in which one concept can substitute for a group of concepts and vice versa. For example, in the signature incompatibilities example given before, this proposed approach can resolve the conflict between the Confirmation and Receipt concepts but it cannot resolve the conflict between the input concepts, as the CreditCard, Amount, and Currency concepts need to be aggregated in order to substitute the Payment concept. To overcome such a limitation, our work in [13, 14] extended CSG graph to capture aggregate conditional substitution semantics of application domain concepts. The new graph is known as the Concepts Substitutability Enhanced Graph (CSEG). CSEG uses the notion of substitution patterns that indicate the mapping types (such as one-to-one, one-to-many, many-to-one, and many-to-many) between application domain concepts with respect to every application domain operation. More details about CSEG are given in Section 4.

RELATED WORK

This section discusses two main related areas for our work. First, we discuss related work in the area of conversation adaptation and then discuss the related work in the area of ontology mapping that shows different approaches for resolving conflicts.

Conversation Adaptation

The problem of synthesizing adapters for incompatible conversations has been studied by many researchers in the area of SOC such as the work described in [2–8, 11] and earlier in the area of component-based software engineering such as the work described in [9]. We can broadly classify these efforts into three categories: manual such as work in [2, 3, 7, 8], semiautomated such as work in [4, 9], and fully automated solutions such as work in [5, 6, 11].

The manual approaches provide users with guidelines to identify conversation incompatibilities and propose templates to resolve identified mismatches. for example, work in [7] tries to mediate between services based on signatures without taking into consideration services behavior, while work in [8] requires adapter specification to be defined manually. On the other hand, work in [3] proposes a method for creating adapters based on mismatch patterns in service composition; however they adopt a syntactic approach for comparing patterns operations, which of course cannot work if different operations sequences or different operation signatures are used.

The semiautomated approaches generate the adapters after receiving some inputs from the users regarding conversation incompatibilities resolution. The fully automated approaches generate the adapters without human intervention provided that conversation models are created according to some restrictions to avoid having signature incompatibilities and protocol deadlocks. Manual and semiautomated approaches are not suitable for dynamic environments due to the following reasons. First, they require experts to analyze the conversation models and to design solutions for incompatibilities resolution, resulting in high financial costs and time barriers for adapter development. This creates obstacles for achieving on-demand customizations and minimizes users' flexibility and agility, especially when users tend to use services for a short term and to change services frequently. Second, the number of services and users in dynamic environments is rapidly growing, which diminishes any chances for having predefined manual customizations policies. Therefore, to have on-demand conversation customizations, adapters should be created automatically. To achieve such a vision, we argue that the middleware should be enabled to automatically create such adapters to avoid any human intervention and to ensure smooth services interoperability. Unfortunately, existing automatic adapter generation approaches are strictly limited [11, 20] as they require no mismatch at the services interface level; otherwise the conversations are considered unadaptable. We argue that such syntactic approaches are not suitable for dynamic environments as service heterogeneity is totally expected in dynamic environments. Hence, conversation incompatibilities should be semantically resolved without any human intervention. Therefore, in this paper, we capture both service conversations and application domain semantics in a machine-understandable format such that we can automatically resolve appearing conflicts without human intervention; more details are given in Sections 5, 6, 7, and 8.

Ontology Mapping

Concepts incompatibilities arise when business systems adopt different application domain ontologies during their interactions. One approach for resolving such incompatibilities is using an intermediate ontology mapping approach that transforms the exchanged concepts into concepts understandable by the interacting systems. Unfortunately, existing approaches for ontology mapping are known for having limited accuracy. This is because such approaches are basically based on generic schematic relations (such as Is-a and Part-of) and ignore the involved usage context as well as the logic of the involved operation.

We argue that the ontology mapping process could be tolerated if the number of ontologies representing a given application domain is small and if there exists a systematic straightforward approach in finding the mappings between semantically related entities. Indeed, in real life, we are expecting the number of ontologies describing a given application domain to be small, as people tend to cluster and unify their understanding. Of course, we are not expecting them to cluster into one group that uses a single ontology; however it is more likely they will cluster into few groups using different ontologies. To fulfil the second requirement, many research efforts have been proposed to provide systematic straightforward approaches for ontology mapping such as [33–37]. A good survey about existing ontology mapping approaches could be found in [22]. For example, work in [33] proposed a language for specifying correspondence rules between data elements adopting a general structure consisting of general ordered labelled trees. Work in [34] developed a translation system for symbolic knowledge. It provides a language to represent complex syntactic transformations and uses syntactic rewriting (via pattern-directed rewrite rules) and semantic rewriting (via partial semantic models and some supported logical inferences) to translate different statements. Its inferences are based on generic taxonomic relationships. Work in [35] provides an ontology mapping approach based on tree structure grammar. They try to combine between internal concept structure information and rules provided by similarity languages. Work in [36] proposed a metric for determining objects similarity using hierarchical domain structure (i.e. Is-a relations) in order to produce more intuitive similarity scores. work in [37] determines the mapping between different models without translating the models into a common language. Such mapping is defined as a set of relationships between expressions over the given model, where syntactical inferences are used to find matching elements. As we can see, existing ontology mapping approaches try to provide a general translation model that can fit in all contexts using generic schematic relations (such as Is-a and Part-of relations), or depending on linguistic similarities to resolve conflicts. We argue that such approaches cannot guarantee high accuracy mapping results in all contexts [10]. Simply because such generic relations and linguistic rules could be sources of ambiguities, which are resulting from the actual domain semantics themselves. For example, the concept Islam could be a name of a religion or a name of a person and could be applied for both males and females. Another example, the Resort concept could be related to the Hotel concept using the *Is-a* relation, however, we cannot substitute the concept Resort by the concept Hotel in all context. Such ambiguities can be resolved only by taking the involved contexts into consideration. Hence, we argue that in order to guarantee the correctness of the mapping results, ontology mappings should be determined in a customized manner according to the usage context as well as the logic of the involved application domain operation (i.e. the transaction needs to be accomplished by interacting systems or users). Next section provides our approach for fulfilling these requirements.

A CONTEXT-SENSITIVE METAONTOLOGY FOR APPLICATIONS DOMAINS

Unlike CSG only capturing bilateral substitution semantics between application domain concepts, CSEG is able to capture the aggregate concept conditional substitution semantics in a context-sensitive manner to allow a concept to be substituted by a group of concepts and vice versa. This is achieved by introducing the notion of substitution patterns. CSEG consists of a collection of segments, such that each segment is corresponding to one of the application domain operations. Each segment consists of a collection of substitution patterns corresponding to the operation input and output concepts. Each substitution pattern consists of a scope, a set of substitution conditions, and a conversion function, as depicted in Figure 6.

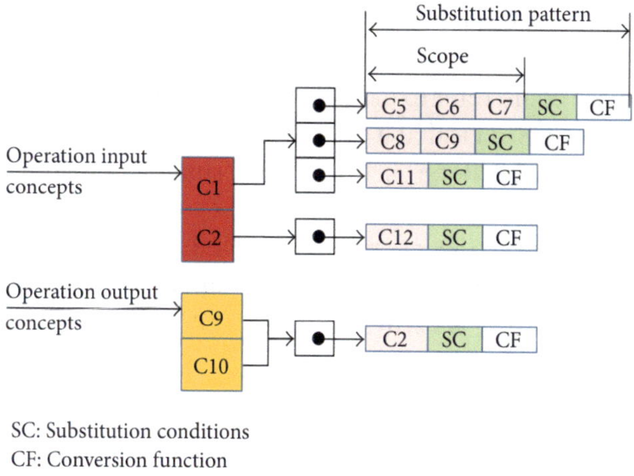

Figure 6. An example for a CSEG segment.

Figure 6 indicates the substitution patterns corresponding to a given operation input and output concepts. For example, the input concept C1 has three substitution patterns. The first pattern indicates that the concepts C5, C6, and C7 can substitute the concept.

A substitution pattern scope is a set of concepts that contains at least one application domain concept. A substitution condition is a condition that must be satisfied by the conversation context in order to consider such substitution as valid. A conversion function indicates the logic needed to convert the scope into the corresponding operation concepts or vice versa. Of course, instead of writing the conversion function code, we could refer to a service or a function that realizes it using its corresponding Uniform Resource Identifier (URI). A substitution pattern could correspond to a subset of concepts. For example, a substitution pattern for a subset of input concepts represents the set of concepts (i.e. the pattern scope) that can substitute such subset of input concepts, while a substitution pattern for a subset of output concepts represents the set of concepts that can be substituted by such subset of output concepts.

Table 2 shows an example of an input and an output substitution patterns for PayOnline operation. The input pattern indicates that CreditCard, Amount, and Currency concepts can be replaced by the Payment concept only if credit card details and the currency are not null and the amount is greater than zero. The output pattern indicates we can substitute the concept Confirmation by the concept Receipt only when conformation is not null. As we can see, substitution patterns are valid only in the contexts satisfying their substitution conditions. Of course instead of writing the conversion function code, we could refer to the URI of its realizing web service. Another advantage of using CSEG is that it systemizes the ontology mapping process, as all that needs to be done is to add the suitable substitution patterns between the ontologies concepts with respect to every domain operation. The mappings between the operations will be automatically determined based on the satisfiability of their pre- and postconditions (details are given later). In the next section, we will show how CSEG substitution patterns are used to resolve concepts incompatibilities.

Table 2. An example for operation substitution patterns.

Operation	Concepts	Scope	Conversion function	Substitution condition
PayOnline	Input: CreditCard Input: Amount Input: Currency	Payment	Payment.Method = Credit Payment.Details = CreditCard.Details Payment.Currency = Currency Payment.CreditAmt = Amount	CreditCard.Details ≠ NULL Amount >0 Currency ≠ NULL
	Output: Receipt	Confirmation	Receipt = Confirmation	Confirmation ≠ NULL

Indeed CSEG could be represented in many different ways differing in their efficiency. However, we prefer to represent it in an XML format as XML is the industrial de facto standard for sharing information. In case the XML file becomes very large, it should be compressed with a query-aware XML compressor and then accessed in its compressed format; more details about this approach could be found in [38]. For example, the substitution patterns depicted in Table 2 could be represented in XML format as shown in Listing 1.

```
<Root>
  <Operation name = "PayOnline">
     <Inputs>
        <Concepts names = {"CreditCard, Amount, Currency"}>
              <SubstitutionPattern>
                 <Scope>
                    <Concepts names = {"Payment"}/>
                 </Scope>
                 <Condition>
                       (CreditCard.Details ≠ NULL) and (Amount >0)
                       and (Currency ≠ NULL)
                 </Condition>
                 <ConversionFunction>
                          "http://example.org/URI/path/convert1.java"
                 </ConversionFunction>
              </SubstitutionPattern>
        </Concepts>
     </Inputs>
     <Outputs>
        <Concepts names = {"Receipt"}>
              <SubstitutionPattern>
                 <Scope>
                    <Concepts names = {"Confirmation"}/>
                 </Scope>
                 <Condition>
                       (Confirmation ≠ NULL)
                 </Condition>
                 <ConversionFunction>
                          "http://example.org/URI/path/convert2.java"
                 </ConversionFunction>
              </SubstitutionPattern>
        </Concepts>
     </Outputs>
  </Operation>
```

-
</Root>

Listing 1. An XML representation for a CSEG segment.

Resolving Concepts Conflicts via Substitutability Semantics

CSEG contains the information indicating which concepts are substitutable with respect to every application domain and also indicates the corresponding conversion functions. Hence, concepts mapping is determined by checking if there exists a sequence of transformations (i.e. substitution patterns) that can be carried out to transform a given concept or a group of concepts into another concept or group of concepts. This is done by checking if there exists a path between the different concepts in the CSEG segment corresponding to the involved application domain operation. Having no path indicates there is no mapping between such concepts according to the logic of the involved operation. We identify the concepts as reachable if such path is found. However, in order to consider reachable concepts as substitutable, we have to make sure that the usage context is not violated by such transformations. This is done by checking if the conditions of the usage context satisfy the substitution conditions defined along the identified path between the concepts. The concepts are considered substitutable only when the usage context satisfies such substitution conditions. Determining condition satisfiability is a tricky process, as conditions could have different scopes (i.e. concepts appearing in the conditions) and yet could be satisfiable; for example, the condition (Capital.Name = Cairo) satisfies the condition (Country.Name = Egypt) in spite of having a different scope. Unfortunately, such cases cannot be resolved by existing condition satisfiability approaches [39, 40] as they are syntactic and require the conditions to have the same scope in order to be examined.

To handle such cases, first we differentiate between the two cases as follows. When satisfiable conditions have the same scope, we identify this case as "*condition direct satisfiability*" which should be determined using existing condition satisfiability approaches. When satisfiable conditions have different scopes, we identify such case as "*condition indirect satisfiability*" which should be determined via generation of intermediate condition, as depicted in Figure 7. The figure indicates that conditions indirect satisfiability implies transforming the first condition into another intermediate condition via a transformation (T) such that the intermediate condition directly satisfies the second condition. Transformation (T) must not violate any condition in the usage context. We determine conditions indirect satisfiability between two different conditions as follows. First, we check if the conditions scopes are reachable. Second, if the scopes are reachable, we use the conversion functions defined along the path to convert the first scope into the second scope and use the obtained values to generate another intermediate condition with the same scope of the second condition. Third, we check if the intermediate condition satisfies the second condition using existing syntactic condition satisfiability approaches. Finally, if the intermediate condition satisfies the second condition, we check if the conditions of the usage context satisfy the substitution conditions defined along the path to accept such transformation. More theoretical details and proofs regarding indirect satisfiability could be found in [10]. As a conversion function could have multiple finite output values, the first condition could be transformed into a finite number of intermediate constraints at a given stage (i.e., a path edge). This forms a finite tree of the possible intermediate constraints that can be obtained from the first condition using the defined finite conversion function. When one of the intermediate constraints of the final stage directly satisfies the second condition, this implies that the first condition can indirectly satisfy the second condition, as indicated in Figure 8. More details about the condition indirect satisfiability approach and the techniques for intermediate conditions generation as well as the corresponding theoretical proofs could be found in [10].

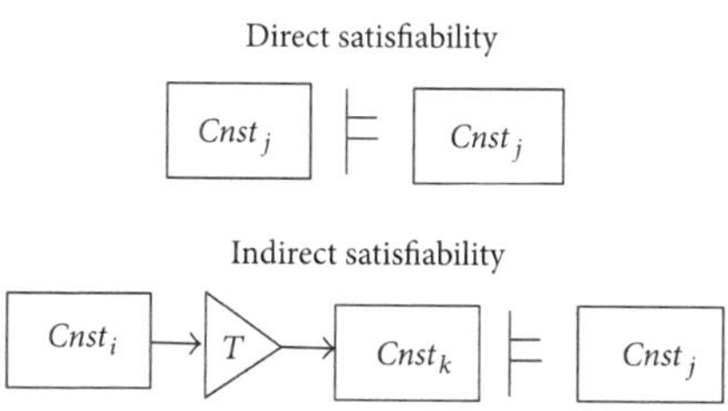

Figure 7. Direct versus indirect condition satisfiability.

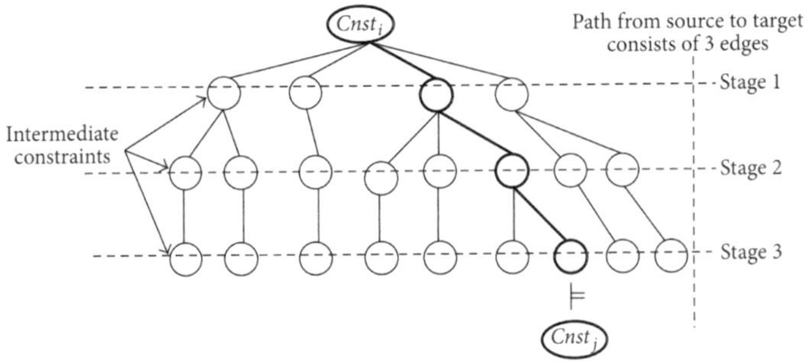

Figure 8. Generated intermediate conditions.

SERVICE CONVERSATION MODEL: G^+ MODEL

Services interactions are captured via the G^+ model [10, 12, 16, 41]. G^+ model captures services goals and interaction contexts as well as the expected interaction scenarios (depicted in Figure 9). A goal is represented by an application domain operation, a scenario is represented by a sequence of application domain operations, and a context is represented by different sets of constraints over application domain concepts (that is pre, post, and capability describing constraints), as in Table 3.

Table 3. Interaction context.

Goal	Freight movement
Preconstraints	{Freight.Det ≠ Null, Origin.Det ≠ Null, Dest.Det ≠ Null, IncoTerm.Type ∈ {FOB, EXW, CIF}}
Desc-Constraints	{Credit.Period = 15, Speciality.Type ⊆ {Motor-Vehicles}}
Postconstraints	{ShippingOrder.Status = Fulfilled, Payment.Status = Received}

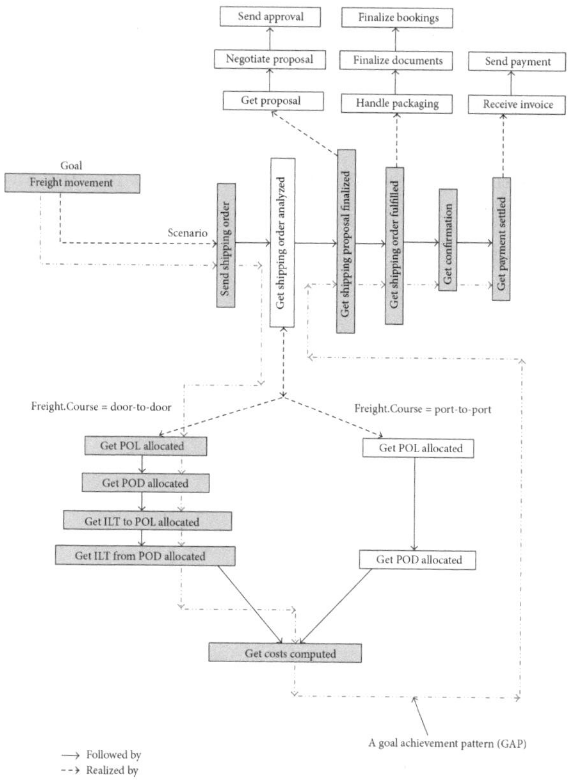

Figure 9. Interaction scenarios.

A Goal Achievement Pattern (GAP) is a global (end-to-end) snapshot of how the service's goal is expected to be accomplished, representing one given way to achieve a goal. A GAP is determined by following the path from the goal node to a leaf operation node, as depicted in Figure 9.

At the point where a branch starts, a group of constraints must be valid in order to visit that branch. This group of constraints acts as a *subcontext* for the GAP. This subcontext will be added to the preconstraints of the context of the G^+ model to form the GAP interaction context, forming what we define as a *conversation context*, and the GAP formulates what we define as a *conversation pattern*. In order to be able to semantically match conversation patterns, we need to generate their corresponding behavior models. A behavior model corresponding to a given conversation pattern is a sequence of conversation states representing the transition point between its operations. The first transition point is the point before invoking the first operation in the pattern, and the final transition point is the point after finishing the execution of the last operation in the pattern. Intermediate transition points are the points located between each pair of consecutive operation. A conversation state is represented by a set of conditions that are guaranteed to be satisfied at the corresponding transition point. For example, the conditions at the first transition point are the preconditions of the conversation context, while the conditions at a given transition point x are the ones constituted from the postconditions of the preceding operations as well as the preconditions of the conversation context that are still satisfied at x. Table 4 shows a simplified example for a sequence of operations and its corresponding state sequence. We propose a new way for representing a behavior state that helps us to improve the matching accuracy. Instead of representing the state as a set of conditions or constraints holding at a given transition point, we differentiate between these constraints based on their effect on the next operation to be executed. As we can see in Table 4, we classify state conditions in two classes: effective conditions and idle conditions. Effective conditions are the minimal subset of the state conditions that satisfies the preconditions of the following operation, while the idle conditions are the maximal subset of the state conditions that are independent from the preconditions of the following operation. This differentiation is important as states will be matched according to their effective conditions only, as including idle conditions in the state matching process just adds unnecessary restrictions as idle conditions have no effect on the invocation of the following operation [10].

Table 4. An example of a conversation pattern and its corresponding state sequence.

Conversation pattern	Preconditions	Postconditions
Conversation context	$C.a = 10$ $C.b = 20$	
OP1	$C.a \neq$ NULL	$C.a < 0$ $C.x = 5$
OP2	$C.b \neq$ NULL $C.x \neq$ NULL	$C.b > 0$
State	Effective conditions	Idle conditions
S_0	$C.a = 10$	$C.b = 20$
S_1	$C.b = 20, C.x = 5$	$C.a < 0$
S_2	$C.a < 0, C.b > 0$	

The first row in Table 4 contains the conversation context. The preconditions of the conversation context are divided into an effective condition ($C.a = 10$) and an idle condition ($C.b = 20$) to form the first state S_0, as only the condition ($C.a = 10$) is satisfying the pre-condition of operation *OP1*. After *OP1* finishes its execution, three conditions are still satisfied ($C.b = 20$), ($C.x = 5$), and ($C.a < 0$), which in turn are divided into effective and idle conditions according to the preconditions of *OP2* to form the state S_1. The process is repeated at every transition point to compute the corresponding state. We consider all the conditions of the final state as effective. Such behavior models could be constructed offline as well as on the fly, and they will be used to determine the mappings between conversation patterns to create the conversation adapter.

SIGNATURE ADAPTATION

This section discusses the proposed approach for signature adaptation. It is based on the context-sensitive conditional concept substitutability approach discussed before to resolve concepts conflicts using CSEG semantics. As a conversation message is formulated according to the vocabulary of the sending service, a chance for signature incompatibility may arise if such a vocabulary is not supported by the receiving service or the receiving service is adopting a different messaging schema. It is fortuitous that a signature incompatibility may be resolved using converters if the operations are substitutable with respect to the involved conversation context [10].

Operations mapping is determined based on their substitutability status. Operations substitutability is determined according to the satisfiability status between their pre- and postconditions, respectively, that an operation *OP1* can be substituted by an operation *OP2* when the preconditions of *OP1* satisfy the preconditions of *OP2* and the postconditions of *OP2* satisfy the postconditions of *OP1*, as indicated in Figure 10. The figure shows that operation *OP2* can substitute operation *OP1* with respect to a given conversation context. *OP2* is adapted to *OP1* by generating an input converter (which converts *OP1* inputs to *OP2* inputs) and an output converter (which converts *OP2* outputs to *OP1* outputs). Converters consist of a set of conversion functions determined according to the mapping types between involved concepts. Operations substitutability is determined according to the satisfiability status between their pre- and postconditions, respectively. An operation *OP1* can be substituted by an operation *OP2* when the preconditions of *OP1* satisfy the preconditions of *OP2* and the postconditions of *OP2* satisfy the postconditions of *OP1*. Operations substitutability is not necessarily bidirectional, as it depends on the satisfiability directions between their conditions. When we have two operations *OP1* and *OP2* with different signatures, we check if the preconditions of *OP1* satisfy the preconditions of *OP2* and the postconditions of *OP2* satisfy the postconditions of *OP1* with respect to the conversation context as discussed above. When such conditions are satisfied, the input and output converters are generated from the conversation functions defined along the identified paths. We summarize the steps needed to generate a converter that transforms a set of concept A to a set of concepts B in Algorithm 1. Generating concepts converters is not a trivial task, as it requires to capture the conversion semantics between application domain concepts, in a context-based finite manner and requires use of these semantic to determine conversion validation with respect to the conversation context. Luckily, concept substitutability graph captures concepts functional substitutability semantics in a context-based manner and provides the conversion semantics and the substitutability constraints that must be satisfied by the conversation context, in order to have a valid conversion. It is important to note that one concept can be converted to another concept in one context, and the same two concepts cannot be converted in other contexts. In order to determine whether two concepts are convertible or not, first we check if the there is a path between the two concepts in the CSEG. If there is no path this means that they cannot be convertible; otherwise, we check the satisfiability of the substitution constraints along the path with respect to the conversation context. If all the constraints are satisfied, this means that the concepts are convertible; otherwise, they are not. Details about this process are in given [10, 16].

Input: Set of concepts A, B, Conversation Context CC, and The CSEG segment S.
Output: Source code of converting A to B with respect to CC.

- **Begin**
- Get Concepts Mapping Matrix Γ between A and B.
- Create an empty class CONVERTER.
- Create a method CONVERT with A as its input, and B as its output.
- **for each** concept c_i in A **do**
  - Get corresponding mapping concept c_j using Γ.
  - **if** There exists a path from c_i to c_j in S **then**
    - **for each** edge $e_{(w,z)}$ in the path **do**
      - Create a method in CONVERTER with same signature and body of the edge conversion function $\Psi_{(w,z)}$.
      - Add $\Psi_{(w,z)}$ invocation code to CONVERT body.
    - **end for**
  - **else**
    - Return Error.
  - **end if**
- **end for**
- Return CONVERTER generated code.
- **End**

Algorithm 1. Converter generator.

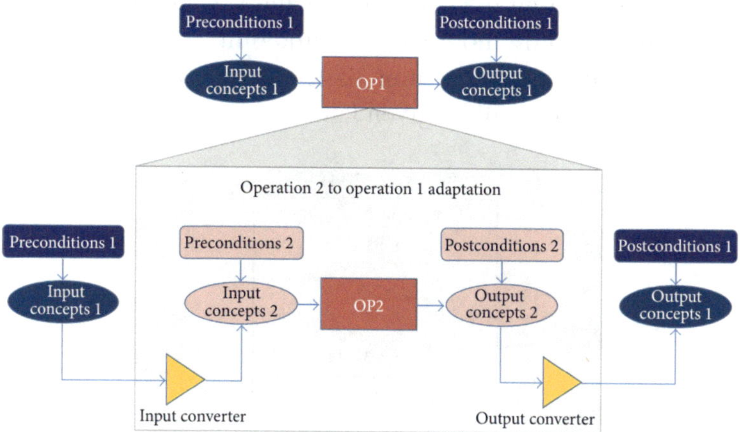

Figure 10. Signature adaptation.

To convert a list of concepts to another list, first we construct a concepts mapping matrix (Γ) between the two lists (one list is represented by the columns, and the other is represented by the rows). A matrix cell has the value 1 if the corresponding concepts are convertible in the direction needed otherwise the cell will have the value 0. When concepts are convertible, we perform the conversion process by invoking the conversion functions defined along with the edges of the path between them. So the invocation code of such conversion functions forms the source code of the needed converter. Steps of generating such converter are indicated in Algorithm 1.

The converter class will have a CONVERT method to be invoked to perform the conversion process. Of course conversion functions along the path are cascaded, so there is no need for adaptation. The converter is represented as a class with different methods corresponding to conversion functions to be invoked. Algorithm 1 requires the converter class to have a CONVERT method, which is invoked to apply the conversions. The algorithm indicates that each element in B should be reachable to a subset of A (i.e., the subset appeared as a scope in a given substitution pattern) and also indicates that the conversation context should satisfy all the substitution conditions defined along the identified path; otherwise such concept substitution is considered invalid and cannot be used. Once substitutions validity is confirmed, the determined concepts mappings are accepted, and the converter is generated. Figure 11 shows an example for a converter consisting of six conversion functions resulting from different types of concept mappings. For example, the conversion function *CF4* is responsible for converting the concepts C6 and C7 into the concept C8. In the next section, we show how the substitutability between two different sequences of operations (conversation patterns) is determined. More details about adapter generation will be given later.

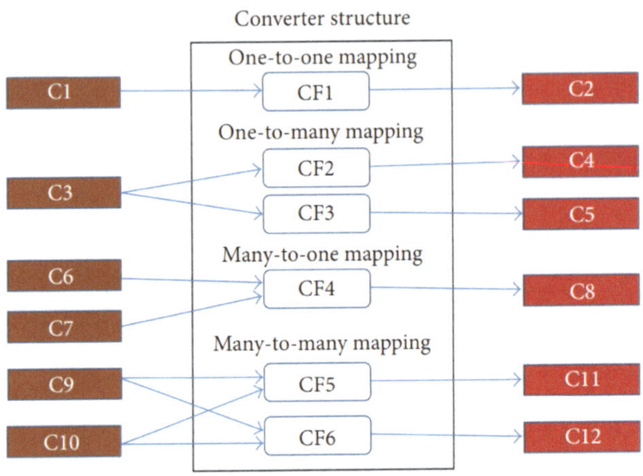

Figure 11. Converter structure.

CONVERSATION PROTOCOL ADAPTATION

One approach for semantically resolving conversation incompatibilities involves the use of the substitutability rule [10, 42] in which two conversation patterns are considered compatible when one pattern can substitute for the other without violating any condition in the corresponding conversation context. In order to determine the substitutability between two conversation patterns, we must check the substitutability of their messages (representing the operations to be performed), which in turn requires checking the substitutability of their input and output concepts. Hence, the first step needed to resolve conversation incompatibilities involves the ability to automatically determine concepts substitutability, as indicated before.

Every service supports a specific number of conversation patterns and requires other services to follow the supported patterns during their interactions. However, protocol incompatibilities could arise when the interacting services expect different ordering for the exchanged message sequences. Protocol incompatibilities may be resolved if there exists a mapping pattern between the operations appeared in the conversation patterns [10]. Conversation adapter structure is decided according to the determined operations mappings, as they specify which messages should be generated by the adapter when a given message or a sequence of messages is received. Operations mappings could be of different types such as one-to-one, one-to-many, many-to-one, and many-to-many mappings and guaranteed to exist if the conversation patterns are substitutable with respect to the conversation context [10]. Hence, to resolve protocol incompatibilities, first we must check the substitutability of the involved conversation patterns, and then find their corresponding operations mappings. Conversation patterns substitutability is determined according to the satisfiability status between their pre and postconditions corresponding to the pre and postconditions of their contexts, respectively, that a conversation pattern *CP1* can be substituted by a conversation pattern *CP2* when the preconditions of *CP1* satisfy the preconditions of *CP2* and the postconditions of *CP2* satisfy the postconditions of *CP1*. Conversation patterns substitutability is not necessarily bidirectional, as it depends on the satisfiability directions between their conditions. To find the operation mappings between two substitutable conversation patterns, we must analyze their corresponding behavior models as operations are matched semantically not syntactically. To find the operation mappings between two substitutable conversation patterns, we must find the mappings between their corresponding behavior states by grouping adjacent states in both models into matching clusters. A state S_x matches a state S_y only when the effective conditions of S_x satisfy the effective conditions of S_y. A cluster CL_x matches another cluster CL_y when the state resulting from merging CL_x states matches the state resulting from merging CL_y states, as depicted in Figure 12.

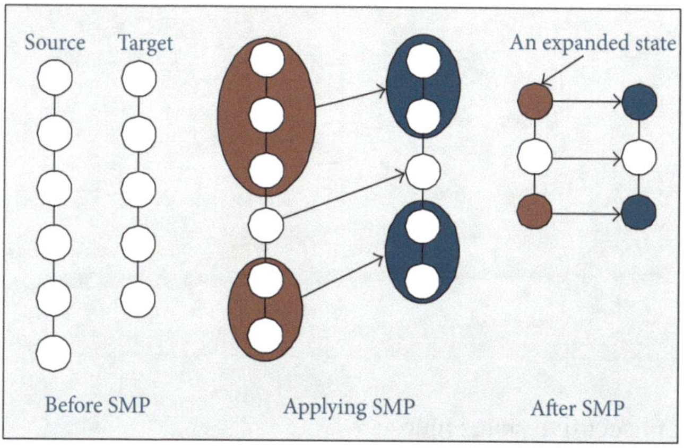

Figure 12. State clustering effect.

The figure shows the initial state sequences, the state clusters, and the final state sequences. Merging two consecutive states S_x, S_{x+1} in a given behavior model to form a new expanded state S_m means that we performed a virtual operation merge between OP_{x+1}, OP_{x+2} to obtain a coarser operation OP_m, as depicted in Figure 13. The figure indicates that the input of OP_m is formulated from the sets of concepts A and B, and its output is formulated from the sets of concepts C and E. As we can see, the set of concepts D does not appear in OP_m signature and consequently will not appear in S_m conditions. Such information hiding provides a chance for having matching states. S_m is computed by reclassifying the effective and idle conditions of S_x into new sets of effective and idle conditions according to the preconditions of OP_m. For example, by merging states S_0, S_1 shown in Table 4, the resulting S_m will have the set $(C.a = 10)$, $(C.b = 20)$ as its effective conditions, and the set $(C.a < 0)$, $(C.b > 0)$ as its idle conditions. As we can see, conditions on $C.x$ do not appear in S_m. We use a Sequence Mediation Procedure (SMP) (discussed in the next subsection) to find such matching clusters. SMP starts by examining the initial states in both sequences, then moves forward and backward along the state sequences until matching clusters are formed and the corresponding operations mappings are determined. The highest level of abstraction that could be reached occurs when all the conversation pattern operations are merged into one operation. As the number of the states is quite small, the backtracking approach does not diminish the performance.

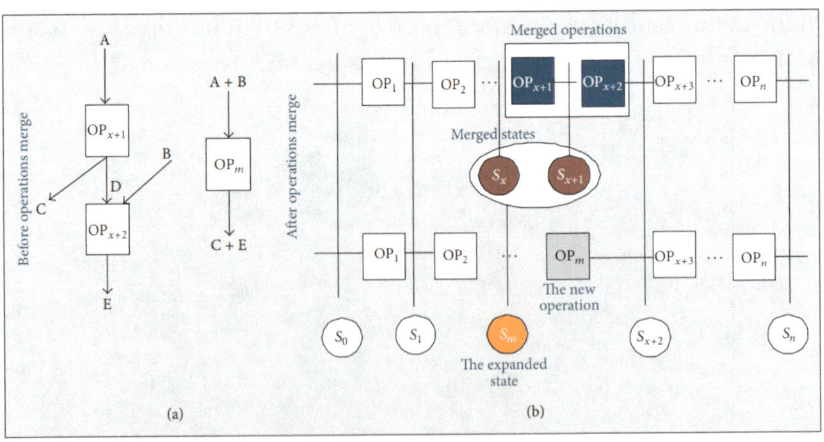

Figure 13. Consecutive states merge.

Conversation Pattern Matching

Sequence Mediator Procedure (SMP) is a procedure used to match different state sequences. Such state sequences are generated from the GAPs (conversation patterns) to be matched. Each transition point x between two consecutive operations Op_x and Op_{x+1} in a given GAP is represented by a behavior state. Such state is captured via constraints active at this transition point x. A constraint at a transition point x is considered effective if it needs to be true in order to invoke Op_{x+1}. A state S_x matches a state S_y when its effective constraints subsume the effective constraints of S_y (theoretical models and proofs could be found in [10]). SMP does not require the state sequences to have the same number of states in order to be matched; however, it applies different state expansion operations to reach to a matching case if possible. When a state is expanded, it could be merged with either its successor states (known as Down Expansion and denoted as $\Downarrow_G$) or its predecessor states (known as Reverse Expansion and denoted as $\Uparrow_G$), where G is the conversation goal, setting the conversation context. SMP uses these different types of state expansions to recluster unmatched state sequences to reach a matching case. This reclustering operation could happen on both state sequences, as indicated in Figure 12. Merging two consecutive states in a given state sequence means that their successor operations are merged to form a new operation, as depicted in Figure 13.

Figure 13 shows that the states S_x and S_{x+1} are merged forming a new state S_m, which is computed as if there is a new operation Op_m in the sequence replacing the operations Op_{x+1} and Op_{x+2}. The input of Op_m is the union between the sets of concepts A and B, its output is the union between the sets of concepts C and E, while the set of concepts D will not appear neither in Op_m input nor in Op_m output.

SMP tries to recluster both state sequences until it reached into an organization that has both sequences matched; if such organization is reached, SMP announces that it found a match and provides the mappings between the resulting clusters. Such mappings are provided in the form of an Operations Mapping Matrix (denoted as Θ) that indicated which operations in a source sequence are mapped to which operations in a target sequence, as indicated in Table 5.

Table 5. Example of a conversation patterns mapping matrix Θ.

Mapping type	CP_x	CP_y
1 to 1	$OP_{(x,1)}$	$OP_{(y,1)}$
1 to many	$OP_{(x,2)}$	$OP_{(y,2)}, OP_{(y,3)}$
Many to 1	$OP_{(x,3)}, OP_{(x,4)}$	$OP_{(y,4)}$
Many to many	$OP_{(x,5)}, OP_{(x,6)}$	$OP_{(y,5)}, OP_{(y,6)}, OP_{(y,7)}$

Once obtaining the operations mapping matrix from SMP, only matched GAPs that require no change in the requested conversation pattern will be chosen, and therefore their corresponding adapters could be generated. SMP starts by examining the first state of the "source target" against the first state of the "target sequence." When the source state matches the target state, SMP applies Algorithm 2 to handle the matching case. When a source state matches a target state, SMP checks the target down expansion to match as many target states as possible with the source state (lines 2 and 3).

```
Input: S[i] (the ith source state), T[j] (the jth target
state), G (the required goal)
Output: Return true when sequences re-clustering is finished.
(1)  Begin
(2)  if (T[j] ⇓G S[i]) then
(3)     Expand T[j]
(4)  end if
(5)  Mark T[j], S[i] as matching peers
(6)  if (j < ‖T[]‖) then
(7)     if (i < ‖S[]‖) then
(8)        Apply Algorithm 4 over (S[i + 1], T[j + 1], G)
(9)     else
(10)       Apply Algorithm 4 over (S[i], T[j + 1], G)
(11)    end if
(12) else
(13)    Return true
(14) end if
(15) End
```

Algorithm 2. SMP matching case handling.

In Algorithm 3, SMP aims to find a matching source cluster for every target state. However, when a source state fails to match a target state, SMP checks if the source state could be down expandable (lines 5–7). If this checking fails too, SMP checks whether the source state could be reverse expanded with respect to the target state (lines 9–11). When a source state cannot be expanded in either directions, SMP tries the successor source states to match the target state using the down and reverse source expansion scenarios (line 16). It stores the unmatched source state for backtracking purposes (line 13). When a target state cannot be matched to any source state, SMP tries reverse expanding the target state to find a match for it (lines 18–20); when that fails this target state is considered unmatched, and the next target state will be examined (lines 22-23). The algorithm continues even if unmatched target state is reached, as this unmatched state could be merged with any of its successors if they are going to be reversely expanded.

```
Input: S[i] (the ith source state), T[j] (the jth target
state), G (the required goal)
Output: Return true when sequences re-clustering is finished.
(1)  Begin
(2)  if (S[i] ⊵G T[j]) then
(3)     Apply Algorithm 3 over (S[i], T[j], G)
(4)  else
(5)     if (S[i] ⇓G T[j]) then
(6)        Expand S[i]
(7)        Apply Algorithm 3 over (S[i], T[j], G)
(8)     else
(9)        if (S[i] ⇑G T[j]) then
(10)          Apply CRO over S[i]
(11)          Apply Algorithm 3 over (S[i], T[j], G)
(12)       else
(13)          BackTrack = S[i]
(14)          if (j < ||T[]||) then
(15)             if (i < ||S[]||) then
(16)                Apply Algorithm 4 over (S[i + 1], T[j], G)
(17)             else
(18)                if (T[j] ⇑G S[i]) then
(19)                   Apply CRO over T[j]
(20)                   Apply Algorithm 3 over (S[i], T[j], G)
(21)                else
(22)                   Mark T[j] as Unmatched
(23)                   Apply Algorithm 4 over (BackTrack,
                       T[j + 1], G)
(24)                end if
(25)             end if
(26)          else
(27)             Merge unmatched source states with their
                 predecessors.
(28)             Return true
(29)          end if
(30)       end if
(31)    end if
(32) end if
(33) End
```

Algorithm 3. Sequence mediator procedure (SMP).

AUTOMATIC ADAPTER GENERATION

Each service has different conversation patterns (generated from its G^+ model) that could use to interact with other services. Such conversation patterns could be matched by one service or by many different services, as depicted in Figure 14.

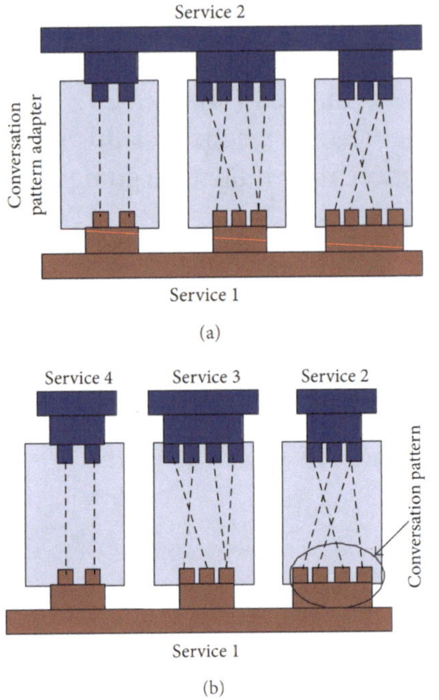

Figure 14. Service conversation patterns adapters.

Figure 14 indicates that each conversation pattern should have its own adapter. Once the required conversation patterns are specified via the management interfaces (as indicated in Section 2.1), the adapter generation process is started. The outcome of the adapter generation process is the source code for the adapter class that consists of the methods to be invoked by the consuming services. The body of these methods consists of the invocation code for the consumed service operations and the invocation code for the corresponding converters. First, we determine the required adapter structure then generate the source code for the adapter and the needed converters. Once the class adapter is generated, it is compiled, and the corresponding WSDL file is generated, in order to expose the adapter class as a service, which could be easily invoked by the consuming service. The details are discussed in the following subsections.

Once two services "decide" to interact with each other, they notify the middleware such that it identifies their substitutable conversation patterns and generates the corresponding conversation adapters. The middleware notifies back the services with the identified substitutable patterns such that each service knows which patterns should be used during the conversation [11]. Once a conversation pattern CP_x is identified as substitutable with a conversation pattern CP_y, the middleware performs the following steps (similar to Algorithm 1) to generate their corresponding conversation adapter, which transforms CP_x incoming messages into CP_y outgoing messages. First, it generates an adapter class with methods corresponding to CP_x operations (incoming messages), such that each method consists of a signature (similar to the signature of the corresponding incoming message) and an empty body (which will be later containining the code for generating the corresponding CP_y outgoing messages). Second, it determines the operations mappings between CP_x and CP_y and then uses these mappings to construct the generation code for the outgoing message. Table 5 provides an example for a CP_x conversation pattern that is substituted by a conversation pattern CP_y, showing the corresponding operations mappings.

Figure 15 shows the corresponding adapter structure. Signature incompatibilities are handled by generating the suitable input and output converters. The outgoing message generation code is constructed as follows.

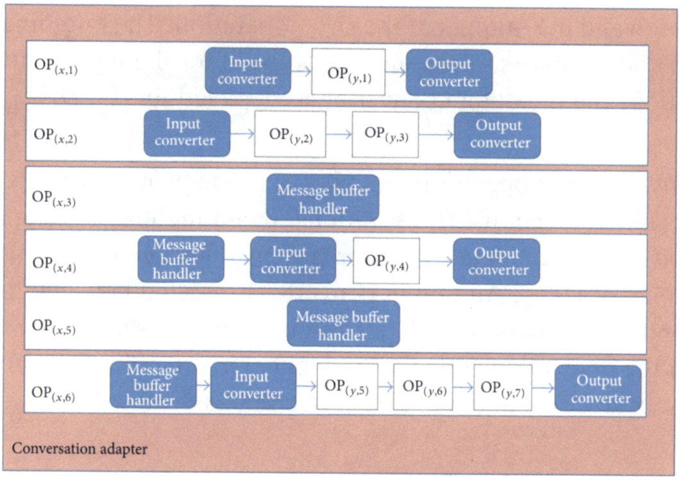

Figure 15. Conversation adapter structure for patterns in Table 5.

In one-to-one operations mappings, one CP_x operation matches one CP_y operation. The input converter is created between the inputs of the CP_x operation and the inputs of CP_y operation. The output converter is created between the outputs of the CP_y operation and the outputs of CP_x operation. The outgoing message generation code consists of the invocation code for the input converter, the CP_y operation, and the output converter, as depicted for Op_{x+1} in Figure 15.

In one-to-many operations mappings, one CP_x operation matches subsequence of CP_y operations. An input converter is created between the inputs of the CP_x operation and the inputs of the OP_{my} operation (resulting from merging the CP_y subsequence). An output converter is created between the outputs of the OP_{my} operation and the outputs of the CP_x operation. The outgoing message generation code consists of the invocation code for the input converter, the CP_y subsequence (multiple messages), and the output converter, as depicted for $_{(x,2)}$ in Figure 15.

In many-to-one operation mapping, a subsequence of CP_x operations matches one CP_y operation. The outgoing message cannot be generated unless all the operations of the CP_x subsequence are received. Hence, before generating the outgoing message, all the incoming messages should be buffered until the last message is received. This is achieved by using a message buffer handler. An input converter is created between the inputs of OP_{mx} (resulting from merging the CP_x subsequence) and the inputs of the CP_x operation. An output converter is created between the outputs of CP_y operation and the outputs of the OP_{mx} operation. The outgoing message generation code consists of the invocation code for the input converter, the CP_y operation, and the output converter, as depicted for $_{(x,3)}$, $OP_{(x,4)}$ in Figure 15.

In many-to-many operation mapping, a subsequence of CP_x operations matches a subsequence of CP_y operations. Incoming messages are buffered as indicated earlier. An input converter is created between the inputs of OP_{mx} and the inputs of OP_{my}. An output converter is created between the outputs of OP_{my} and the outputs of OP_{mx}. The outgoing message generation code consists of the invocation code for the input converter, the CP_y subsequence (multiple messages), and the output converter, as depicted for $_{(x,5)}$, $OP_{(x,6)}$ in Figure 15.

Once the adapter class is successfully generated, the middleware can reroute the conversation messages to the adapter service (corresponding to the generated class) to perform the needed conversation customizations. Invoking operations from existing services is a straightforward simple task, however generating the inputs and outputs converters is not, as we need to find the mappings between the concepts and their conversion functions. Steps of generating such adapter class are indicated in Algorithm 4. Algorithm 4 simply starts by creating an empty class then adds methods to this class with the same signatures of the consuming service conversation pattern. For each created method, it gets the sequence of operations realizing the method with the help of the operation mapping matrix (Θ). Then, it creates the concepts converters by calling the ConverterGenerator function (depicted in Algorithm 1) with the proper parameters. Finally, it adds the converter generated code to the adapter if no error resulted during the generation.

```
Input: CPx (consuming GAP), CPy, Operations Mapping
Matrix Θ.
Output: Source code of the adapter between CPx and
CPy.
  Begin
  Create an empty class ADAPTER.
  for each operation Op(x,i) in GAPx do
     Create a method in ADAPTER with same signature
     as Op(x,i).
     Get realizing operations subsequence OPSeq from
     CPy using Θ.
     Call ConverterGenerator (Op(x,i) Inputs, OPSeq Inputs).
     // Algorithm 2
     if Converter Code generated successfully then
        Add Inputs Converter Code to method body.
        Add invocation Code for OPSeq to method body.
        Call ConverterGenerator (OPSeq Output, Op(x,i)
        Outputs).
        if Converter Code generated successfully then
           Add Outputs Converter Code to method body.
        else
           Return Error.
        end if
     else
        Return Error.
     end if
  end for
  Return ADAPTER generated code.
  End
```

Algorithm 4. Adapter automatic generator.

In case the algorithm returns error, this means conversation adaptation cannot be performed; therefore, these services cannot talk to each other on the fly, and a manual adapter needs to be created to enable such conversation.

EXPERIMENTS

This section provides simulation experiments used for verifying the proposed approaches. First, we start by the verifying experiments for the proposed signature adaptation approach; then we introduce the verifying experiments for the proposed conversation adaptation approach.

Signature Adaptation

To verify the proposed signature adaptation using conditional ontology mapping approach, we use a simulation approach to compare between the proposed approach and the generic mapping approach that adopts only Is-a relations to match signature concepts (both input and output concepts). The used comparison metric is the F-measure metric.

F-measure metric combines between the retrieval precision and recall metrics and is used as an indicator for accuracy that approaches with higher values which means that they are more accurate. F-measure is computed as $(2 * Precision * Recall)/(Recall + Precision)$. The experiment starts by generating two random sets of independent concepts (representing two different ontologies). One set will be used as the original dataset, and the second one will be used as a query set. For each concept in the query set, we randomly generate an Is-a relation to a corresponding concept in the original dataset (i.e., mapping using Is-a relation). For each pair of concepts having an Is-a relation, we generate a corresponding substitution pattern in the CSEG. For simplicity, the substitution pattern is generated as follows. The scope is equal to the original dataset concept. The substitution condition is generated as greater than condition with a randomly generated integer number (e.g., C1 > 10). The conversion function is just an equality function (e.g., C1 = C2). From the generated set of concepts, we generate a random signature (i.e., a random operation) by randomly choosing a set of input concept and a set of output concepts. For each generated signature in the query set, we generate a corresponding context. For simplicity, the context will consist of one equality condition with a randomly generated integer number (e.g., C1 = 20). Hence, not all the substitution patterns defined in the CSEG will be valid according to the generated contexts. We submit the query set to the two approaches to find matches in the original dataset, and based on the retrieved concepts the F-measure is computed. Figure 16 depicts the results. As we can see, the generic approach ignores the contexts and retrieves the whole original dataset as answers, which results in low F-measure values, while the proposed approach succeed to reach 100%.

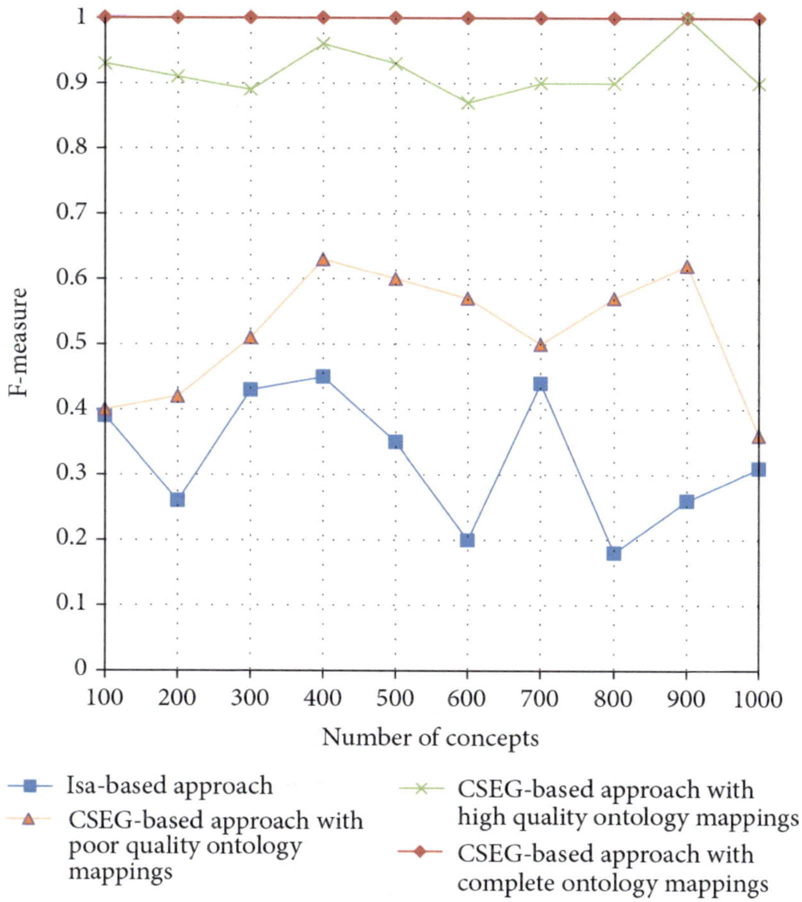

Figure 16. Signature adaptation approaches comparison.

However, this result could be misleading, as the experiment is done with complete CSEG patterns. In practice, an ontology designer may skip some substitution patterns when defining CSEG patterns. Therefore, the proposed approach will not be able to resolve the cases with missing patterns. In other words, the accuracy of the proposed approach mainly depends on the quality of the defined ontology mappings. To show such effect, we repeated the experiment except that we store only a portion of the generated substitution patterns. A high-quality ontology mapping means that up to 25% of the generated patterns are missing. A low-quality ontology mapping means that from 50% to 80% of the generated patterns are missing. Then, we compute the F-measure values for each case. Results are depicted in Figure 16. As we can see, when low-quality mappings are used, the proposed approach accuracy is negatively affected. The worst case complexity of the proposed approach is $O(n * m^{\|p\|})$, where n is the average number of substitution patterns of domain operations, m is average number of possible outputs generated from conversion functions, and $\|p\|$ is the length of the path p linking between mapped concepts (details could be found in [10]). The factor $m^{\|p\|}$ is the cost endured to find a sequence of generated intermediate conditions to indirectly match two conditions. However, in practice, n, m, and p are expected to be small; hence, we argue that the performance of the proposed approach is acceptable.

Conversation Adaptation

Currently, there is no standard datasets for service conversations. Hence, to verify the proposed approach for automated adapter generation, we follow a simulation approach similar to the one used in [10]. The proposed simulation approach compares three approaches for automated adapter generation. The first approach is a syntactic approach that requires no changes at the services interface level of the operations. It cannot resolve any semantic differences. We use this approach as a benchmark for our works. The second approach is our approach proposed in [11] that uses bilateral concept substitution to resolve signatures incompatibilities. We use this approach to show the effect of not supporting concept aggregation. The third approach is the approach proposed in this paper that uses aggregate concept conditional substitution semantics to resolve signature incompatibilities. The used comparison metric is the adaptation recall metric. It is similar to the retrieval recall metric and is computed as the percentage of the number of adapted conversation patterns (i.e., the ones that have a successfully generated conversation adapter) with respect to the actual number of the adaptable conversation patterns in the dataset.

The experiment starts by generating a random set of independent conversation patterns, for which each pattern has a unique operation, and each operation has different input and output concepts. A query set is generated as a copy of the original set. The query set is submitted to the three adaptation approaches in order to generate the adapters between the query set patterns and the original set patterns. As the two sets are identical and the conversation patterns are independent, each pattern in the query set will have only one substitutable pattern in the original set (i.e., its copy). The second phase of the experiment involves the gradual mutation of the query set and submission of the mutated query set to the three approaches, and then we check the number of adapters generated by each approach to compute the adaptation recall metric. The mutation process starts by mutating 10% of the query set and then continues increasing the percentage by 10% until the query set is completely mutated. The value of 10% is an arbitrary percentage chosen to show the effect of semantic mutations on the approach. At each step, the adaptation recall metric is computed for the three approaches. The mutation process is performed by changing the signatures of the operations with completely new ones. Then the corresponding substitution patterns are added between the old concepts and the new concepts in the CSEG. The number of concepts in a substitution pattern is randomly chosen between 1 (to ensure having cases of bilateral substitution) and 5 (an arbitrary number for concept aggregation). For simplicity, conversion functions are generated by assigning the old values of the concepts to the new values of the concepts, and the substitution conditions are generated as not null conditions.

The experiment results are depicted in Figure 17(a). The figure shows that the syntactic approach could not handle any mutated cases, as it cannot resolve signature incompatibilities. Hence, its corresponding adaptation recall values drops proportionally to the mutation percentage. The bilateral substitution approach only solved the cases with substitution patterns having one concept in their scopes, while it could not solve the cases with substitution patterns having more than one concept in their scopes (i.e., cases representing concept aggregation). Hence, its corresponding adaptation recall values are higher than the values of the syntactic approach (as it solved bilateral substitution cases) and lower than the values of the proposed approach (as it could not resolve cases require concept aggregation). On the other hand, the proposed approach managed to generate adapters for all the mutated cases, providing a stable adaptation recall value of one.

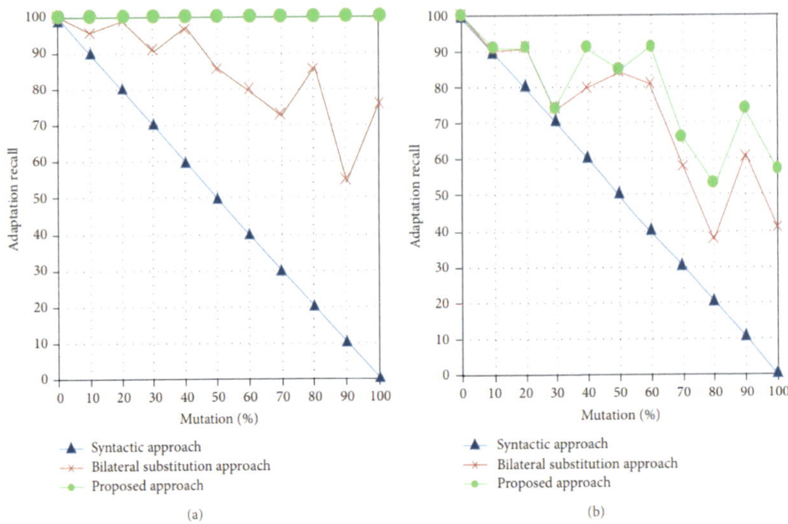

Figure 17. Conversation adaptation approaches comparison: (a) with complete substitution patterns (b) with missing substitution patterns.

However, these results could be misleading, as the experiment is performed with complete CSEG patterns. In practice, an ontology designer may skip some substitution patterns when defining CSEG patterns, depending on his/her domain knowledge and modelling skills. Therefore, the proposed approach will not be able to resolve the cases with missing patterns. In other words, the accuracy of the proposed approach mainly depends on the quality of the defined ontology mappings. To show such effects, we repeated the previous experiment except that we store only a random portion (0%–100%) of the generated substitution patterns. The results are depicted in Figure 17(b). The figure shows that the proposed approach could not resolve all the mutation cases due to missing substitution patterns; however, it succeeds in adapting more cases than the other approaches.

The worst case complexity of the proposed approach is (n^3), where n is the number of operations in a conversation pattern (a theoretical proof can be found in [10]). In practice, n is expected to be small; hence, we argue that the performance of the proposed approach is acceptable, especially when compared to the time needed for manually developing conversation adapters (which could require several days). We will focus our future research efforts to optimize the proposed algorithms and apply them to real-life application domains, which require involvement of application-domain experts to precisely define the needed CSEG.

CASE STUDY

Given a service S_1 and S_2 with GAPs depicted in Table 6. In order to find wether these GAPs are matching or not, we have to extract the behavior models of each GAP. Let us assume the operations definitions as in Tables 7 and 8. Hence, extracted behavior models will be as listed in Table 9. Assuming that we have a CSG segment as depicted in Table 10, and applying the SMP procedure, we will find the matching behavior models states as indicated in Table 11. We can see from the table that S_1 operations *Send—Shipping—Order*, *Get—POL—Allocated*, *Get—POD—Allocated*, *and Get—Costs—Computed* are matching the operation *Send—Cargo—Details* of S_2. Hence, the corresponding adapter method is created; accordingly, the rest of the adapter methods is created by the mappings given in Table 11.

Table 6. An example of two matching GAPs

	S_2 GAP	S_1 GAP
Precon-straints	{Cargo.Det = 1000 cars, Cargo.POL = Mel-bourne-Australia, Cargo.POD = Alexan-dria-Egypt, Cargo.Course = Port-To-Port, IncoTerm.Type = CIF}	{Freight.Det $\neq$ Null, Origin.Det $\neq$ Null, Dest.Det $\neq$ Null, Freight.Course = Port-To-Port, IncoTerm.Type $\in$ {FOB, EXW, CIF}}
Desc-Con-straints	{Payment.type = Credit, Speciality.Type = Motor-Vehicles}	{Credit.Period = 15, Speciality.Type $\subseteq$ {Motor-Vehicles, Dangerous-Cargo}}
Postcon-straints	{Cargo.Status = Ac-complished}	{ShippingOrder.Status = Fulfilled, Payment.Status = Re-ceived}
Goal	Cargo transportation	Freight movement
Operation sequence	(1) Send-Cargo-Details	(1) Send-Shipping-Order
	(2) Get-Offer	(2) Get-POL-Allocated
	(3) Negotiate-Offer	(3) Get-POD-Allocated
	(4) Accept-Offer	(4) Get-Costs-Computed
	(5) Execute-Offer	(5) Get-Proposal
	(6) Send-Payment	(6) Negotiate-Proposal
		(7) Send-Approval

		(8) Handle-Packaging
		(9) Finalize-Documents
		(10) Finalize-Bookings
		(11) Get-Confirmation
		(12) Receive-Invoice
		(13) Send-Payment

Table 7. Part of the ontology operations' definitions adopted by S_2.

Operation	**Preconstraints**	**Postconstraints**
Send-Cargo-Details	{Cargo.Det ≠ Null, Cargo.POL ≠ Null, Cargo.POD ≠ Null, IncoTerm.Type ≠ Null}	{Cargo.Status = Received}
Get-Offer	{Cargo.Status = Received, Cargo.Course ≠ Null}	{Offer.Status = Sent}
Negotiate-Offer	{Offer.Status = Sent}	{Offer.Status = Approved}
Accept-Offer	{Offer.Status = Approved}	{Offer.Status = Accepted}
Execute-Offer	{Offer.Status = Accepted}	{Offer.Status = Executed}
Send-payment	{Offer.Status = Executed}	{Cargo.Status = Accomplished}

Table 8. Part of the ontology operations' definitions adopted by S_1.

Operation	**Preconstraints**	**Postconstraints**
Send-Shipping-Order	{Freight.Det ≠ Null, Origin.Det ≠ Null, Dest.Det ≠ Null, Freight.Course ≠ Null, IncoTerm.Type ≠ Null}	{ShippingOrder.Status = Created}
Get-Shipping-Order-Analyzed	{ShippingOrder.Status = Created}	{ShippingOrder.Status = Analyzed}
Get-POL-Allocated	{ShippingOrder.Status = Created}	{POL.Status = Allocated}
Get-POD-Allocated	{POL.Status = Allocated}	{POL.Status = Allocated, POD.Status = Allocated}
Get-ILT-To-POL-Allocated	{POL.Status = Allocated}	ILT.ToStatus = Allocated

Get-ILT-From-POD-Allocated	{POD.Status = Allocated}	ILT.FromStatus = Allocated
Get-Costs-Computed	{POL.Status = Allocated, POD.Status = Allocated}	{ShippingOrder.Status = Analyzed}
Get-Shipping-Proposal-Finalized	{ShippingOrder.Status = Analyzed}	{ShippingOrder.Status = Approved}
Get-Proposal	{ShippingOrder.Status = Analyzed}	{Proposal.Status = Sent}
Negotiate-Proposal	{Proposal.Status = Sent}	{Proposal.Status = Approved}
Send-Proposal	{Proposal.Status = Approved}	{ShippingOrder.Status = Approved}
Get-Shipping-Order-Fulfilled	{ShippingOrder.Status = Approved}	{ShippingOrder.Status = Executed}
Handle-Packaging	{ShippingOrder.Status = Approved}	{Packaging.Status = Accomplished}
Finalize-Documents	{Packaging.Status = Accomplished}	{Documentation.Status = Accomplished}
Finalize-Bookings	{Documentation.Status = Accomplished}	{ShippingOrder.Status = Executed}
Get-Confirmation	{ShippingOrder.Status = Executed}	{ShippingOrder.Status = Confirmed}
Get-Payment-Settled	{ShippingOrder.Status = Confirmed}	{ShippingOrder.Status = Fulfilled, Payment.Status = Received}
Receive-Invoice	{ShippingOrder.Status = Confirmed}	{ShippingOrder.Status = Pending}
Send-Payment	{ShippingOrder.Status = Pending}	{ShippingOrder.Status = Fulfilled, Payment.Status = Received}

Table 9. S_1 and S_2 behavior models.

S_1 Behavior model	
S_0	⟨{*Freight.Det* ≠ Null, *Origin.Det* ≠ Null, *Dest.Det* ≠ Null, *Freight.Course* = Port-to-Port, IncoTerm.Type ∈ {FOB, EXW, CIF}}, {} ⟩
S_1	⟨{*ShippingOrder.Status* = Created}, {} ⟩
S_2	⟨{*POL.Status* = Allocated}, {} ⟩

S_3	⟨{*POL.Status* = Allocated, *POD.Status* = Allocated}, {} ⟩
S_4	⟨{*ShippingOrder.Status* = Analyzed}, {} ⟩
S_5	⟨{*Proposal.Status* = Sent}, {} ⟩
S_6	⟨{*Proposal.Status* = Approved}, {} ⟩
S_7	⟨{*ShippingOrder.Status* = Approved}, {} ⟩
S_8	⟨{*Packaging.Status* = Accomplished}, {} ⟩
S_9	⟨{*Documentation.Status* = Accomplished}, {} ⟩
S_{10}	⟨{*ShippingOrder.Status* = Executed}, {} ⟩
S_{11}	⟨{*ShippingOrder.Status* = Confirmed}, {} ⟩
S_{12}	⟨{*ShippingOrder.Status* = Pending}, {} ⟩
S_{13}	{*ShippingOrder.Status* = Fulfilled,
	Payment.Status = Received}, {} ⟩
S_2 Behavior model	
S_0	⟨{*Cargo.Det* = 1000 Cars, *Cargo.POL* = Melbourne-Australia, *Cargo.POD* = Alexandria-Egypt, IncoTerm.Type = FOB}, {*Cargo.Course* = Port-to-Port}⟩
S_1	⟨{*Cargo.Course* = Port-to-Port, *Cargo.Status* = Received}, {} ⟩
S_2	⟨{*Offer.Status* = Sent}, {} ⟩
S_3	⟨{*Offer.Status* = Approved}, {} ⟩
S_4	⟨{*Offer.Status* = Accepted}, {} ⟩
S_5	⟨{*Offer.Status* = Executed}, {} ⟩
S_6	⟨{*Cargo.Status* = Accomplished}, {} ⟩

Table 10. CSG segment for CargoTransportation operation.

Source	Destination	Conversion Fn	Substitution Cond.
Cargo.Det	Freight.Det	Freight.Det = Cargo.Det	
Freight.Det	Cargo.Det	Cargo.Det = Freight.Det	
Cargo.POL	Origin.Det	Origin.Det = Cargo.POL	
Origin.Det	Cargo.POL	Cargo.POL = Origin.Det	
Cargo.POD	Dest.Det	Dest.Det = Cargo.POD	
Dest.Det	Cargo.POD	Cargo.POD = Dest.Det	
Cargo.Type	Freight.Type	Freight.Type = Cargo.Type	
Freight.Type	Cargo.Type	Cargo.Type = Freight.Type	

Credit.Period	Payment.Type	IF (Credit.Period > 0) THEN
		Payment.Type = Credit
		ELSE
		Payment.Type = Cash
		END IF
			Credit.Period ≥ 0

Payment.Type	Credit.Period	IF (Payment.Type = Credit) THEN
		Credit.Period ∈ {15, 30, 45, 60}
		ELSE
		Credit.Period = 0
		END IF
			Payment.Type ∈ {Credit, Cash}

Order.Stat	Cargo.Stat	SWITCH (Order.Stat) CASE Fulfilled: Cargo.Stat = Done CASE Created: Cargo.Stat = Received END CASE	Order.Stat ∈ {Fulfilled, Created}
Cargo.Stat	Order.Stat	SWITCH (Cargo.Stat) CASE Done: Order.Stat = Fulfilled CASE Received: Order.Stat = Created END CASE	Cargo.Stat ∈ {Done, Received}
Proposal.Stat	Offer.Stat	Offer.Stat = Proposal.Stat	Proposal.Stat ∈ {Sent,Approved}
Offer.Stat	Proposal.Stat	Proposal.Stat = Offer.Stat	Offer.Stat ∈ {Sent, Approved}

Order.Stat	Offer.Stat	IF (Order.Stat = Approved) THEN
		Offer.Stat = Accepted
		ELSE
		Offer.Stat = Executed
		END IF
			Order.Stat ∈ {Approved, Executed}

Offer.Stat	Order.Stat	IF (Offer.Stat = Accepted) THEN
		Order.Stat = Approved
		ELSE
		Order.Stat = Executed
		END IF
			Offer.Stat ∈ {Accepted, Executed}

Payment.Stat	Cargo.Stat	IF (Payment.Stat = Received) THEN
		Cargo.Stat = Done
		END IF
			Payment.Stat = Received

Cargo.Stat	Payment.Stat	IF (Cargo.Stat = Done) THEN
		Payment.Stat = Received
		END IF
			Cargo.Stat = Done

Table 11. Matching behavior models using SMP.

S_1 behavior model	S_2 behavior model
S_0, S_1, S_2, S_3, S_4	S_0, S_1
S_5	S_2
S_6	S_3
S_7, S_8, S_9	S_4
S_{10}, S_{11}, S_{12}	S_5
S_{13}	S_6

CONCLUSION

In this paper, we have proposed an automated approach for generating service conversation adapters on the fly in dynamic smart environments, where services interact with each other in seamless transparent manner without human intervention. The proposed approach customizes service conversations in a context-sensitive manner by resolving conversation conflicts (signature and/or protocol) using aggregate concept conditional substitution semantics captured by the proposed concepts substitutability extended graph (CSEG) that required to be a part of the adopted application domain ontology. We illustrated how such semantics are used to resolve signature and protocol incompatibilities. We provided the algorithms needed for automatic adapter generation and presented the verifying simulation experiments. Finally, we indicated how the adapter structure is determined and provided the algorithms needed for adapter source code generation. The proposed approach enables services in dynamic environments to smoothly interact with one another without having semantic interoperability concerns, thus increasing the chances for service reuse, and consequently improving the efficiency of dynamic environments. We believe that the proposed approach helps in improving business agility and responsiveness and of course resembles an important step toward achieving the IoS vision.

REFERENCES

1. M. Papazoglou and D. Georgakopoulos, "Service oriented computing," *Communications of the ACM*, vol. 46, no. 10, pp. 24–28, 2003.
2. M. Dumas, M. Spork, and K. Wang, "Adapt or perish: algebra and visual notation for service interface adaptation," in *Business Process Management*, vol. 4102 of *Lecture Notes in Computer Science*, pp. 65–80, 2006.
3. B. Benatallah, F. Casati, D. Grigori, H. R. Motahari Nezhad, and F. Toumani, "Developing adapters for web services integration," in *Proceedings of the 17th International Conference on Advanced Information Systems Engineering (CAiSE '05)*, pp. 415–429, June 2005.
4. H. R. Motahari Nezhad, B. Benatallah, A. Martens, F. Curbera, and F. Casati, "Semi-automated adaptation of service interactions," in *Proceedings of the 16th International World Wide Web Conference (WWW '07)*, pp. 993–1002, May 2007.
5. R. Mateescu, P. Poizat, and G. Salaün, "Behavioral adaptation of component compositions based on process algebra encodings," in *Proceedings of the 22nd IEEE/ACM International Conference on Automated Software Engineering (ASE '07)*, pp. 385–388, November 2007.
6. A. Brogi and R. Popescu, "Automated generation of BPEL adapters," in *Proceedings of the 4th International Conference on Service-Oriented Computing (ICSOC '06)*, vol. 4294 of *Lecture Notes in Computer Science*, pp. 27–39, 2006.
7. J. Hau, W. Lee, and S. Newhouse, "The ICENI semantic service adaptation framework," in *UK e-Science All Hands Meeting*, pp. 79–86, 2003.
8. A. Brogi and R. Popescu, "Service adaptation through trace inspection," *International Journal of Business Process Integration and Management*, vol. 2, no. 1, pp. 9–16, 2007.
9. D. M. Yellin and R. E. Strom, "Protocol specifications and component adaptors," *ACM Transactions on Programming Languages and Systems*, vol. 19, no. 2, pp. 292–333, 1997.
10. I. Elgedawy, Z. Tari, and J. A. Thom, "Correctness-aware high-level functional matching approaches for semantic Web services," *ACM Transactions on the Web*, vol. 2, no. 2, article 12, 2008.

11. I. Elgedawy, "Automatic generation for web services conversations adapters," in *Proceedings of the 24th International Symposium on Computer and Information Sciences (ISCIS '09)*, pp. 616–621, Guzelyurt, Turkey, September 2009.
12. I. Elgedawy, Z. Tari, and M. Winikoff, "Exact functional context matching for Web services," in *Proceedings of the Second International Conference on Service Oriented Computing (ICSOC '04)*, pp. 143–152, New York, NY, USA, November 2004.
13. I. Elgedawy, "A context-sensitive approach for ontology mapping using concepts substitution semantics," in *Proceedings of the 25th International Symposium on Computer and Information Sciences (ISCIS '10)*, vol. 62 of *Lecture Notes in Electrical Engineering*, pp. 323–328, London, UK, 2010.
14. I. Elgedawy, "Conditional ontology mapping," in *Proceedings of the 36th IEEE International Conference on Computer Software and Applications (COMPSAC '12), the 7th IEEE International Workshop on Engineering Semantic Agent Systems (ESAS '12)*, Izmir, Turkey, 2012.
15. I. Elgedawy, Z. Tari, and M. Winikoff, "Scenario matching using functional substitutability in web services," in *Proceedings of the 5th International Conference on Web Information Systems Engineering (WISE '04)*, Brisbane, Australia, 2004.
16. I. Elgedawy, Z. Tari, and M. Winikoff, "Exact functional context matching for Web services," in *Proceedings of the 2nd International Conference on Service Oriented Computing (ICSOC '04)*, pp. 143–152, Amsterdam, Netherlands, November 2004.
17. F. Casati, E. Shan, U. Dayal, and M.-C. Shan, "Business—oriented management of Web services," *Communications of the ACM*, vol. 46, no. 10, pp. 55–60, 2003.
18. M. P. Papazoglou and W.-J. van den Heuvel, "Web services management: a survey," *IEEE Internet Computing*, vol. 9, no. 6, pp. 58–64, 2005.
19. W3C, "Web service choreography interface," 2002, http://www.w3.org/TR/wsci/.
20. M. Dumas, B. Benatallah, and H. R. M. Nezhad, "Web service protocols: compatibility and adaptation," *IEEE Data Engineering Bulletin*, vol. 31, no. 3, pp. 40–44, 2008.

21. M. Nagarajan, K. Verma, A. P. Sheth, J. Miller, and J. Lathem, "Semantic interoperability of Web services—challenges and experiences," in *Proceedings of the 4th IEEE International Conference on Web Services (ICWS '06)*, pp. 373–380, September 2006.
22. Y. Kalfoglou and M. Schorlemmer, "Ontology mapping: the state of the art," *Knowledge Engineering Review*, vol. 18, no. 1, pp. 1–31, 2003.
23. N. Shadbolt, W. Hall, and T. Berners-Lee, "The semantic web revisited," *IEEE Intelligent Systems*, vol. 21, no. 3, pp. 96–101, 2006.
24. B. C. Grau, I. Horrocks, B. Motik, B. Parsia, P. Patel-Schneider, and U. Sattler, "OWL 2: the next step for OWL," *Web Semantics*, vol. 6, no. 4, pp. 309–322, 2008.
25. D. Roman, U. Keller, and H. Lausen, "Web service modeling ontology (WSMO)," Feburary 2005, http://www.wsmo.org/TR/d2/v1.1/20050210/.
26. "OWL-Services-Coalition, OWL-S: semantic markup for web services," 2003, http://www.daml.org/services/owl-s/1.0/owl-s.pdf.
27. J. Kopecký, T. Vitvar, C. Bournez, and J. Farrell, "SAWSDL: semantic annotations for WSDL and XML schema," *IEEE Internet Computing*, vol. 11, no. 6, pp. 60–67, 2007.
28. M. Kova, J. Bentahar, Z. Maamar, and H. Yahyaoui, "A formal verification approach of conversations in composite web services using NuSMV," in *Proceedings of the Conference on New Trends in Software Methodologies, Tools and Techniques*, 2009.
29. L. Ardissono, A. Goy, and G. Petrone, "Enabling conversations with web services," in *Proceedings of the 2nd International Joint Conference on Autonomous Agents and Multiagent Systems (AAMAS '03)*, pp. 819–826, July 2003.
30. M. T. Kone, A. Shimazu, and T. Nakajima, "The state of the art in agent communication languages," *Knowledge and Information Systems*, vol. 2, no. 3, 2000.
31. M. B. Juric, *Business Process Execution Language for Web Services BPEL and BPEL4WS*, Packt Publishing, Birmingham, UK, 2nd edition, 2006.
32. V. Kashyap and A. Sheth, "Semantic and schematic similarities between database objects: a context-based approach," *The VLDB Journal*, vol. 5, no. 4, pp. 276–304, 1996.

33. S. Abiteboul, S. Cluet, and T. Milo, "Correspondence and translation for heterogeneous data," *Theoretical Computer Science*, vol. 275, no. 1-2, pp. 179–213, 2002.
34. H. Chalupksy, "Ontomorph: a translation system for symbolic knowledge," in *Proceedings of the 17th International Conference on Knowledge Representation and Reasoning*, Breckenridge, Colo, USA, 2000.
35. S. Li, H. Hu, and X. Hu, "An ontology mapping method based on tree structure," in *Proceedings of the 2nd International Conference on Semantics Knowledge and Grid (SKG '06)*, November 2006.
36. P. Ganesan, H. Garcia-Molina, and J. Widom, "Exploiting hierarchical domain structure to compute similarity," *ACM Transactions on Information Systems*, vol. 21, no. 1, pp. 64–93, 2003.
37. J. Madhavan, P. A. Bernstein, P. Domingos, and A. Y. Halevy, "Representing and reasoning about mappings between domain models," in *Proceedings of the 18th National Conference on Artificial Intelligence (AAAI '02)*, pp. 80–86, August 2002.
38. I. Elgedawy, B. Srivastava, and S. Mittal, "Exploring queriability of encrypted and compressed XML data," in *Proceedings of the 24th International Symposium on Computer and Information Sciences (ISCIS '09)*, pp. 141–146, Guzelyurt, Turkey, September 2009.
39. J. Pearson and P. Jeavons, *A survey of tractable constraint satisfaction problems* CSD-TR-97-15, Oxford University, Computing Laboratory, Oxford, UK, 1997, http://citeseerx.ist.psu.edu/viewdoc/summary?doi=10.1.1.43.9045.
40. P. G. Jeavons and M. C. Cooper, "Tractable constraints on ordered domains," *Artificial Intelligence*, vol. 79, no. 2, pp. 327–339, 1995.
41. I. Elgedawy, "A conceptual framework for web services semantic discovery," in *Proceedings of On The Move (OTM) to Meaningful Internet Systems*, Catania, Italy, 2003.
42. Y. Taher, D. Benslimane, M.-C. Fauvet, and Z. Maamar, "Towards an approach for web services substitution," in *Proceedings of the 10th International Database Engineering and Applications Symposium (IDEAS '06)*, pp. 166–173, December 2006.

INDEX

I

J

K

L

M

N

O

P

W

X

Z

Printed in the United States
by Baker & Taylor Publisher Services